Th

OCEANOGRAPHY OF ASIAN MARGINAL SEAS

Elsevier Oceanography Series, 54

OCEANOGRAPHY OF ASIAN MARGINAL SEAS

Edited by

K. TAKANO

Institute of Biological Sciences
University of Tsukuba
Tsukuba, Ibaraki 305, Japan

ELSEVIER

Amsterdam — Oxford — New York — Tokyo   1991

ELSEVIER SCIENCE PUBLISHERS B.V.
Sara Burgerhartstraat 25
P.O. Box 211, 1000 AE Amsterdam, The Netherlands

Distributors for the United States and Canada:

ELSEVIER SCIENCE PUBLISHING COMPANY INC.
655, Avenue of the Americas
New York, NY 10010, U.S.A.

```
         Library of Congress Cataloging-in-Publication Data


Oceanography of Asian marginal seas / edited by K. Takano.
      p.   cm. -- (Elsevier oceanography series ; 54)
    Includes bibliographical references and index.
    ISBN 0-444-88805-5
    1. Oceanography--North Pacific Ocean--Congress.   I. Takano,
  Kenzō, 1929-     II. Series.
  GC791.O24  1991
  551.46'55--dc20                                        91-8897
                                                            CIP
```

ISBN 0-444-88805-5

© Elsevier Science Publishers B.V., 1991

PREFACE

The fifth JECSS Workshop was held in Kangnung, Korea on 18–22 September 1989.

JECSS, an acronym for Japan and East China Seas Study, was initiated by Prof. Ichiye of Texas A&M University in 1981 as an international forum by, and for, those who have interest in oceanography, especially in physical oceanography, of the Japan Sea, East China Sea and adjacent waters. It is a forum not only for exchange of information, views and data, but also for discussion and planning of joint studies.

The workshop was convened by Profs. T. Ichiye (Texas A&M University) and K. Takano (University of Tsukuba), organized by Profs. K. Kim (Seoul National University), Byung Ho Choi (Sung Kyun Kwan University) and Dr. Heung-Jae Lie (Korea Ocean Research and Development Institute). It was associated with the Regional Committee for the WESTPAC of the Intergovernmental Oceanographic Commission (IOC), supported by the Research Institute of Oceanography of Seoul National University, and sponsored by the Oceanological Society of Korea, the Korean Oceanógraphic Commission and the American Geophysical Union.

The JECSS workshops have been held every two years. The numbers of participants are listed below.

	China	Japan	Korea	Philippines	Thailand	UK	USA	USSR
			Number of participants					
1st June 1981	1	27	6				7	
2nd April 1983	6	27	7				6	
3rd May 1985	8	30	7				6	
4th September 1987	15	33	9				7	
5th September 1989	13	30	55	1	1	1	9	4

Compared to the previous four JECSS workshops held at Tsukuba in Japan the fifth workshop is worthy of special mention: through much effort of the local organizers, colleagues could attend the workshop from the People's Republic of China and the USSR which have no diplomatic relation with Korea, and from the Philippines, Thailand, and the UK for the first time. Participation from many countries on Asian marginal seas is of crucial importance to JECSS activity.

Seventy-four papers were presented at the workshop. This Proceedings volume contains thirty-one papers including five papers from Chinese scientists who could not attend the meeting because of difficulties with formalities.

We would like to thank the anonymous reviewers of the submitted papers for their comments and suggestions.

Authors and titles of the papers which were presented at the workshop but do not appear in this volume are listed on the next pages.

The Proceedings of the first to fourth workshops were published in a special issue of La Mer, 20 (1982), of la Société franco–japonaise d'océanographie, Elsevier Oceanography Series, 39 (1984), Progress in Oceanography, 17 (1986), pp. 1–399, and Progress in Oceanography, 21 (1989), pp. 227–536.

Editors of the Proceedings

Kenzo Takano (Editor-in-chief, University of Tsukuba)
Kuh Kim (Seoul National University)
Ya Hsueh (florida State University)
Hsien-Wen Li (National Taiwan Ocean University)
Fagao Zhang (Institute of Oceanology, Academia Sinica)
Kazuo Kawatate (Kyushu University)

List of papers which do not appear in this volume

CONTENTS

X

OBSERVATIONAL CHARACTERISTICS OF INTERNAL TEMPERATURE FLUCTUATIONS IN THE MID-LATITUDE NORTH PACIFIC

Jae-Yul Yun[1,2], James M.Price[3] and Lorenz Magaard[1]
1 Department of Oceanography, University of Hawaii, Honolulu, Hawaii 96822, U.S.A.
2 Department of Oceanography, Korea Naval Academy, Jinhae, Kyungnam, Korea
3 Hawaii Institute of Geophysics, University of Hawaii, Honolulu, Hawaii, 96822, U.S.A.

ABSTRACT

Using the TRANSPAC XBT data at a depth of 300 m, the regional variability of energy, time scales, length scales, and phase propagation of internal temperature fluctuations in the mid-latitude North Pacific is examined. The results show that the regional variability in the eastern half of the basin is substantially different from that in the western half.

In the eastern North Pacific, the energy of the internal temperature fluctuation is very low and fairly uniformly distributed. Time scales and meridional length scales are distributed over broad ranges, zonal length scales are relatively small, and the direction of phase propagation is almost due west. At the eastern boundary, the opposite tendency in time and length scale distribution holds.

In contrast, the energy in the western North Pacific is high, particularly along the main axis of the Kuroshio Extension Current (KEC), and decreases toward the east. It also decreases toward the north and south. Time scales are small near the western boundary and increase eastward. Both zonal and meridional length scales are large near the western boundary and decrease eastward. Phase propagation along the KEC appeared to be eastward, while that in the outer regions north and south of the KEC seemed to be westward with poleward components to the north and south, respectively.

INTRODUCTION

Since the development of the TRANSPAC XBT program (White and Bernstein, 1979), the number of studies on the variability of energy, time scales, length scales and phase propagation of the internal temperature fluctuations has substantially increased for the mid-latitude North Pacific. It is well known from many studies that the potential energy or the variance of internal temperature fluctuations is much higher in the western North Pacific, especially along the axis of the Kuroshio current system, than in the eastern North Pacific with a relatively sudden change at about 170°W (Roden, 1977; White, 1977; Wilson and Dugan, 1978; Kang and Magaard, 1980; Bernstein and White, 1981; White, 1982; Harrison et al., 1983; Mizuno and White, 1984).

Concerning the time scales the results of earlier studies suggest that there are predominant interannual fluctuations (White and Walker, 1974; Price and Magaard, 1980; White, 1983) and near-annual ones with periods ranging from several months to two years (Bernstein and White, 1974; Emery and Magaard, 1976; Bernstein and White, 1981; White, 1982) in the North Pacific. This is also clearly shown in Magaard's (1983) composite model spectrum for the area of 20-25°N, 175-130°W.

The most commonly observed wavelength of the internal temperature fluctuations both in the zonal and in the meridional direction is reported to be in a range of 400- 1000 km in the mid-latitude North Pacific (Roden, 1977; Wilson and Dugan, 1978; Kang and Magaard, 1980; Harrison et al., 1981). The phase propagation in the mid-latitude North Pacific, according to the results of earlier studies (Kang and Magaard, 1980; Bernstein and White, 1981; White, 1982), is basically westward with a speed of 1-4 cm/sec. However, Mizuno and White (1984), using the same data set as the present one, recently found that the phase propagation is eastward along the axis of the Kuroshio Extension Current (KEC).

Although the basic features of the observational characteristics in the mid-latitude North Pacific seem to be known from earlier studies, the regional variabilities of the characteristics have yet to be studied rather systematically. The purpose of this paper is to investigate the geographical distribution of energy, time scales, length scales and phase propagation of the temperature fluctuations at a depth of 300 m in the mid-latitude North Pacific. In this paper we utilize several standard analysis methods such as spectrum analysis, autocorrelation analysis, time-longitude and time-latitude contours, and harmonic analysis without discussing the methodology in detail.

THE DATA

In this paper we use seasonally averaged temperature data at 300 m depth, mostly from the TRANSPAC XBT program and partly from the Japanese Oceanographic Data Center for the data in the region near Japan. This data set has 0.5° latitude by 0.5° longitude grid coverage from the coast of Japan to the coast of California between 30 and 45°N and extends, in time, from summer 1976 to spring 1980. In the course of the edition of this data set each grid point has been interpolated first by fitting a trend surface to the nearest eight surrounding observations and then selecting the value at the grid point from this surface. This data set was provided by Dr. Warren White of the Scripps Institution of Oceanography. It has been used by White (1982) in the eastern North Pacific (180-120°W) and by Mizuno and White (1984) in the western North Pacific (130°E-170°W). For more detailed descriptions of the data set the reader is referred to their papers.

MEAN AND STANDARD DEVIATION

The mean temperature profiles at a depth of 300 m (Fig. 1) indicate that the

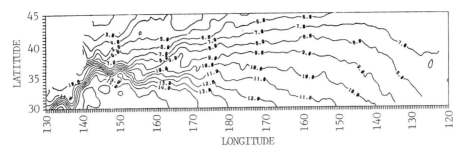

Fig. 1. Mean temperature at 300 m depth

KEC appears to be bifurcate at about 152°E and to become two nonzonal branches, the northeastward (NE) branch (along the isotherms of 6-8°C) and the southeastward (SE) branch (along the isotherms of 11-14°C). This bifurcation of the mean flow may be triggered by the Shatsky Rise as is inferred from its location near the point of the bifurcation. The profiles also show that the mean temperature gradients decrease from about 3°C/100 km at 150°E to about 1°C/100 km at 165°E along the SE branch and to about 0.3°C/100 km at 160°W indicating the eastward decrease in geostrophic velocity.

The standard deviation of the temperatures (Fig. 2) shows a drastic change at

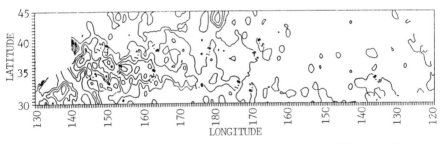

Fig. 2. Standard deviation of the temperature at 300 m depth.

173°W along the main axis of the KEC suggesting that the vigorous current-related fluctuations terminate there. West of 173°W the standard deviation is mostly larger than 0.5°C and especially large along the SE branch of the KEC. East of 173°W the standard deviation is mostly less than 0.5°C with a fairly uniform distribution. Along the SE branch, the standard deviation shows a distribution of highs and lows with intervals of 300-750 km between the neighboring maxima. These intervals may be considered as wavelengths of the fully developed finite amplitude instability of the KEC.

The shape of the 1°C contour line of the standard deviation indicates that the energy sources are apparently located along the axis of the KEC and that the energy decreases outward (particularly toward the east, north and south) from the sources. In the far-fields away from the KEC the 1°C contour line of the standard deviation indicates a radiation of energy from the east toward the west, probably as stable Rossby waves. If areas with values of the standard deviation less than 0.5°C are considered as areas of background fluctuations, the meridional length scale of decay from the current-related source region to the northern region of the background fluctuations in the far-fields is about 1000 km.

TIME SCALES

Frequency spectrum analysis

Frequency spectra have been estimated using the direct Fourier transform method and a spatial smoothing instead of a smoothing in the frequency domain, because of the small number of data (16 points) in time. The Fourier coefficients have been calculated from the time series at each grid point. These coefficients are then averaged over each of the six subareas (Fig. 3),

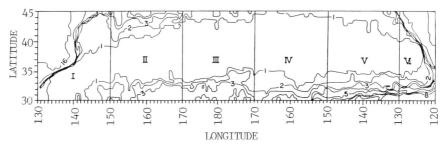

Fig. 3. Distribution of subareas I-VI and contours indicating the numbers of the missing data point in time. Inside the label 1 there is no missing data point in time

over the area west of 170°W, over the area east of 170°W, and over the entire study area. The frequency spectra are then calculated from the spatially averaged coefficients. Therefore, a total of 9 spectra have been produced. In this analysis the grid points with gappy data have been excluded in determining the spectra. Thus, the actual area covered by this analysis is within the contour line labled 1 in Fig. 3. The same estimation procedure is repeated after removing the linear trend in each time series to determine to what degree the linear trend affects the spectrum.

Since the data are not independent from a grid point to another, the effective number of degree of freedom has to be calculated to assign the confidence interval for each spectrum. Using the formula given by Emery and Magaard (1976) the effective number of degree of freedom is calculated for subarea I which has the smallest number of grid points (354) and it is 100. Since all the other subareas contain more number of grid points than subarea I does, the number of degree of freedom of 100 is used safely to assign 95%

confidence interval for each spectrum In Figs. 4 and 5 the broken lines

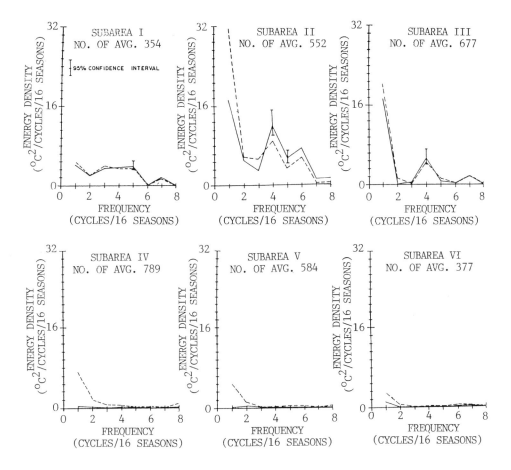

Fig. 4. Spatially smoothed power spectra in each subarea. Solid (broken) lines indicate the spectra without (with) the linear trends.

indicate the spatially smoothed spectra with the linear trends included and the solid lines indicate those with the linear trends removed.

In subarea I (near the western boundary), a considerable amount of energy is found only in a frequency range of 0.75–1.25 cycles/year (cpy). In this region the removal of the linear trends makes no difference for the shape of the spectra indicating a lack of very low frequency fluctuations. In subareas II and III (the interior western North Pacific) the energy levels are high (the highest in subarea II) at all frequencies. In the spectra, distinct peaks are shown at 1 cpy (15% of the total energy) and 1.5 cpy (9%) in subarea II and at 1 cpy (16%) and 1.75 cpy (6%) in subarea III. When the linear trends are removed, both peaks in subarea II become more distinct (from 15% to 23% for 1 cpy and from 9% to 14% of the total energy for 1.5 cpy) while both peaks in subarea III show little change (from 16% to 21% for 1 cpy and remained at 6% for 1.75 cpy). The spectra illustrate that the linear trends account for a larger portion of the energy in subarea II than in subarea III, while the energy at the lowest frequency after the removal of the linear trend is higher in subarea III (67%) than in subarea II (33%).

In subareas IV and V (the interior eastern North Pacific) the energy level is very low at all frequencies compared to that in subareas II and III, as is anticipated from the distribution of the standard deviation. The energy at 1 cpy is not as distinct as that in subareas II and III. In these subareas the spectra look quite different depending on whether the linear trends are included or not. When the linear trends are included, the energy level decreases drastically from the lowest frequency (62% in subarea II and 58% in subarea III) to the next lowest (14% in subarea II and 15% in subarea III) and then slowly decreases toward the higher frequencies. When the linear trends are removed, the energy at 1 cpy (16%) in subarea IV and at 0.5 cpy (32%) and 1 cpy (15%) in subarea V become important. In subarea VI (near the eastern boundary) the linear trend and the lowest frequency account for a large portion of total energy but the portion is not as large as that in subareas IV and V. In this subarea the energy level is relatively high at 1.5 cpy (10 and 11% respectively with and without the linear trends) and 1.75 cpy (5 and 9%) as well as at 1 cpy (6 and 8%). This indicates a large increase in very small time scale fluctuations toward the eastern boundary.

Hence, it is very clear that the spectra in subareas IV and V are distinct from those in subareas II and III. It is certain that long-term fluctuations in the eastern North Pacific are different from those in the western North Pacific with respect to relavant time scale and the amount of energy. We speculate that the high energy at very low frequency in subareas IV and V may indicate the existence of the same low frequency signal as shown in Magaard's (1983) model spectrum.

The spectra also indicate a different distribution of the small time scales between the two regions. The most dominant periods are annual (23 and 34% respectively with and without the linear trends) in the interior western North Pacific (Fig. 5a) and annual (5 and 18%), biannual (15 and 20%) and semiannual (6 and 11%) in the interior eastern North Pacific (Fig. 5b). The average spectra over the entire study area (Fig. 5c) show the annual period as being the most dominant (13 and 34%).

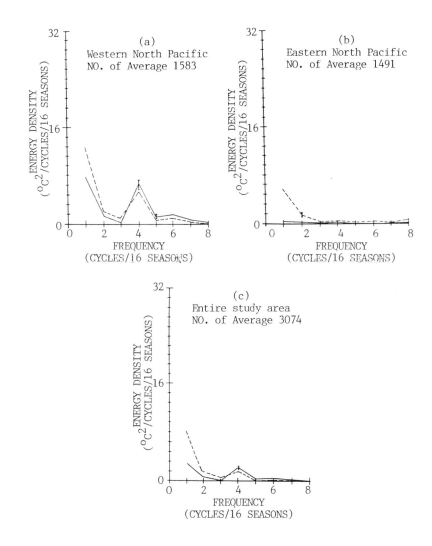

Fig. 5. Spatially smoothed power spectra in the western half, the eastern half, and the entire study area. Solid (broken) lines indicate the spectra without (with) the linear trends.

Autocorrelation analysis

In order to study the time scales by a different method, an autocorrelation analysis method has been applied. This method determines the first zero-crossing time lag (hereafter called ZXTL) at each grid point. In this analysis the time series have been used without any filtering and the only grid points that have more than 12 consecutive data points are included in the analysis. For the present data, autocorrelations beyond the ZXTL are found to be drastically reduced at most of the grid points. Therefore, one can consider the ZXTL as decorrelation time scales.

Table 1 displays the total number of estimated ZXTLs in each subarea and

TABLE 1

Total number of estimated first zero-crossing time lags in each subarea and their distribution (in percentage of the total number) in 7 categories of time lags.

Longitude band (Total est.)	130-150E (693)	150-170E (989)	170E-170W (852)	170-150W (910)	150-130W (808)	140-120W (255)
Zero-crossing(months)	I	II	III	IV	V	VI
< 3	48	36	35	14	29	58
3-6	30	37	34	17	16	23
6-9	11	11	12	15	15	10
9-12	8	7	10	13	13	7
12-15	1	5	7	19	13	2
15-18	0	3	2	17	9	0
> 18	1	1	0	5	4	0

their distribution (in percentage of the total number) in 7 categories of time lags. The most commonly occuring ZXTL is less than three months. A comparable decorrelation time scale of about 2 months has been obtained by Bernstein and White (1981) from about 2 years of the TRANSPAC XBT data in the western North Pacific. The next most commonly occuring ZXTLs are 3 to 6 months.

The ZXTLs of less than three months occur most commonly in subareas I and VI (near both boundaries) and it is especially so in subarea VI. The ZXTLs are generally more broadly distributed in the interior than in the boundary region. In the interior region the ZXTLs are distributed over a narrower range in the western than in the eastern North Pacific. As one proceeds from subarea I to subarea IV, the distribution of the ZXTLs becomes broader with a gradual increase in number of larger time lags. However, the small ZXTLs of less than 3 months and 3 to 6 months are still dominant in subareas II and III. The tendency of the distribution of the ZXTLs broadening toward subarea III is indicated by the percentage of the total number of ZXTLs in the first two categories of time lag (less than 3 months and 3 to 6 months) decreasing from 78% in subarea I to 73% in subarea II and then to 69% in subarea III. The ZXTLs of 6 months or larger at 155-170°E as shown in Fig. 6 may be a consequence of the Shatsky Rise, since the topography can scatter energy toward larger time scales (Rhines and Bretherton, 1974).

The ZXTLs are distributed over the broadest range in subarea IV. From subarea IV to subarea VI the number of small ZXTLs increase substantially, although the distribution tends to be broad still in subarea V. The percentage of time lags in the first two categories increase from only 31% in subarea IV

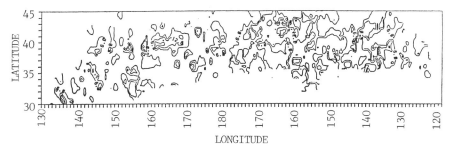

Fig. 6. Geographical distribution of the first zero-crossing time lags (in months) of the temperature at 300 m depth.

to 45% in subarea V and then to 81% in subarea VI. The percentages of time lags larger than 9 months are 10, 16, 19, 54, 39, and 9% in sunareas I to VI, respectively. This again indicates a possible existence of long-term fluctuations in the eastern North Pacific as discussed in the frequency spectrum analysis. However, the confidence interval of the autocorrelation function will be much larger at these larger ZXTLs than at the smaller ones.

LENGTH SCALES

Wavenumber spectrum analysis

Wavenumber spectra are estimated from the space series of the temperature data using the standard Blackman-Turkey method. For the zonal wavenumber spectra in the western North Pacific, the space series covering the distance from near Japan to the date line are used after removal of the spatial linear trends. The average record length is about 45 degrees in the zonal direction; the sampling rate is 0.5°. In order to obtain a composite zonal wavenumber spectra at latitudes, 30, 31,....45, individual spectra are estimated for each season at latitudes 1° and $(1+0.5)^{\circ}$. Then for each latitude 1°, 32 individual spectra (2 spatial points x 16 time points) are averaged at each wavenumber. The effective number of degree of freedom for the composite spectrum is estimated using the formula given in Emery and Magaard (1976) and is about 60. This effective number of degree of freedom is used to assign the 95% confidence interval to the composite spectra. The composite wavenumber spectra at several selected latitudes (33, 36, 39 and 44°N in the western North Pacific and 32, 35, 39 and 44°N in the eastern North Pacific) are shown in Figs. 7a and 7b.

For the zonal wavenumber spectra the energy levels are high at $33-36^{\circ}$N (the maximum is at $35-36^{\circ}$N) at most wavenumbers and low at all wavenumbers at $42-45^{\circ}$N (the minimum is at 42°N) in the western North Pacific (Fig. 7a). In the eastern North Pacific the energy levels are nearly two orders of magnitude smaller than the maximum energy in the western North Pacific. In the eastern

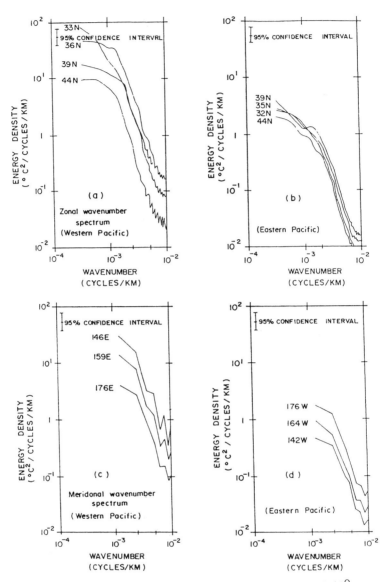

Fig. 7. Zonal wavenumber spectra (a) at 33, 36, 39 and 44°N in the western North Pacific and (b) at 32, 35, 39 and 44°N in the eastern North Pacific. Meridional wavenumber spectra (c) at 146, 159 and 176°E, and (d) at 176, 164 and 142°W.

North Pacific there is very little difference in energy levels among different latitudes. The energy levels are slightly higher at 39°N and slightly lower at 45°N in the eastern North Pacific.

For the zonal wavenumber spectra the spectral slopes change at wavelengths of 600–750 km in the western North Pacific and at wavelengths of 420–500 km in the eastern North Pacific. The difference in the wavelengths at which the slope change occurs reflects the difference in dominant length scales. The slope changes are more gradual in the eastern than in the western North Pacific. In the western North Pacific the spectral slopes at 35-45°N are -2.7 to -3.2 at wavelengths of 100-750 km and nearly flat at wavelengths larger than 750 km. However, those at 31-35°N are -2.7 to -3.0 at wavelengths of 100-750 km and -1 to -1.2 at wavelengths larger than 750 km indicating a broader distribution of length scales at 31-35°N than at 35-45°N. At 31-32°N there are also marginal peaks at a wavelength of about 750 km. The spectral slopes in the eastern North Pacific are -2.6 to -3.0 at wavelengths of 100-500 km. Hence, they are gentler than those in the western North Pacific. The gentler slopes in the eastern North Pacific indicate an increase in number of very small scale fluctuations. The slopes at wavelengths larger than 500 km are about -1 (slightly flatter at 31-34°N) in the eastern North Pacific. The steeper slopes at this wavelength range indicate a more gradual change in length scale distribution. At 38-42°N a marginal peak is found at wavelengths 600-750 km.

Composite meridional wavenumber spectra are estimated using the meridional space series covering 15°. Each composite spectrum is obtained by averaging spectra from 2.5° longitude (5 grid points) and 16 seasons. The effective number of degrees of freedom for these composite spectra is about 55. The spectra at several selected longitudes (146, 159, 176°E, 176, 164 and 142°W) are shown in Figs. 7c and 7d. In the meridional wavenumber spectra the highest energy is found at 138.5-146°E at all wavenumbers. The energy levels decrease quite monotonically toward the east at all wavenumbers. The various spectra in the eastern North Pacific have about the same shape and energy level at each wavenumber. The energy minima are found at 151.5, 149, 141.5 and 129°W.

For these meridional spectra the spectral slopes change at a wavelength of 400 km. The spectral slopes are -2.5 to -3.4 over wavelengths of 100-400 km with steeper slopes in the western than in the eastern North Pacific indicating more frequent occurence of small scale fluctuations in the latter region. The spectral slopes at wavelengths larger than 400 km are -0.4 to -1.0 with subareas I, IV, II, III, V, and VI in a sequence of descending magnitude of the slope.

Autocorrelation analysis

The autocorrelation analysis method is used to study the length scale distribution in both zonal and meridional directions in each subarea. In this method the zonal first zero-crossing space lags (hereafter called ZZXSL) are computed in each subarea from the zonal space series of the temperature fluctuations. For the same reason as described in the autocorrelation analysis of the time series, the ZZXSLs can be interpreted as decorrelation length scales. Table 2 shows the total number of estimated ZZXSL in each subarea and

TABLE 2

Total number of estimated zonal first zero-crossing space lags in each subarea and their distribution (in percentage of the total number) in 9 categories of space lags.

Longitude band (Total est.)	130-150E (159)	150-170E (451)	170E-170W (438)	170-150W (444)	150-130W (446)	140-120W (326)
Zero-crossing (km)	I	II	III	IV	V	VI
50-100	0	0	0	1	2	1
100-150	9	17	27	31	25	18
150-200	27	36	31	36	34	30
200-250	18	15	21	11	18	19
250-300	18	14	12	10	10	14
300-350	23	8	3	4	6	11
350-400	3	5	4	2	3	5
400-450	2	4	2	2	2	2
450-500	0	0	1	1	1	1

their distribution (in percentage of the total number) in 9 categories of space lags. In this analysis only the space series that have more than 30 consecutive data points are included.

The ZZXSLs are generally large and they are distributed over a broad range in subareas I and VI. In the interior North Pacific the ZZXSLs are generally small and they are distributed over a narrow range. The ZZXSLs are distributed over the broadest range in subarea VI, although the space lags of 150-200 km are still dominant (30%) among all the categories. The ZZXSLs which have the maximum percentage are the smallest and they are distributed over the narrowest range in subarea IV. The percentages of the total number of estimated ZZXSLs in the categories of 100-150 km and 150-200 km are 36, 53, 58, 67, 59, and 48% in subareas I-VI, respectively. As one proceeds away from subarea IV toward both boundaries, the number of large ZZXSLs increase gradually. The percentages of the ZZXSLs larger than 300 km are 28, 17, 10, 9, 15, and 19% in subareas I-VI, respectively. For the entire study area the most commonly occuring ZZXSLs are 150-200 km.

Meridional first zero-crossing space lag (hereafter called MZXSL) is estimated at each longitudinal grid point and for each season. In this analysis only the meridional space series that have more than 20 consecutive data points have been included. Table 3 shows the total number of estimated MZXSLs in each subarea and their distribution (in percentage of the total number) in 8 categories of space lags, In general the MZXSLs are smaller than the ZZXSLs.

The MZXSLs are distributed over a narrow range in subareas I and VI but the most commonly occuring MZXSLs are different between the two subareas. They are 200-250 km (46%) in subarea I while 100-150 km (50%) in subarea VI. In the

TABLE 3

Total number of estimated meridional first zero-crossing space lags in each subarea and their distribution (in percentage of the total number) in 8 categories of space lags.

Longitude band (Total est.)	140-150E (279)	150-170E (625)	170E-170W (640)	170-150W (640)	150-130W (631)	130-120W (164)
Zero-crossing (km)	I	II	III	IV	V	VI
50-100	1	1	2	3	7	12
100-150	21	28	38	21	39	50
150-200	25	39	31	28	28	27
200-250	46	24	19	13	12	9
250-300	7	7	8	17	10	1
300-350	0	1	1	14	2	1
350-400	0	0	1	4	1	0
400-450	0	0	0	0	1	0

latter subarea the percentage at 50-100 km is also large with 12%. In the interior region they are distributed more broadly in subareas IV and V than in subareas II and III. The broadest distribution is shown in subarea IV. As one proceeds from subarea I to subarea III, the most commonly occuring MZXSLs change from 250-300 km (46%) in subarea I, to 150-200 km (39%) in subarea II, and then to 100-150 km (38%) in subarea III. From subarea III to subarea IV the distribution of the MZXSLs becomes broad quite suddenly with an increase in number of large space lags. In subarea V the distribution of the MZXSLs is over a narrower range than in subarea IV and the commonly occuring MZXSL is again 100-150 km (39%). The percentages of the MZXSLs in the first three categories (the space lags of 50-100, 100-150, and 150-200 km) are 47, 68, 71, 52, 74, and 89% and those in the range larger than 250 km are 7, 8, 10, 35, 14, and 2% in subareas I-VI, respectively.

PHASE PROPAGATION

Time-longitude and time-latitude contours

Time-longitude matrices at each full degree of latitude are contoured to examine the direction of zonal phase propagation. Figs. 8a-h exhibit the contours at each odd-numbered latitude. In the figures an upward slope of contour toward the west indicates westward phase propagation. The zonal phase propagation at 44-45°N (Fig. 8a for the contour at 45°N) does not show any preferred direction as there is no consistent slope. However, at 43°N (Fig. 8b) there is a tendency of upward slopes to the west intermittently over short time periods and locally over short distances. Westward phase propagation is relatively evident at 37-42°N except in the regions near both eastern and western boundaries (Figs. 8c-e at 41, 39 and 37°N, respectively). The westward phase propagation is most pronounced at 39°N with speeds of about 2

14

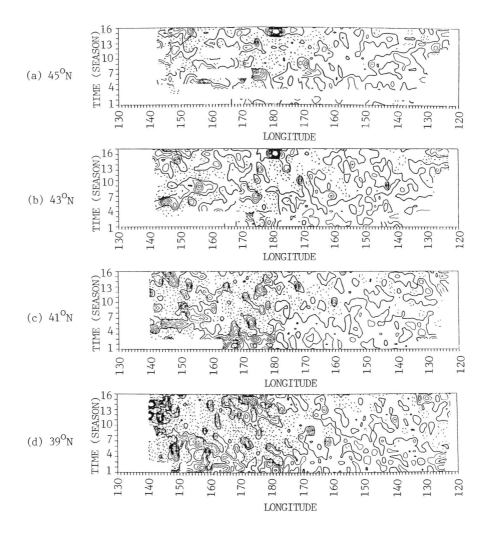

Fig. 8. Time-longitude plots of the temperature (at 300 m depth) variability at each odd-numbered latitude. The contour interval is 0.5°C.

cm/sec. At 37-38°N it becomes less evident. West of 160°E the direction of phase propagation is not obvious even at 37-45°N.

At 34-38°N (Figs. 8e and f at 37 and 35°N, respectively) there appears to be eastward phase propagation in the western North Pacific in some seasons, and it shows most clearly at 35-36°N. The eastward phase speed is about 3 cm/sec for the first few years at 160°E-170°W. However, even in the region where phase propagation appears to be eastward, there is some indication of westward phase propagation at least for a short period of time. In the eastern North

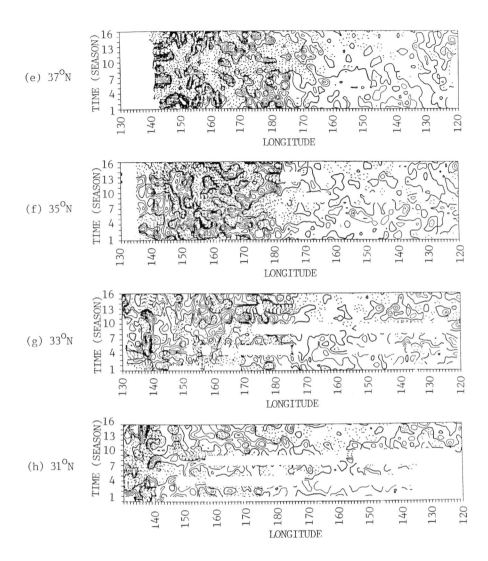

(e) 37°N

(f) 35°N

(g) 33°N

(h) 31°N

Fig. 8. (continued)

Pacific westward phase propagation appears to occur even at 35-36°N.

Recalling that this latitudinal band is basically located along the southeastward branch of the KEC west of 165°E and located in the region between the two branches at 165-179°W, eastward phase speed indicates the disturbances being carried along the axis of the current (Hansen, 1970; Mizuno and White, 1984). At 30-33°N eastward phase propagation tends to disappear and westward phase propagation tends to appear as one prodeeds southward (Fig. 8h at 31°N). The phase speed seems to be higher toward the south, for example, about 4.5 cm/sec between 165°E and 168°W for the last one year period at 31°N.

Time-latitude matrices are contoured in Fig. 9 to examine the direction of

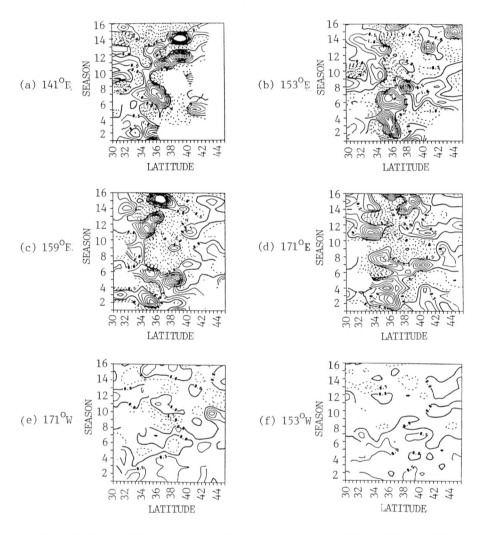

Fig. 9. Time-latitude plots of the temperature (at 300 depth) variability at several selected longitudes. The contour interval is 0.5°C.

meridional phase propagation. The slopes of the contours indicate phase propagation from the axis of the KEC to the north and south. The northward and southward phase propagation respectively in the regions north and south of the axis of the KEC are relatively evident at 141-171°E. A typical phase speed both to the north and south is approximately 1-2 cm/sec, although the estimates are often available over short period of time only.

Harmonic analysis

A harmonic analysis method is employed to investigate phase propagation of the temperature fluctuations for the annual period. Percentages of the total variance explained by the annual harmonics are also obtained. In order to reduce the error bar the original half degree grid data have been averaged over each 1° latitude by 1° longitude before applying the method. The percentages of the total variance explained by the annual harmonics (raw bandwidth of 10.7-13.7 months) are mostly 20-40% in the western North Pacific and 10-20% in the eastern North Pacific. This is an agreement with the results of the time scale analysis.

The zonal phase propagation for the annual harmonics is shown in Figs. 10a-h. In the figures westward phase propagation is indicated by an upward

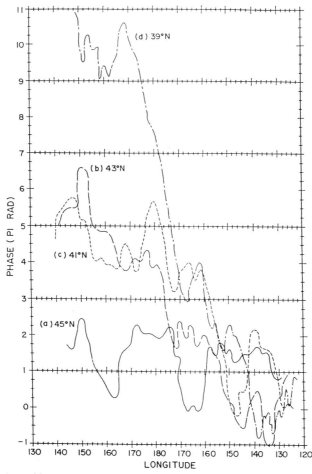

Fig. 10. Phase of the annual harmonics along zones at latitudes of (a) 45°, (b) 43°, (c) 41°, (d) 39°, (e) 37°, (f) 35°, (g) 33°, and (h) 31°N.

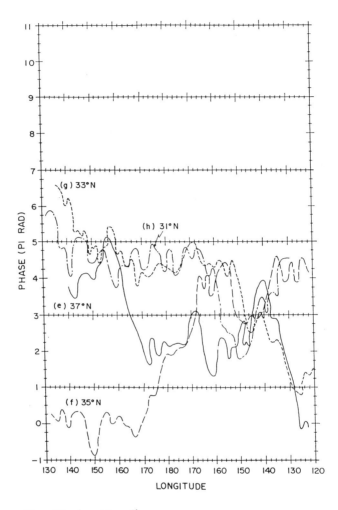

Fig. 10. (continued)

slope to the west. This analysis shows generally the same phase propagation characteristics as discussed in the time-longitude matrices. The westward phase speed at 43°N is roughly 2.1 cm/sec for longitudes between 180 and 168°W. At 39°N the westward phase propagation is very obvious at 150°E-140°W, and the phase speed between 167°E and 167°W is about 1.9 cm/sec. At 35 and 36°N eastward phase propagation occurs between 165°E and 170°W with a speed of roughly 3.8 cm/sec (Fig. 10f at 35°N). As one proceeds away from 35°N toward the south, westward phase propagation becomes relatively apparent again. At 31-33°N the westward phase speeds are roughly 2.5-3.2 cm/sec.

The meridional phases are shown in Fig. 11. It turns out to be possible to categorize areas with similar distributions of meridional phases while the

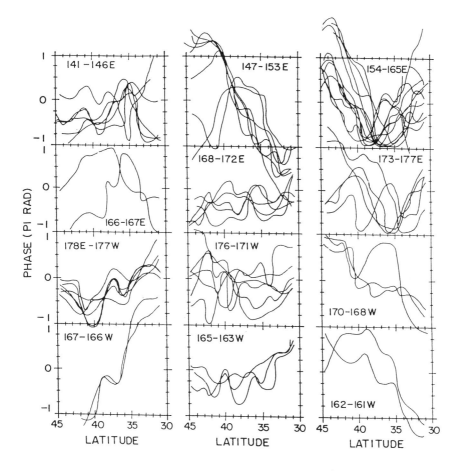

Fig. 11. Phase of the annual harmonics along meridians at various longitudes. The range of longitudes within a box is selected to group similarly looking curves.

distribution changes wildly from an area to the next. The meridional phases for 154-165°E show a considerably persistent pattern in which the relative phase minima exist around 36°N and the phases increase from the region of the phase minima toward the north and south. The phase speed is approximately 2.3 cm/sec both to the north and south. This tendency of phase propagation to the far-fields from about 36°N seems to prevail only in the western North Pacific. In some cases there are two relative phase minima indicating that the energy sources are located along the two branches of the current. East of 165°E, meridional phase propagation often appears to be alternating northward and southward.

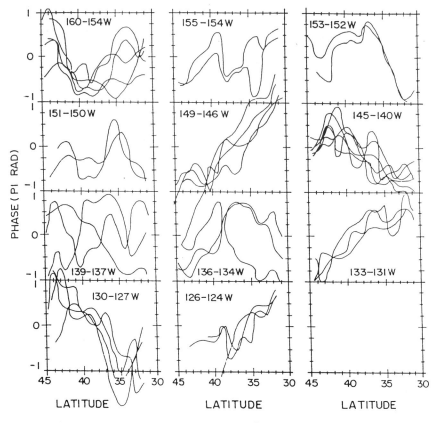

Fig. 11. (continued)

The distribution of the phase of the annual harmonics over the entire study area is displayed in Fig. 12. In this figure only the 120° phase lines are

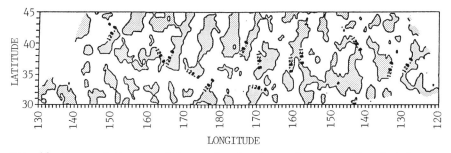

Fig. 12. Phase of the annual harmonics in the study area. The shaded region indicates phase less than 120°.

contoured in order to see a simple phase pattern. The figure shows fairly

regular intervals of the phase lines, except in the regions along the axes of
the KEC. The figure appears to indicate that the phase propagates to the east
along the axis of the KEC, to the northwest in the region north of the axis, and
to the southwest in the region south of the axis in the western North Pacific.
In the entire eastern North Pacific the phase propagates to the west.

SUMMARY AND CONCLUSIONS

The energy levels are much higher in the western (the highest in subarea II)
than in the eastern North Pacific at all frequencies with a drastic change
occuring at 173°W. The annual peak is highly distinct in the western North
Pacific but in the eastern North Pacific it becomes distinct when the linear
trends are removed. The linear trend accounts for a large portion of the total
energy in the eastern North Pacific indicating a possible existence of strong
interannual fluctuations. For the zonal wavenumber spectra the energy levels
are highest at $35-36^{\circ}$N and lowest at 42°N at most wavenumbers in the western
North Pacific. They are highest at 39°N and lowest at 45°N in the eastern
North Pacific, but with only a small difference. For the meridional wavenumber
spectra the energy levels are highest at $138.5-146^{\circ}$E and lowest at 151.5, 149,
141.5 and 129°W at all wavenumbers.

In subarea I (near the western boundary) the time scales are small and both
zonal and meridional length scales are large in general. In this subarea the
time and meridional length scales are distributed over a narrow range but the
zonal length scales are distributed over a broad range. The dominant period,
zonal wavelength and meridional wavelength are shorter than a year, ranges of
600-800 km and 1200-1400 km and a range of 800-1000 km, respectively.

In subarea VI (near the eastern boundary) the time and meridional length
scales are very small but the zonal length scales are generally large compared
to those in subareas IV and V. In this subarea the time scales are distributed
over a narrow range and the zonal length scales are distributed over a broad
range as in subarea I. However, the meridional length scales are distributed
over a narrow range unlike those in subarea I. The dominant period, zonal
wavelength and meridional wavelength are shorter than a year, a range of 600-800
km and a range of 400-600 km, respectively.

Subareas II and III (the interior western North Pacific) are regions of
gradual transition for the time series from a narrow range of distribution of
predominantly short scales in subarea I to a broad range with an increase in
number of large scales in subarea IV. The gradual transition also occurs for
the zonal length scales from a broad distribution with relatively large scales
in subarea I to a narrow distribution of predominantly small scales in subarea
IV. The meridional length scales are distributed over a narrow range in the
western North Pacific but the dominant length scales changes gradually from

large scales in subarea I to small scales in subarea III. The dominant period, zonal wavelength and meridional wavelength are respectively near-annual, a range of 600-800 km, and ranges of 600-800 km in subarea II and 400-600 km in subarea III.

In subareas IV and V (the interior eastern North Pacific) the time and meridional length scales are distributed over broad ranges with increase in number of large scales compared to those in subareas II and III. However, the zonal length scales are distributed over a narrow range of predominantly small scales in these subareas. In subarea IV the distributions of the time and meridional length scales are the broadest while that of the zonal length scales are over the narrowest range (especially at 38-42°N) among all the subareas. In general, the zonal length scales are smaller than the meridional ones in this subarea. In these subareas there appears to be no single dominant time scale, but the near-annual, biannual period and interannual fluctuations may be similarly important. The dominant wavelengths are in a range of 400-800 km in both directions. From subarea IV toward subarea VI the number of small time and meridional length scales and the number of large zonal length scales tend to become large.

Phase propagation is to the east along the axis of the KEC, to the northwest in the region north of the axis and to the southwest in the region south of the axis in the western North Pacific. It is to the west in the eastern North Pacific. The zonal phase speeds are about 3-4 cm/sec eastward along the axis of the KEC and 2-4 cm/sec westward in the region away from the KEC with a slight increase toward the south. The meridional phase speeds are about 2 cm/sec to the north and south in the western North Pacific.

REFERENCES

Bernstein, R., and W. White, 1974. Time and length scales of baroclinic eddies in the central North Pacific. J. Phys. Oceanogr., 4: 613-624.
Bernstein, R., and W. White, 1981. Stationary and traveling mesoscale perturbations in the Kuroshio Extension current. J. Phys. Oceanogr., 11: 692-704.
Emery, W. J., and L. Magaard, 1976. Baroclinic Rossby waves as inferred from temperature fluctuations in the eastern Pacific. J. Mar. Res., 34: 365-385.
Hansen, D. V., 1970. Gulf Stream meanders between Cape Hatteras and the Grand Banks. Deep-sea Res., 17: 495-511.
Harrison, D. E., W. J. Emery, J. P. Dugan, and B.-C. Li, 1983. Mid-latitude mesoscale temperature variability in six multiship XBT surveys. J. Phys. Oceanogr., 13: 648-662.
Kang, Y. Q., and L. Magaard, 1980. Annual baroclinic Rossby waves in the central North Pacific. J. Phys. Oceanogr., 10: 1159-1167.
Magaard, L., 1983. On the potential energy of baroclinic Rossby waves in the North Pacific. J. Phys. Oceanogr., 7: 41-49.
Mizuno, K., and W. B. White, 1984. Annual and interannual variability in the Kuroshio current system. J. Phys. Oceanogr., 13: 1847-1867.
Price, J. M., and L. Magaard, 1980. Rossby wave analysis of the baroclinic

potential energy in the upper 500 meters of the North Pacific. J. Mar. Res., 38: 249-264.

Roden, G. I., 1977. On the long-wave disturbances of dynamic height in the North Pacific. J. Phys. Oceanogr., 7: 41-49.

White, W. B., 1977. Secular variability in the baroclinic structure of the interior North Pacific from 1950-1970. J. Mar. Res., 35: 3, 587-607.

White, W. B., 1982. Traveling wave-like mesoscale perturbations in the North Pacific. J. Phys. Oceanogr., 12: 231-243.

White, W. B., 1983. Westward propagation of short term climatic anomalies in the western North Pacific Ocean from 1964-1974. J. Mar. Res., 41: 113-125.

White, W. B., and L. Bernstein, 1979. Design of an oceanographic network in the mid-latitude North Pacific. J. Phys. Oceanogr., 9: 592-606.

White, W. B., and A. E. Walker, 1974. Time and depth scales of anomalous subsurface temperature at Ocean Weather Station P, N, and V in the North Pacific. J. Geophys. Res., 1979: 4517-4522.

Wilson, W. S., and J. P. Dugan, 1978. Mesoscale thermal variability in the vicinity of the Kuroshio Extension. J. Phys. Oceanogr., 8: 537-540.

TIDAL COMPUTATION OF THE EAST CHINA SEA, THE YELLOW SEA AND THE EAST SEA

Sok Kuh KANG, Sang-Ryong LEE* and Ki-Dai YUM

Coastal Engineering Laboratory, Korea Ocean Research and Development Institute
*Department of Marine Science, Pusan National University, Korea

ABSTRACT

The M_2 tidal phenomena in the entire surrounding seas of Korea Peninsula have been investigated under the condition of one model area, majorly to understand which factors govern the amphidromic system in each sub-area. In this study two-dimensional numerical model based upon an implicit scheme has been used. Due to the large depth differences of each sea in the modelled region the tide generating force is included as well as other inertia terms. The model region toward south and south-eastern boundaries is extended to Ryukyu Islands to utilize the measured data.

The computed M_2 co-tidal and co-amplitude lines in the Yellow Sea and the East China Sea are in good agreement with those of existing tidal charts. The four amphidromic points known to exist in the Yellow Sea and the East China Sea as well as two amphidromic points in the East Sea (Japan Sea) are successfully reproduced. The results of computation also show that the tide generating forces play greatly different roles in the each sub-sea model region. The tidal amplitude due to the tide generating force explains several percent of the tide generating force in the Yellow Sea and the East China Sea. It is shown that the realistic amphidromic system in the East Sea can not be explained without the tide generating forces. It is inferred that the reflected wave at the head region of Tartary Bay is more reinforced by the tide generating force. Also the role of the driving force at the head of Tartary Bay for the amphidromic system is discussed.

An analysis for tidal volume fluxes and volume transport is made for the five straits existing within model area to assess the possible contribution to tidal regime in the linked sub-model area. And the tidal volume fluxes through five straits in the model region are compared with those calculated by simple analytical method (Defant, 1961) and those calculated by Odamaki (1989b).

1. INTRODUCTION

In this paper the M_2 tidal regimes of the East China Sea, the Yellow Sea and the East Sea (Japan Sea) are investigated under one model area by using a two-dimensional depth averaged numerical model based upon an implicit scheme. The model domain covers the whole region of the Yellow Sea and the East China Sea extending to the southeast as far as Ryukyu Islands, to the south as far as Taiwan Strait as well as covering the East Sea connected by the Korea Strait (see Figure 1).

The depths used for the model were obtained from SYNBAPS II data set from MIAS and US chart No. 529. As shown Figure 2, the mean depth of each sea in the model area is very different from each other. It is also noted that there exist abrupt changes of depth at several regions. From the Ryukyu Islands, the ocean bottom rises rapidly from 2,000 m to 200 m and the depths of the entire north-western model area are less than 100m. The East Sea is a semi-closed sea connected with other seas by four straits : the Korea Strait between Korea and Japan, the Tsugaru Strait between Honshu and Hokkaido, the Soya Strait between Sakhalin and Hokkaido and the

26

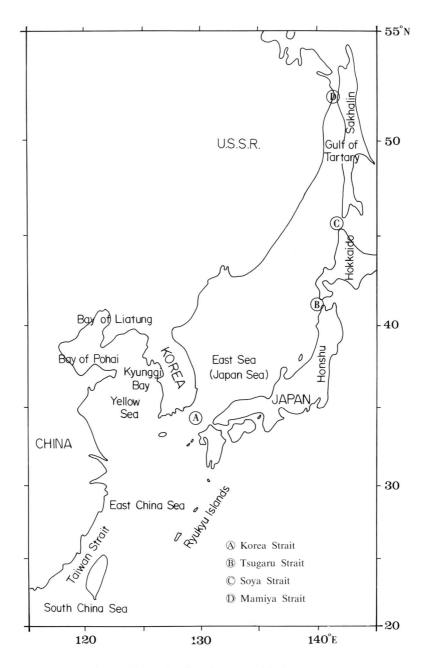

Figure. 1. Map showing the geographical names.

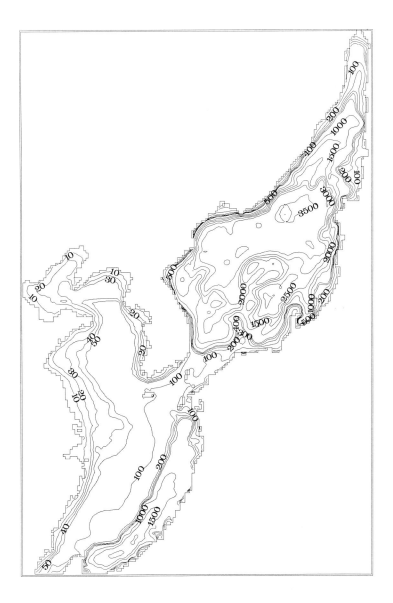

Figure. 2. Bathymetric map of the modelled area. Numerals show depth in meters.

Mamiya Strait between Sakhalin and the continent.

The most comprehensive work on the tides and the tidal currents in these boundary seas was done by Ogura (1933), who has edited the co-tidal M_2 and co-range (M_2+S_2) charts in these seas using the method of Proudman and Doodson (in Defant(1961)), based upon a considerable number of tidal measurements. And recently the co-tidal and co-range charts of semidiurnal and diurnal tides in the East China Sea, the Yellow Sea and the East Sea are redrawn with the new tide and tidal current data by Nishida (1980). The M_2 tidal charts are shown in Figure 3(a) and (b) following Nishida. According to the tidal chart of M_2 tide there exist four amphidromic points in the Yellow Sea and the East China Sea and two amphidromic points in the East Sea. It is worth noting that the large tidal ranges up to 6 m along the west coast of Korea occur while those of the East Sea are less than 0.2 m in the central part of the sea. In this paper the tidal charts by Nishida are used for the comparison between model-generated tidal charts and existing ones because Nishida's charts have an explicit expression of co-amplitude chart of M_2 while that by Ogura (1933) is for (M_2+S_2) co-range chart. Recently Fang (1986) also edited the tidal charts for M_2 tide based upon the observed tide on coast and islands. Especially the differences from those of Ogura and Nishida appear in the locations of the amphidromic points occurring in Liatung and Pohai bays.

As noted by Defant (1961), the tidal phenomenon of the entire East China Sea is almost exclusively conditioned by those water-masses which penetrate through the canals between the Ryukyu Islands within a half tidal period and then flow out again within the following half tidal period. Defant also described that the tides of this boundary seas (the East China Sea and the Yellow Sea) are essentially of Pacific origin as a result of qualitative analyses of water transport during 6 h of the M_2 tide through the canals of the Ryukyu Islands ($350\ km^3$), the northern entrance of the Taiwan Strait ($130\ km^3$) and the Korea Starit ($20\ km^3$). In this paper the qualitative assessment of water transport during 6 h of the M_2 tide through Taiwan Strait and Korea Strait as well as three other straits in the East Sea will be given based upon the computed results of the numerical model.

Besides the works mentioned above there have been several analytical or numerical studies to investigate the tides of the Yellow Sea and the East China Sea or the East Sea. An (1977) developed a Yellow Sea numerical model with the open boundary being established on a straight line from southern tip of Korea to Shanghai of China. He gave some explanation for the high tidal ranges in Kyŏnggi Bay in terms of its shallowness and a resonant oscillation of 10 hour period. He also commented that the amplitude due to the tide generating potential is very small (less than about 3 percent) compared to that of the boundary condition in the Yellow Sea.

An extensive study by numerical model for the tidal phenomena in the Yellow Sea and the East China Sea was carried out by Choi (1980). In his study he employed the modelling techniques based upon an explicit scheme. The model area in his study extends to the south as far as Nothern Taiwan (Formosa) and seaward as far as the edge of the continental shelf with the grid resolution 1/5° in latitude and 1/4° in longitude. In his study general amphidromic system in the Yellow Sea were reasonably reproduced while in the Liatung Bay it appears as deamphidromic pattern. Through the comparison of the computed tidal constants with the observed ones of M_2 tide he showed that the differences are approximately 10 percent in amplitude and 10° in phase.

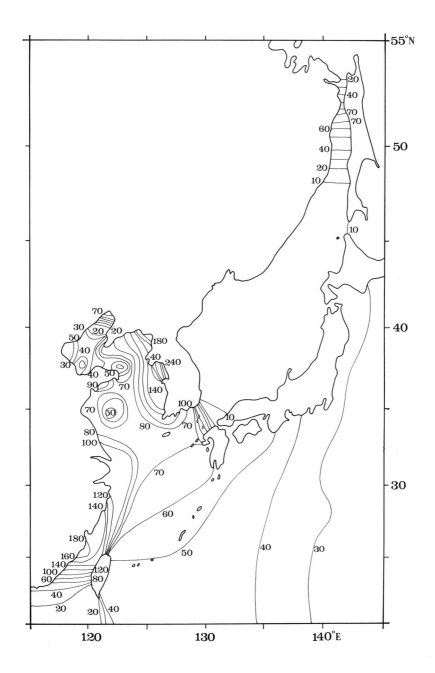

Figure. 3. Tidal charts for the M_2 tide in the seas around the Korea Peninsular (Nishida, 1980)
(a) Co-amplitude chart(amplitude : cm).

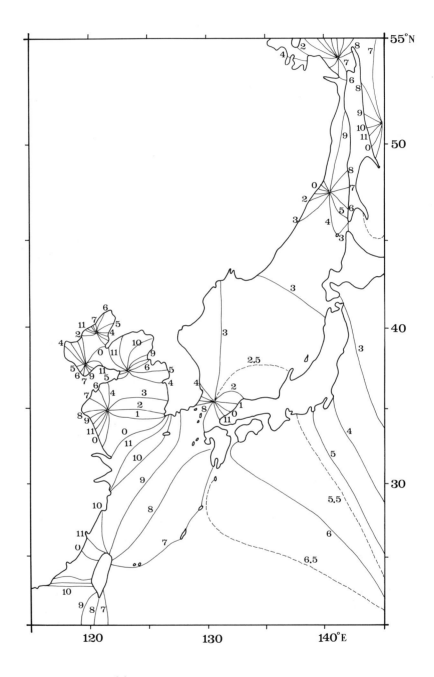

Figure. 3(b). Co-tidal chart of M_2 referred to 135° E.

But as the open boundaries in the study are located at the edge of the continental shelf, he had to use the tidal charts edited by Ogura (1933) for the open boundary input. He also does not include the tide generating force in governing equations. This tidal model has been further refined to resolve the flow over the continental shelf in more detail with the resolution of 1/15° in latitude and 1/12° in longitude in another two-dimensional study (Choi, 1988).

Kang (1984) has investigated the co-oscillating tides in the Yellow Sea south of Shantung Peninsula analytically by a superposition of Kelvin and Poincare waves. He explained that the large tidal range of Kyŏnggi Bay is due to the modifications of Poincare and Kelvin waves by Ongjin Peninsular. He also showed that the asymmetry of amphidromic system of M_2 tides arises primarily due to the partial penetration of tidal energy through the opening at the bay head. He also expected that the tidal energy dissipation by bottom friction might contribute to the asymmetry of the amphidromic system. But for mathematical simplicity the frictional effect to investigate the asymmetry was excluded in his analyses.

For the tide of the entire East Sea a hydrodynamical theory has been given by Ogura on the basis of the canal theory. Recently Odamaki (1989a) re-edited co-tidal chart in the Korea Strait on the basis of intensive observed data and the amphidromic point of the M_2 are similar to Ogura's result while the amphidromic points of the K_1 and O_1 tides are remarkably shifted toward the Korean coast compared to Ogura's charts. Odamaki (1989b) also investigated the generation of tides in the East Sea with relation to the tidal volume fluxes at the attached straits, which are estimated with the observed tidal current data. In his study he separated the tides into the co-oscillating tides induced by tidal volume fluxes and the independent tide by tide-generating force using an one-dimensional tidal model. Through his study he estimated that the amplitude of the independent tide in the Tartary Bay is the same as that of the co-oscillating tide attributed to the Korea Strait. But due to the limitation of the one dimensional study the effect of tide generating force on the amphidromic system could not be realistically investigated and the contribution of the Mamiya Strait to the tidal regime of the East Sea was ignored by letting the strait closed.

In this study the M_2 tide in the entire surrounding seas of Korea Peninsula is investigated to understand which factors determine the amphidromic system. To meet the requirement that nearly same accuracy should keep over the largely different depth diifference in model area, the two dimensional numerical model employing implicit scheme is used in this study. And as the model region is including both the shallow and deep water depths the tide generating force is introduced to see what different effect it has on each sub-model area. The model region toward south and south-eastern boundaries is also extended to Ryukyu Islands for the model to utilize the measured data along the islands and to avoid possible error that might occur when using the edited tidal charts. Also the four straits surrounding the East Sea are set open to study possible contribution to the tidal regime in the East Sea.

2. A TWO-DIMENSIONAL TIDAL MODEL

2.1 *Governing equations*

The governing equations are described under the spherical polar coordinate system to consider the curvature of the earth and the variation of Coriolis force with latitude and the variation of the generating forces with latitude as well as longitude. The usual depth averaged two-dimensional equations of shallow water waves can be alternatively represented in terms of depth-integrated velocities as follows :

$$\frac{\partial U}{\partial t} + \frac{gh}{R\,\cos\phi}\,\frac{\partial}{\partial\chi}(\eta - (1 + k_2 - h_2')E) = \frac{UV\tan\phi}{Rh} + 2\omega\sin\phi V - \frac{k_b U\sqrt{U^2 + V^2}}{h^2} \qquad (1)$$

$$\frac{\partial V}{\partial t} + \frac{gh}{R}\,\frac{\partial}{\partial\phi}(\eta - (1 + k_2 - h_2')E) = \frac{V^2\tan\phi}{Rh} - 2\omega\sin\phi U - \frac{k_b V\sqrt{U^2 + V^2}}{h^2} \qquad (2)$$

$$R\cos\phi\,\frac{\partial\eta}{\partial t} + \frac{\partial U}{\partial\chi} + \frac{\partial(V\cos\phi)}{\partial\phi} = 0 \qquad (3)$$

where χ is longitude, ϕ latitude, h water depth, η surface elevation, R radius of the earth, $k_b = g/C^2$, g gravity acceleration, C=Chezy coefficient with uniform value or $(h + \eta)^{1/6}/n$ with n constant. The term $(1 + k_2 - h_2')E$ in (1) and (2) represents tide generating forces. The modification factor by earth tide, $(1 + k_2 - h_2')$ is set to 0.69 following Pingree and Griffiths (1987), and k_2 and h_2' are Love numbers.

And $E = He\times\cos2\theta\times\cos(\omega t - PL + \omega S)$, in which He, ω, P, L and S denote the amplitude of the equilibrium tide for the M_2, angular frequency, species number (2 for M_2 component), longitude, latitude and standard time, respectively.

In this study the convection terms are ignored as the main purpose of this study is mainly to investigate the amphidromic system. The bottom friction term has been denoted as quadratic friction form following the general applicability of the quadratic friction law for semi-diurnal tide modelling(Pingree and Griffiths, 1987).

2.2 *Numerical method and computation*

i) *Numerical method.* The governing equations were solved using the linearized Abbott (1979) type scheme with four component equation of 2-dimensional equations which is an implicit scheme and has second order accuracy in space and time. In the computation the two equations that is, x-momentum and continuity equation are alogrithmically connected in a first stage in such a way as to advance U directly from $n\delta t$ to $(n+1)\delta t$ while advancing η only from $n\delta t$ to $(n+1/2)\delta t$. The second stage connects y-momentum equation and continuity equation algorithmically to advance V directly from $n\delta t$ to $(n+1)\delta t$ while advancing η from $(n+1/2)\delta t$ to $(n+1)\delta t$. In each stage the equations are solved using the double sweep method. The more detailed description has been given by Abbott (1979). Due to the large differences of depth of each sea in the model area and the existence of a deep water depth in the East Sea the proper choice of time interval may be important and from this point this scheme is efficient in choosing the time interval (δt) than the explicit scheme which should satisfy the Courant-Friedrichs-Lewy condition.

Letting mean depths of the East Sea and the Yellow Sea 2,500 m and 500 m, repectively, the Courant number ratio is about 9 for the chosen time interval and grid size. The characteristics of the scheme used in the region with basins of greatly different mean depths should show nearly same accuracy for the large different Courant number. The characteristics of Abbott type scheme has been thoroughly investigated by Sobey (1970), Abbott (1979) and Abbott et al. (1981).In Figure

4 the phase portrait of the scheme used in this study is shown following Abbott (1979).

The amplification factor is 1 for all wave number and the celerity ratio is not unity, which is functions of the direction of propagation relative to grid line. This figure shows phase portraits in phase shifts as function of numbers of points per wave length when propagation occurs at angle $\lambda = 45°$ to the grid lines, x or y. Also it is shown that in order to use the capabilities of the scheme for working at the high Courant number, something in the range of 20-100 points per wave length is required for the main wave component. As shown in Figures 3(b) and 5 (grid system), the typical grid points per wave length in the Pohai Bay and the East Sea are about 80 and several hundreds, respectively. And the chosen time interval (δt) is about 175 s and the mean interval of grid space between h and U (or V) is about 7,000 m. The mean depth in the Pohai Bay and the East Sea is about 20 m and 2,500 m, respectively, so that the Courant number is 0.5 and 5.5, respectively. It is therefore expected that for such ranges of Courant number the scheme used in this study is expected to be sufficient enough to reproduce the accurate amplitude and phase of M_2 tide from the theoretical point of view.

ii) *computation*. The initial and boundary conditions required for the solution of equations (1) to (3) are described in what follows. At $t = 0$, $U(\chi, \theta, t)$, $V(\chi, \theta, t)$ and $\eta(\chi, \theta, t) = 0$ are specified at every points. At a land boundary, the component of the flow normal to the boundary is zero. The open boundaries consist of five sections ; section along the Ryukyu Islands, section along the Taiwan Strait and three sections in the East Sea, that is, the Tsugaru Strait, the Soya Strait and the Mamiya Strait. Along an open boundary, elevation is specified as a function of time. For the M_2 tidal constituent the open boundary values are given by

$$\eta(\chi, \theta, t) = H_{M2}(\chi, \theta) \cos(\omega_{M2}t - g_{M2}(\chi, \theta)) \tag{4}$$

where g_{M2} is local phase lag referred to 135° E and H_{M2}, ω_{M2} are amplitude and angular frequency of M_2 component, respectively. The values of tidal constants (H_{M2} and g_{M2}) are directly obtained from measured data not from the tidal chart. By using the measured data as bourdary condition with some interpolatin we can avoid the possible error that could occur when we use tidal chart. But in this study the inner disturbances due to tide-generating force were not permitted to radiate through the open boundaries under the assumption that the effect is limitted around the boundary. The grid resolution is $1/8°(2\delta\chi)$ in longitude and $1/6°(2\delta\phi)$ in latitude. the time interval is 174.6647s with which one M_2 period corresponds to 256 time steps. To check the convergence of numerical solution 13 cycle run was made. The grid system and the locations of 12 checking points for elevation and five strait sections for the compution of tidal volume fluxes are indicated in Figure 5.

3. RESULTS AND ANALYSIS

A series of numerical experiments were carried out to simulate the M_2 tidal distribution in order to reach reasonable agreemet with tidal charts newly edited from additional observations (Nishida, 1980). The various bottom friction parameters were used, that is, with the spatially uniform Chezy value or nonuniform Chezy value with respect to water depth at each grid point. At this stage the overall reproducation of M_2 tidal distribution with uniform frictional factor is shown due to its relatively more realistic reproduction of tidal charts than using a nonuniform parameter.

In Figure 6 the computed elevations at 12 grid points are shown to confirm the fact that

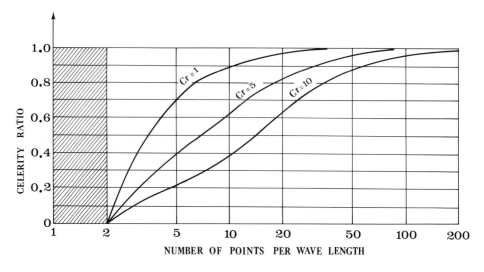

Figure. 4. Phase portrait for propagation along direction at 45° to grid lines for linearized Abbott
type scheme(Abbott(1979)).

Table. 1. Amplitude of tidal volume flux through straits in the model region.

(unit : × $10^6 m^3/s (=1\ Sv)$)

Strait	Amplitude
Korea	5.34
Tsugaru	1.93
Soya	0.54
Tartary	0.09
Taiwan	2.64

Table. 2. Inflow and Outflow of volume transport in the model region during one M_2 cycle.

(unit : × $10^{10} m^3/s$)

Strait		Inflow	Outflow
Korea	Eastern channel	3.714	3.677
	Western channel	3.922	3.936
	Sum	7.633	7.608
	Tsugaru	2.875	2.638
	Soya	0.891	0.684
	Tartary	0.140	0.122
Sum in the East Sea boundary		11.539	11.052
	Taiwan	3.828	3.687

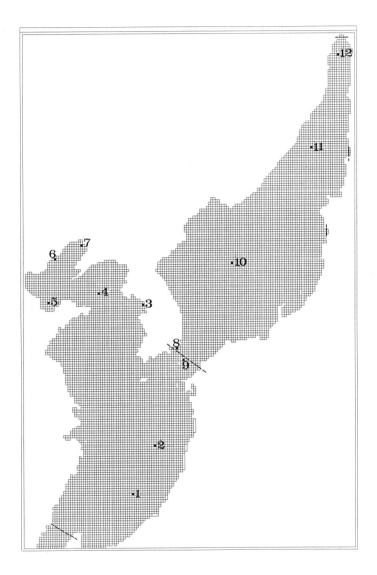

Figure. 5. The location of the checking points(1-12) for numerical soultion's convergence and sections
for computing the tidal volume fluxes and volume transport through five straits.

the computed results reach stationary state around 7th cycle after the start of calculation from the initial state and to see the different amplifications at various sub-sea points. The short period oscillations superposed on the major oscillation are noted at points 1 and 2 which are located in continental slope. These are thought to be due to the generation of short period component associated with the abrupt change of bottom topography rather than numerical instability. The large tidal amplitude at the point 3 of Kyŏnggi Bay occurs as expected. At the points 4, 5, 6, 7 located Seohan, Pohai, Liatung bays relatively different converged results are shown especially with the strong nonlinear characteristic occurring in grid point 6. The elevations at points 9, 10, 11, 12 in the East Sea are shown to converge around 7th cycle while in point 11 the nonlinear characteristic is shown. The results were analysed for M_2 constituent by the Fourier analysis.

3.1 *Comparison of model-generated tidal charts with existing tidal charts*

Figure 7(a) and 7(b) show the model-generated co-amplitude and co-tidal charts of the M_2 tide. The results have been simulated with inclusion of tide generating forces with friction parameter $k_b = 0.0025$. The general patterns of the computed charts are in good agreement with the existing ones in Figure 3(a) and 3(b). The four amphidromic points in the Yellow Sea and the East China Sea as well as two amphidromic points in the East Sea have been reasonably reproduced. Also 8 hour co-tidal line as well as co-range line located nearly along the continental slope has been fairly well simulated without any serious oscillation. The large tidal ranges in the Kyŏnggi Bay west coast of Korea are also reasonably generated. The low tidal ranges in the almost entire area of the Eas Sea due to the deep water depth are also apparent in the model results. But some deviations of co-tidal hour in the bays of Liatung and Pohai are shown as well as a little northward movement of co-amplitude line of 10 cm in the Tartary Bay of the East Sea.

For the locations of amphidromic points occurring in the bays of Liatung and Pohai the computed ones move more westward or northwestward, compared with those of Nishida (1980) or Ogura (1933), but appearing more consistent with the results by Fang (1986). Choi's fine grid model study (1988) also show that the amphidromic point in Liatung Bay appears in deamphidromic pattern while the amphidromic point in Pohai Bay moves northwestward similarly in this study. From these observational and numerical results the more thorough analysis based upon measurement seems to be required to confirm where and how the amphidromic system exists.

A little severe dissipation of tidal amplitude at Tartary Bay appearing as a little northward movement of co-amplitude line of 10 cm and slightly westward movement of amphidromic point in Korea Strait are thought to be partly due to the relatively high bottom dissipation by using uniform friction parameter over the whole model area irrespective of the difference in water depth between the seas. In fact, when considering the bottom friction factor as variable of depth (the form of $(depth)^{1/6}$), the friction factor ratio is 2 to 1 for the depths 50 m and 250 m for the Yellow Sea and the East Sea, respectively. In fact, the response for the less friction factor is also studied and, as expected, such effects appear though not shown in this study.

The ocean current, a branch of Kuroshio, in the East Sea is also expected to play a role to the movement of amphidromic system located at the Korea Strait through wave current interaction.

3.2 *The effect of tide generating forces*

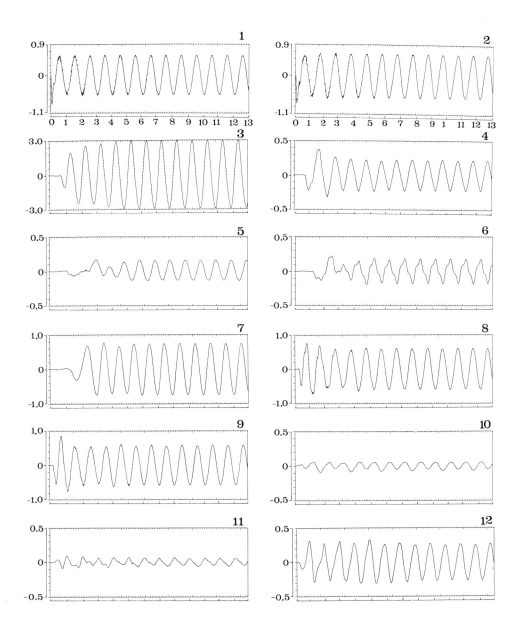

Figure. 6. Temporal variation of the computed surface elevation in meters at 12 points for 13 cycles of M_2 tide(time unit in tidal period).

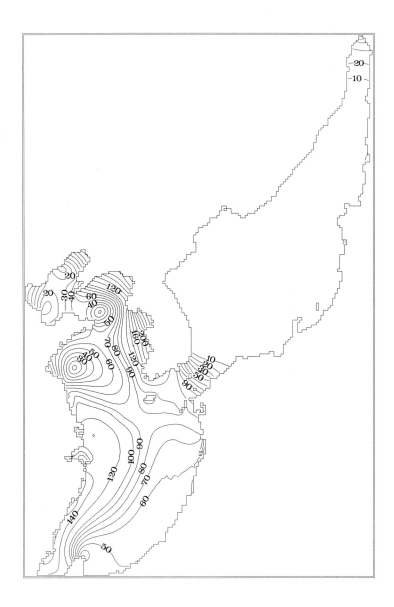

Figure. 7. Model-generated tidal charts with tide generating force included. (a) co-amplitude chart (amplitude : cm).

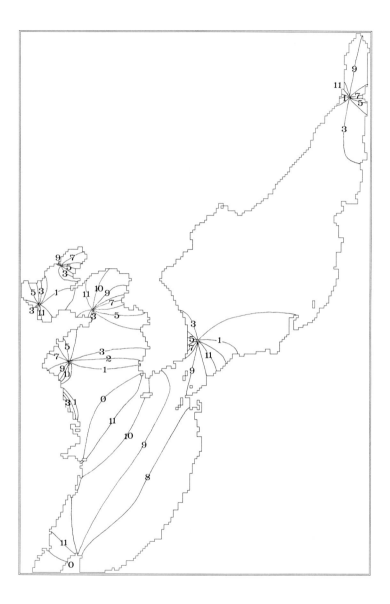

Figure. 7(b). Co-tidal chart referred to 135° E.

Figure 8(a) and 8(b) show the computed co-tidal co-amplitude lines without tide generating forces with an uniform friction parameter (k_b=0.0025). The difference of co-tidal lines between Figure 7(b) and 8(b) is obvious in the results of the East Sea where without tide-generating force the amphidromic point in the Tartary Bay disappears while the amphidromic point just above the Korea Strait moves unrealistically toward northwest direction compared with the results including tide generating force and with existing tidal chart (Nishida (1980)). In case of co-amplitude lines (Figure 7(a) and 8(a)) any significant differences in the Yellow Sea and the East China Sea are not noticeable probably due to minor changes while some differences at the Tartary Bay of the East Sea are shown.

To see the effect of tide generating forces upon the tidal dynamics of the modelled area the pure contribution of independent tide due to tide generating force has been computed by analysis of the differences of elevations with tide generating force and without it have been compared. Figure 9(a) and (b) show the model generated co-amplitude and co-tidal lines due to the independent tide by tide generating force. As shown in Figure 9(b), the independent tide in the Yellow Sea shows the pattern that the equilibrium tidal wave propagates from south and south-east direction to nothern direction. The amphidromic points by independent tide also appear at the nearly same locations in the existing tidal chart except the amphidromic point located in the Gulf of Pohai.

In the East Sea the independent tide propagates from central area both to north-eastern direction and south-western directions. Figure 9(a) shows the computed co-amplitude lines by the independent tide in the model area. In the Yellow Sea and the East China Sea the large amplitudes up to 11 cm due to the tide generating force appear in the west coast of Korea as in the co-oscillating tide, but in the Tartary Bay of the East Sea the amplitudes due to the independent tide increase gradually and reach up to about 25 cm, which is almost the same order of magnitude of tide flowing into the East Sea through Korea Strait, as also discussed by Odamaki (1989b). The independent tide in the distribution of amplitudes accounts for about 1-4% in the Yellow Sea and the East China Sea while in the East Sea it explains more than 10% over the nearly whole area, and in the Tartary Bay up to about several tens of percent.

Through these analysis the independent tide has been shown to be very important over the whole area in the East Sea. Especially it is worth noting that the amphidromic point located in the Tartary Bay does not appear without the independent tide even though the bay head is open. Therefore it could be inferred that the reflected Kelvin wave at the head area of the Tartary Bay is more reinforced by tide generating force than the reflected wave formed only by the interaction of incoming wave through Korea Strait with the waves imposed on the open boundaries of the Mamiya Strait.

3.3 Tidal volume fluxes through five straits

The tidal volume fluxes through five straits (see Figure 5 for the location) in modelled region have been computed and the time variations of tidal volume fluxes through each strait are plotted in Figure 10, which shows that the volume fluxes passing through the Taiwan, the Korea, the Mamiya straits reach stationary state during 7 or 8th cycle while in the Tsugaru and the Soya straits they reach stationary state beyond 10th cycle. The amplitude of tidal volume fluxes and the volume transports during one M_2 tidal cycle have been listed in Table 1 and Table 2. The amplitude of

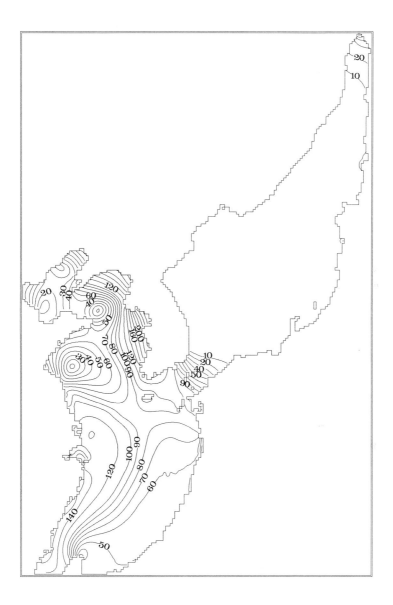

Figure. 8. Model-generated tidal charts with tide generating force excluded. (a) co-amplitude chart (amplitude : cm).

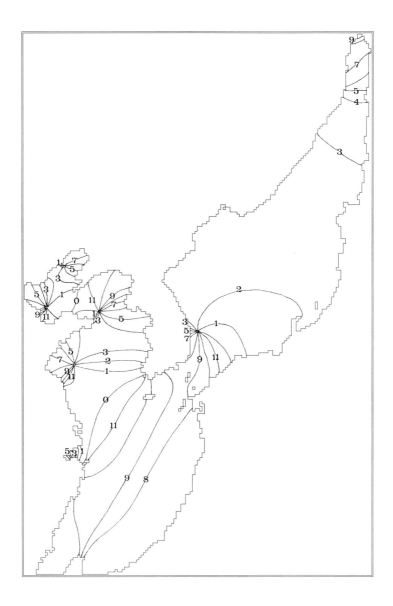

Figure. 8(b). Co-tidal chart referred to 135° E.

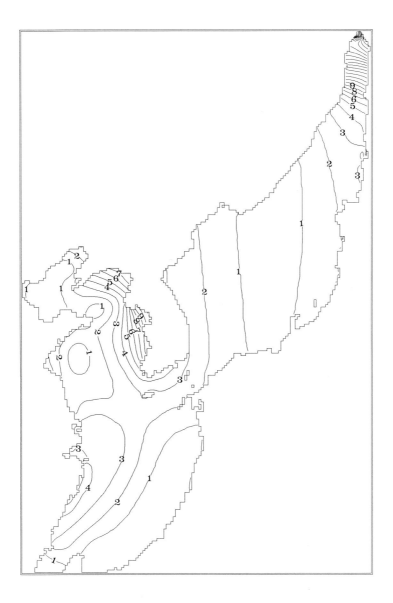

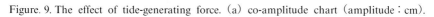

Figure. 9. The effect of tide-generating force. (a) co-amplitude chart (amplitude : cm).

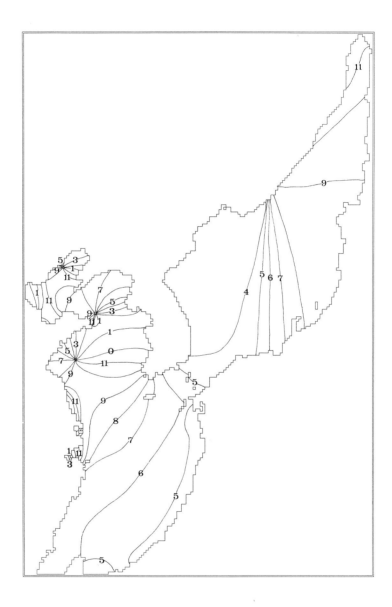

Figure. 9(b). Co-tidal chart referred to 135° E.

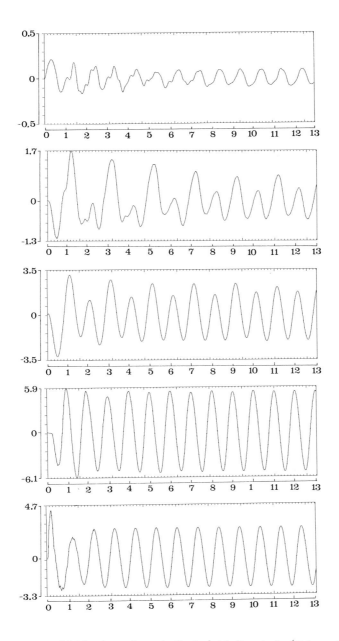

Figure. 10. Convergence of tidal volume fluxes in Sv throutgh five straits (Taiwan, Korea, Tsugaru, Soya, Tartary Straits in order of bottom to top. Time unit in tidal period).

tidal volume flux at the Taiwan Strait is about 2.64 Sv ($=10^6$ m^3/s) and the volume transport during half period of M_2 is about 38 km^3, but the ratio of each strait to total volume flux are similar to those estimated by Odamaki.

The inflow volume transport into the East Sea for about half M_2 tidal cycle through the Korea, the Mamiya straits are 76.3 km^3, 1.4 km^3 while those through the Soya and the Tsugaru straits are 8.9 km^3 and 28.8 km^3. So it could be said that the co-oscillating tide through the Korea Strait is dominant for the M_2 tide, as noted by Defant (1961). The contribution by the Tsugaru Strait also seems to be influential considering the amount of volume transport. The amounts of inflow and outflow through the Korea Strait have been computed to be almost the same as each other. In contrast to the Taiwan Strait the volume transport through the Korea Strait is about 5 times as large as that computed by Defant(1961). One of the reason of this difference is expected to be partly due to the possible difference of the location of section where the tidal amplitudes change rapidly from place to place for computing volume transport.

4. CONCLUSION AND DISCUSSIONS

Through this study it has been shown that the tidal phenomena in the entire seas surrounding Korea Peninsular could be investigated under one oscillating system of the whole seas using a two-dimensional depth averaged numerical model. In this study southern part of open boundaries were extended to Ryukyu Islands for the boundary value to be obtained from the observed data. Due to the large mean depth difference between each sea the tide generating force has been included in the computation. To overcome the problem related to time interval limitation by CFL condition in case of using the explicit method the implicit scheme has been successfully employed in this study.

The computed M_2 co-amplitude lines in the Yellow Sea and the East China Sea are in good agreement with those of existing tidal charts. The four amphidromic points known to exist in the Yellow Sea and the East China Sea as well as two amphidromic points in the East Sea are successfully reproduced with inclusion of tide generating forces. Even along the continental slope where rapid depth changes occur the computed results have been founded to be in good agreement with the observed results. The results of computation also show that tide generating forces play greatly different roles in the each sub-sea of the model region. The tidal amplitude due to the tide generating force in the Yellow Sea and East China Sea explains several percent of the real observed amplitude while the tidal regime in the East Sea is greatly influenced by the tide generating force. It might be inferred that the reflected waves at the head area of the Tartary Bay is rather reinforced by the tide generating force.

In the Tartary Bay of the East Sea the experiment letting the head as open boundaries was found to give more reasonable results for the proper simulation of the amphidromic system than letting the head closed, even though the tidal volume fluxes through the Mamiya Strait is very small compared with those passing through other straits, as shown in the section 3.3. In case a bay with its head opening is too narrow and too deep for Poincare modes to propagate semi-diurnal energy, Hendershott and Speranza (1971) showed that the co-oscillating tide must be primarily a superposition of oppositely travelling waves and that an asymmetry of amphidromic system in such a bay results from partial absorption of incident power flux at the bay head. In case of the

Tartary Bay, in contrast to the Yellow Sea investigated by Kang (1984), it has a step in depth along the bay from 1,000 m to 100 m and the amphidromic point is located on the continental slope, which step as well as bottom friction make it difficult to investigate the amphidromic system analytically. The critical period of the first Poincare wave, T_c is as follows (see Kang (1984)).

$$T_c = \frac{2\pi}{\left\{ f^2 + \frac{gHm^2\pi^2}{B^2} \right\}^{1/2}} \tag{5}$$

where f, g, H, m, B are Coriolis parameter, gravity acceleration, depth, cross channel mode number of Poincare wave and bay width, respectively. For depth of 100 m, 500 m, 1,000 m and width of 150 km the critical periods are 2.6, 1.2, 0.8 hr, respectively. So, even though considering the depth slope, it could be said that the co-oscillanting tides in the Tartary Bay consist of Kelvin waves. The possible asymmetry in the geometry condition like the Tartary Bay is first expected to occur due to a partial absorption of incident power flux at the bay head (Hendershott and Speranza (1971)) and the energy dissipation by bottom friction (Rienecker and Teubner (1980)). To check the role of head opening another experiment has been made with the head closed. The computed results even with tide generating force showed that amphidromic points within realistic range of bottom friction factor, seems to move inside the land. When considering that the head opening has the effect of letting amphidromic point move westward in the nothern hemisphere it is thought that the driving force at the head of the Tartary Bay plays some role in the amphidromic system as well as the tide generating force, as discussed in section 3.2.

For the bottom friction the quadratic form was found to give general agreements with observed ones even th some evidence of a little excessive dissipation appears in the results of the East Sea area, wh.ch was also shown through simple analysis. Further studies for more accurate reproduction of tidal amplitude and phase seem to be needed with relation to bottom friction.

In this study the nonlinear phenomena in the modelled area have not been investigated as well as not including the study on the tidal regime of diurnal component. There topics will be considered in the next study associated with a series of studies for investigating the tidal regime in these seas.

ACKNOWLEDGEMENTS

The authors would like to thank Drs. H. J. Lie (Korea Ocean Research and Development Inst.), Y. H. Seung (Inha Univ.) and Y. Q. Kang (Pusan Fishery Univ.) for their reading manuscripts and valuable comments. We appreciate Dr. B. H. Choi for review and helpful comments. Thanks are also due to Mr. W. D. Paik for drawing figures. The work was partially funded by the Ministry of Science and Technology and the computations by Cray2s was supported by System Engineering Research Institute, KIST.

REFERENCES

Abbott, M.B., 1979. Computational Hydraulics : Element of the Theory of Free-Surface Flows, Pitman, London.

Abbott, M.B., A. McCoWan, and I. R. Warren, 1981. Numerical modelling of free-surface flows and coastal waters, Ed. H.B. Fisher, Academic Press, New York.

An, H.S. 1977. A numerical experiment of the M_2 tide in the Yellow Sea. Jour. of the Oceanogra. Soc. of Japan, 33 : 103-110.

Defant, A. 1961. Physical Oceanography, Vol. 2.

Choi, B.H. 1980. A tidal model of the Yellow Sea and the Eastern China Sea. KORDI Rep. 80-02, Korea Ocean Research and Development Institute, Seoul 72 pp.

Choi, B.H. 1988. A fine grid two-dimensional M_2 tidal model of the East China Sea. Jour. of Korean Assoc. of Hydrol. Sc., 21(2) : 183-192.

Fang, G. 1986. Tides and tidal currents in the marginal seas adjacent to China. Presented at tenth international symposium on earth tide. Madrid.

Hendershott, M.C. and A. Speranza. 1971. Co-oscillating tides in long, narrow bays ; the Taylor problem revisited. Deep-Sea Res., 18 : 959-980.

Kang, Y.Q. 1984. An analytical model of tidal waves in the Yellow Sea. Jour. Mar. Res., 42 : 473-483.

Nishida, H. 1980. Improved tidal charts for the western part of the north Pacific Ocean, Rep. of Hydrogra. Res., No. 15.

Odamaki, M. 1989a. Tides and tidal currents in the Tsusima Strait. Jour. of the Oceanogra. Soc. of Japan, 45 : 65-82.

Odamaki, M. 1989b. Co-oscillating and independent tides of the Japan Sea. Jour. of the Oceanogra. Soc. of Japan, 45 : 217-232.

Ogura, S. 1933. The tides in the Seas adjacent to Japan., Hydrogr. Bull. Dep. Imp. Jap. Navy, 7, 1-189.

Pingree, R.D. and D.K. Griffiths. 1987. Tidal friction for semidiurnal tides. Continental Shelf Res., 7(10) : 1181-1209.

Rienecker, M.M. and M.D. Teubner. 1980. A note on frictional effects in Taylor's problem. Jour. Mar. Res., 38 : 183-191.

Sobey, R.J. 1970. Finite difference schemes compared for wave-deformation characteristics ets., Tech. Memor. No. 32, US Army Corps of Engineers, Coastal Eng. Res. Center, Washington, D.C.

NONLINEAR ROSSBY WAVES IN THE INERTIAL BOUNDARY CURRENT AND THEIR POSSIBLE RELATION TO THE VARIABILITY OF THE KUROSHIO

Qinyu Liu and Zenghao Qin
Institute of Physical Oceanography, Ocean University of Qingdao, Qingdao 266003 (China)

ABSTRACT
Based on a qualitative theory of ordinary differential equations, the stability characteristics of nonlinear barotropic Rossby waves propagating in the western inertial boundary current of the ocean is discussed with geostrophic momentum approximation. A criterion is obtained about the occurrence of the bifurcation in the inertial boundry current, which may be applied to the variability of the Kuroshio.

INTRODUCTION

In recent decades, remarkable advances and achievements have been made in the study of the large scale ocean waves and their instabilities,especially of the linear barotropic and baroclinic Rossby waves. Some phenomena closely related to the large scale ocean waves have been clarified. Pedlosky(1979) has systemati- cally studied the linear Rossby wave instability under various circumstances. Kang et al.(1982) have studied it in a non-zonal shearing ocean current. However, very few authors have dealt with the nonlinear Rossby waves and their instabilities in the oceans. This is because not only the nonlinear problem is mathe- matically difficult to deal with but also the nonlinear terms can be neglected in most cases in large-scale ocean motions.

Within the inertial boundary current zone in the western boundary of the ocean, e.g.,the Kuroshio and Gulf Stream, the current velocity is usually one order of magnitude greater than that in other regions, and in some processes nonlinear terms in the momentum equations can not be omitted. Because the western boundary region is mostly over the continental shelf and the sea-floor topography is quite complicated , the interactions between the currents and those between waves can not be neg- lected. The inertial boundary current results from the combin- ation of the westward-propagating Rossby waves obstructed by the western boundary and the beta-effect (Pedlosky,1979). Hence, the study of the nonlinear Rossby waves in this region is helpful to understand the dynamical mechanism of inertial boundary currents such as the Kuroshio and Gulf Stream.

Liu et al. (1983, 1987) have studied nonlinear waves both in the atmosphere and ocean using stability theory of ordinary differential equations. However, their studies ignore the bottom topography, basic current and frictional effects. The present authors (Liu and Qin,1990) have studied the topographically trapped waves and the nonlinear waves in the tropical ocean and atmosphere, which are, however, not applicable for the non-linear waves in the inertial boundary currents.

An attempt is made to study nonlinear Rossby waves in the inertial boundary current and to explain to some degree the Kuroshio variability.

2 NONLINEAR BAROTROPIC ROSSBY WAVES

In order to generalize the discussion of the inertial boundary current, the local Cartesian coordinates, x, y, z are chosen : the y-axis coincides with the straight coastline and x-axis lies perpendicular to it with positive direction pointing off-shore and z-axis goes upward pointing to the local zenith. The plane oxy coincides with undisturbed sea surface and η (x, y, t) is the free surface elevation above the equilibrium level. Assuming the bottom friction to be linear for the velocity component with bottom resistance coefficient r and denoting the constant basic inertial boundary current by $\overline{V}$ satisfying

$\overline{V}=(g/f) \partial \eta/\partial x > 0$, we get the nonlinear equations for barotropic

motion of an inviscid, homogeneous and incom-pressible water overlying the uneven sea-bed z = -h(x,y) :

$$(\partial/\partial t + u\partial/\partial x + v\partial/\partial y + \overline{V}\partial/\partial y) u - fv = -g\partial\eta/\partial x - ru \qquad (1)$$

$$(\partial/\partial t + u\partial/\partial x + v\partial/\partial y + \overline{V}\partial/\partial y) v + fu = -g\partial\eta/\partial y - rv \qquad (2)$$

$$(\partial/\partial t + \overline{V}\partial/\partial y) + (u\partial/\partial x + v\partial/\partial y)(\overline{\eta} + h + \eta)$$

$$+ (\overline{\eta} + h + \eta)(\partial u/\partial x + \partial v/\partial y) = 0 \qquad (3)$$

where u and v are the depth-averaged components of the horizontal velocity in the x- and y-directions, respectively.

The Coriolis parameter f is assumed to be $f = f_0 + \beta_x x + \beta_y y$, where $\beta_x = \beta \sin \psi$, $\beta_y = \beta \cos \psi$, f_0 is the Coriolis parameter and β is its northward gradient at a prescribed latitute and ψ is the counterclockwise angle of the x-axis from the east. The other symbols in Eqs (1)-(3) are usual.

The geostrophic momentum approximation (Hoskins,1975) is made to filter out the inertia-gravity waves which is inherent in the model and simultaneously to retain the nonlinear effect unchanged, i.e.

$$\partial v/\partial x - \partial u/\partial y = (g/f)(\partial^2 \eta/\partial x^2 + \partial^2 \eta/\partial y^2)$$

$$\beta_y v + \beta_x u = (g/f)(\beta_y \partial \eta/\partial x - \beta_x \partial \eta/\partial y)$$

so that Eqs. (1) and (2) can be combined into the following vorticity equation

$$(\partial/\partial t + u\partial/\partial x + v\partial/\partial y + \overline{V}\partial/\partial y + r)(\partial^2 \eta/\partial x^2 + \partial^2 \eta/\partial y^2) + \beta_y \partial \eta/\partial x$$

$$-\beta_x \partial \eta/\partial y + (f_0^2/g)(\partial u/\partial x + \partial v/\partial y) = 0 \qquad (4)$$

The term $(\partial v/\partial x - \partial u/\partial y)(\partial u/\partial x + \partial v/\partial y)$ is negligibly small in

comparison with term $f_0(\partial u/\partial x + \partial v/\partial y)$, so that it is omitted.

For Rossby wave of wave numbers k and l in x- and y-directions and angular frequency ω , solutions of Eqs. (3) and (4) are of form

$$u=U(\vartheta), \quad v=V(\vartheta), \quad =H(\vartheta) \qquad (5)$$

where $\vartheta = kx+ly-\omega t$, $\omega \neq l\overline{V}$, and the parameter $\overline{V}$ is assumed to be independent on ϑ. Eqs. (3) and (4) are then transformed to

$$(-\omega + kU + lV + l\overline{V})(k^2 + l^2)H''' + (\beta_y k - \beta_x l)H' + r(k^2 + l^2)H''$$

$$+ (f_0^2/g)(kU + lV)' = 0 \qquad (6)$$

$$(-\omega + 1\overline{V})H' + [(kU + 1V)(H + h + \overline{\eta})]' = 0 \qquad (7)$$

Here, the superscript denotes the differentiation with respect to θ. Integrating Eq. (7) and putting the integral constant to be zero, we have

$$(kU + 1V) = H(\omega - 1\overline{V})/(H + h + \overline{\eta}) \qquad (8)$$

Inserting Eq. (8) into (6) results in

$$H''' = PH'' + QH' = F(H,H',H'') \qquad (9)$$

where

$$P = r(H + h + \overline{\eta})/[(\omega - 1\overline{V})(h + \overline{\eta}), \quad Q = P(\beta_y k - \beta_x 1)/(k^2 + 1^2)r$$

$$+ (f_o^2/g)/[(k^2 + 1^2)(H + h + \overline{\eta})] \qquad (10)$$

The zero solution or equilibrium points of the ordinary differential equation (9) are $H = H' = H'' = 0$, which is equivalent to the average inertial boundary current with the corresponding uneven sea surface $\overline{\eta}$ and without perturbed u, v and η.

F(H, H', H'') is an analytical function if the conditions $\omega \neq 1\overline{V}$, and $H \neq -(h + \overline{\eta})$ hold good. Eq.(9) can be replaced by its equivalent set of equations:

$$dH/d\theta = H'$$

$$dH'/d\theta = H'' \qquad (11)$$

$$dH/d\theta = P(H) H'' + Q(H) H'$$

Consequently, there are some equilibrium points $(H_o, 0, 0)$ in the phase plane in which H_o depends upon the values of P and Q. The zero solution must be the equilibrium point no matter what the values of P and Q will be.

Expanding Eq. (9) in Taylor's series near the point (0, 0, 0) leads to

$$H''' = Q_1 H' + Q_2 H'' + Q_3 HH' + Q_4 HH'' + \ldots \ldots \qquad (11)'$$

where $Q_1 = (\beta_y k - \beta_x l)/[(\omega - l\bar{V})(k^2 + l^2)] + (f_o/g)/[(k^2 + l^2)(h + \bar{\eta})]$

$, Q_2 = r/(\omega - l\bar{V}), \quad Q_3 = (\beta_y k - \beta_x l)/[(\omega - l\bar{V})(k^2 + l^2)(h + \bar{\eta})]$

$- f_o^2/[g(k^2 + l^2)(h + \bar{\eta})]$

$$Q_4 = r/[(\omega - l\bar{V})(h + \bar{\eta})]$$

Eq.(11) is a nonlinear ordinary differential equation, which is difficult to solve. If the nonlinear terms higher than the third order and the friction terms are neglected, Eq. (11) can be replaced by the equation:

$$H''' = Q_1 H' + Q_3 HH' \qquad (12)$$

which is a KdV equation provided that the horizontal variations of h and are $\bar{\eta}$ neglected. Its solution is an elliptical cosinoidal function (Liu, 1983) and its frequency is related not only to the wave length, ocean depth, basic current and coast orientation, but also to the wave amplitude.

When the friction is taken into account, the derivation of the solution would be quite difficult. Now let's use the stability theory of the ordinary differential equation to explore the characteristics of the solution of Eq. (9).

3 STABILITY AND BIFURCATION

Eq.(9) can be approximated by a nonlinear equation

$$H''' = Q_1 H' + Q_2 H'' + Q_3 HH' + Q_4 HH'' \qquad (13)$$

The stability in the vicinity of the zero solution of Eq. (13)

within the phase space (H, H', H'') can be studied from the stability theory of the ordinary differential equation.

First of all, let's discuss the stability of the linearized Eq. (13)

$$H''' = Q_1 H' + Q_2 H'' \qquad (14)$$

Its corresponding characteristic equation is

$$\lambda^2 - A\lambda + B = 0 \qquad (15)$$

where $A = Q_2$, $B = -Q_1$.

Set:

$$h_c = (f_0/g)(\omega - 1\bar{V})/(\beta_x 1 - \beta_y k) - \bar{\eta} , \qquad (16)$$

$$\bar{V}c = \omega/1, \quad M = -Q_1/Q_2^2 , \qquad (17)$$

From the characteristic root of Eq.(15) and the expressions for Q1 and Q2, the stability behaviour in the vicinity of the zero solution of Eq (14) can be determined as shown in Table 1.

Table 1

Stability behaviour of equilibrium point (zero solution)

r	h	$\bar{V}$	M	stability behaviour of equilibium point
r>0	h>h_c	$\bar{v} > \bar{v}_c$	M>1/4	unstable focal point
			M=1/4	unstable degenerate nodal point
			0<M<1/4	unstable nodal point
		$\bar{V} < \bar{V}_c$	M>1/4	stable focal point
			M=1/4	stable degenerate nodal point
			0<M<1/4	stable nodal point
	h>h_c	$\bar{V} > \bar{V}_c$	M<0	unstable saddle point
		$\bar{V} < \bar{V}_c$	M<0	stable saddle point
r=0	h>h_c		M>0	unstable saddle point
	h<h_c	$\bar{V} \neq \bar{V}_c$	M<0	centre
r>0	h=h_c	$\bar{V} > \bar{V}_c$	M=0	unstable singular nodal point
		$\bar{V} < \bar{V}_c$	M=0	stable singular nodal point

It is seen from Table 1 that the equilibrium point of the approximate linear equation (14) becomes the central point only when $r = 0$ and $h < h_c$, i.e., for $r = 0$, $h < h_c$ only . The stability of the third-order differential equation (13) can not be determined yet. In other circumstances, the zero solution stability of Eq. (13) as in Table 1 is obtained (Zhang, 1981).

Eq. (13) is transformed into a set of ordinary differential equations:

$$dX_1 / d\theta = \quad QX_3 + Q_3 X_2 X_3$$

$$dX_2 / d\theta = \quad X_3 \qquad\qquad\qquad\qquad (18)$$

$$dX_3 / d\theta = \quad X_1$$

where $X_1 = H''$, $X_2 = H$ and $X_3 = H'$. When $r = 0$, Eqs.(18) satisfies the tubular center theorem (Zhang, 1981), and therefore the solution of Eq.(13) through any point in a certain domain of zero solution is explained as a periodical one, and this equiliburium point is the central point. This confirms the validity of the conclusion that Eq.(13) have a solution of elliptical cosinoidal function for $r = 0$. Therefore the stability of the nonlinear wave derived from Eq. (13) in the vicinity of zero solution is described by Table 1.

The above discussion refers to the Rossby wave of specific frequency and number. In fact, the incident and reflected Rossby wave appear simultaneously in the inertial boundary current region. For example, the incident wave can be expressed as $U_i = -\text{Re}\{A\exp[i(k_i x + ly - \omega t)]\}$ while the reflected wave is of a similar form $U_r = \text{Re}\{A\exp[i(k_r x + ly - \omega t)]\}$, if there is no energy loss. The boundary requirement leads to a definite relationship between the wave number ki of the incident wave and the wave number k_r of the reflected wave which can be derived theoretically without any difficulty in linear cases. In general, $k_r > k_i$ holds true for the western inertial boundary currents(Pedlosky, 1979).

It is shown from the above consideration that the nonlinear Rossby wave in the $\overline{V}$ zone of the inertial boundary current possesses stability features as follows.

(a) There is a critical mean current $\overline{V}c = \omega/l$ for the given coast orientation and bottom topography. When the mean current velocity changes from $\overline{V}c$, the stability of the zero solution for the single wave also changes, i.e., the stable (unstable) equilibrium point changes to the unstable (stable) equilibrium point . The bifurcation value $\overline{V}c$ is determined from the ratio of the wave frequency to the wave number along the coast. The wave propagating along the coast of higher frequency and of relatively long wavelenth is rather stable, and vice versa.

(b) When the mean velocity $\overline{V}$ remains unchanged, a critical value h_c for topography exits. When a small change occurs in the sea-floor topography around this critical value, the equi-librium point changes for the single wave and the saddle-nodal bifurcation and the saddle-focal bifurcation may occur. The bifurcation value h_c is related to the coast orientation, wave frequency and wave number.

If the bottom friction is neglected (i.e. r = 0),we have

$$h_c = (f_0/g)(\omega - l\overline{V})/(g\beta_y k) - \overline{\eta} \text{ for the western inertial boundary}$$

currents.

Consequently, two conclusions may be reached.

(i) If the mean current speed is greater than its critical value, the reflected wave tends to be unstable in comparison with the incident wave of the same wave frequency because the x-component wave number of the reflected wave k_r is generally greater than that (k_i) of the incident wave.

(ii) If the mean current speed is less than its critical value, the reflected wave is tends to be stable with a periodic solution.

(c) Friction not only dissipates the wave energy, but also considerably distorts the periodicity of the wave. Table 1 shows that when friction exits, (amplitude attenuating) stable and (amplitude increasing) unstable nonperiodic wave occur, while the periodic wave with equilibrium point at the centre does not occur.

(d) The unstable Rossby wave occur provided that the mean

current speed is so large that the frequency and wave number of the individual Rossby wave in the inertial boundary current satisfy the condition $\overline{V} > \overline{V}c$.

4 EXPLANATION OF VARIATIONS OF THE KUROSHIO

Kuroshio is a well-known western boundary current. Its motion is considered to be quasi-geostrophic (Pedlosky, 1979), and mainly Rossby waves. Although lots of studies have been done on variations of the Kuroshio, their processes are not yet clear. By using above theoretical results, an attempt is made to have some insight into them.

4.1 Difference between the Kuroshio in the East China Sea and the Kuroshio to the south of Japan

Observations show that the Kuroshio in the East China Sea has no significant meander except small ones in autumn and winter over the sea to the northeast of Taiwan Province. Its motion is controlled strongly by the bottom topography. However , the Kuroshio to the south of Japan shows significant and stationary meander which slightly changes with months, seasons, and years(Guan,1979). This difference may be related to the following factors.

(i) Coast orientation. In the East China Sea, the coastline is oriented approximately in the north-south direction. The main axis of the Kuroshio is basically on the eastern side of the 400 m isobath and along the 500m isobath. Therefore $|\beta_y k| > |\beta_x \ell|$ .However, to the south of Japan, the orientation of the coastline is mainly east-west, more likely to be zonal, thus $|\beta_y k| < |\beta_x \ell|$. This difference could result in different topographic bifurcation h in these two regions. Even if other conditions are similar, the nonlinear Rossby waves can have different stability behaviours and equilibrium states due to different coastline orientation.

(ii) Topography. The Kuroshio is situated at the outer edge of the continental shelf in the East China Sea where water depth varies greatly. On its right side is the bulgy continental slope (Guan, 1979) and the distant Ryukyu Islands exist as its barrier. Therefore, in this region h>>H, and the approxi-mation $H + h + \overline{\eta} = h + \overline{\eta}$ is valid in P and Q of Eq.(9). This means that the nonlinear waves can be regarded as linear waves. The streamlines of the quasi-geostrophic linear Rossby wave are almost parallel to the isobathes (Pedlosky, 1979). Thus the current flows steadily along the isobathes. The Kuroshio axis undergoes no significant changes. Only at shallower waters on both sides of the Kuroshio, the nonlinear effect becomes important. Currents which do not flow along the isobathes may occur, as a counter-currenrt on the eastern side of the Kuroshio and a warm current in winter and spring on its western side.

However, to the south of Japan, the bottom is sharply sloped , and there is no barrier on the southern side of the Kuroshio. Therefore, in some circumstances the Kuroshio could be forced to change its path.

(iii) Scale of disturbance in the bottom topography and coast. In the Kuroshio region in the East China Sea, the scales of disturbances resulting from the sea-bottom topography and coastline are very small, so that long Rossby waves are hardly developed. Disturbances could be possibly brought about from the equatorial current, wind stress, atmospheric temperature and other factors but they are too weak to produce variations in the velocity and path of the Kuroshio as observed. However, the Kuroshio to the south of Japan may cause nonlinear Rossby waves through the disturbances of large scales from the presence of Kyushu Island. These waves could make different behaviours of the Kuroshio in the East China Sea and south of Japan.

4.2 The Kuroshio meander

In the past 50 years or more there are 6 occasions of the Kuroshio meander to the south of Japan(Guan,1983; Sun, et. al., 1989; METEOROLOGICAL AGENCY MARINE WEATHER OBSERVATIONS by JMA) i,e.,1939-1944, 1953-1955, 1959-1963, 1975-1980, 1981-1984, and recently October 1986-May 1988. In most cases, the time for forming the meander is short but the duration for its attenu-ation is long. This is called bimodal path of the Kuroshio (Guan,1981). From the nonlinear Rossy wave theory , this bimodality can be considered as two equilibrium states of the waves. As mentioned above, the nonlinear Rossby waves in the inertial boundary current have more than two equilibrium states only when paramenter μ is positive. For the Kuroshio,

β_x and $\beta_x l - \beta_x^2$ are always negative. Hence , the requirement for the appearance of the two equilibrium states is $\bar{V} > \bar{V}c = \omega/l,$

which is coincident with observations. Guan (1983) has pointed out: " When the annual average of the surface wind stress vorticity near Hawaiian Islands keeps increasing for several years or increases rapidly to its peak, the Kuroshio in the East China Sea is also intensified and reaches its peak. During the peak period, the Kuroshio to the south of Japan shows a large meander ". This phenomenon can be explained as follows: for the Rossby waves caused by the disturbances of Kyushu Island, $\bar{V}c$ is a constant. When the mean flow is weak, μ is negative. Therefore there is only one stable equilibrium state which is the permenent inertial boundary current.However, the Rossby wave will have two equilibrium states if the atmospheric circulation and oceanic circulation intensify the Kuroshio in the East China Sea where $\bar{V} > \widehat{V}c$. In the region south of Japan,the Rossby waves may shift one equilibrium state to two equilibrium states, so that there occurs a large meander whose scale depends on the magnitude of $\bar{V}$.

5 CONCLUDING REMARKS

Assuming a basic current along an arbitrary coastline we transformed the governing equations into a single nonlinear ordinary equation for Rossby wave amplitude, whose stability in the immediate vicinity of the zero solution in the phase space was derived with the qualitative theory of the ordinary differential equation. The solution and stability characteristics for the single nonlinear barotropic Rossby wave in the inertial boundary currents are summarized as follows.

The solution having a periodic elliptical cosinoidal function for the amplitude of nonlinear barotropic Rossby waves is possible only when the bottom friction is negligible, the bottom is flat and a uniform basic current are postulated. the wave frequency depends not only on the wave length, water depth , basic current and orientation of coastline, but also on the wave amplitude.

The bottom friction tends not only to dissipate the wave energy but also to distort the wave periodicity.

The stability behaviour of the equilibrium point might be changed to give rise to various bifurcations depending on the bottom topography, coast orientation and beta-effect.

For the specified coast orientation, bottom topography, the magnitude of the basic current velocity, wave frequency

and wave length could not only affect the stability of zero
solution but also determine whether there are two equilibrium
states.

By use of the stability criteria thus obtained, an attempt
was made to explain the meander of the Kuroshio. The dynamical
difference of the Kuroshio between the East China Sea and the
south of Japan is attributed to the difference in the bottom
topography, orientation and scales of the coastline. The
bimodal path of the Kuroshio may be associated with the
presence of two equilibrium states which satisfy the condition
that the basic current velocity is greater than the phase
velocities of Rossby waves propagating parallel to the
coastline.

The serious shortcoming of the present study should be that
the basic current is constant in both space and time. The
result would be different with a basic current variable with
space and / or time. The present study is , nevertheless,
successful in suggesting a possible process of the Kuroshio
meandering in terms of the equilibrium states of the nonlinear
Rossby waves, though we could not estimate to what extent
our oversimplification affects the behaviour of the nonlinear
Rossby waves in the western boundary region.

Acknowledgement. This study is supported by the National
Science Foundation of The People's Republic of China. This
support is gratefully acknowledged. Appreciation is given to
Mr. Duan Yihong and Mr. Ma De-hua who timely typed the
manuscript.

6 REFERENCES

Guan, B.X., 1979. Some results from the study of the variation
    of the Kuroshio in the East China Sea. Oceanologia Et
    Limnologia Sinica, 10: 297-306.
Guan, B.X., 1983. The main results of survey and study on the
    Kuroshio in the East China Sea and to the East of Taiwan.
    Acta Oceanologica Sinica, 5: 133-145.
Hoskins, B.J., 1975. The geostrophic momentum appraxmation and
    the semi-geostrophic equations. Jour. Atmos. Sci., 32:
    233-242.
Kang, Y.Q., 1982. On stable and unstable Rossby waves in
    non-zonal oceanic shear flow. Jour. Phys. Oceanography, 12:
    528-537.
Liao, K.R., 1985. The tubular centre theorem of the ordinary
    differential equation. Acta Mathematca, 2: 174-181.
Liu, S.K. and Liu, S.D., 1983. Nonlinear waves of the fluid on
    the earth. Scientia Sinica, Series B, 3: 297-306.
Liu, S.K. and Liu, S.D., 1987. Nonlinear waves with semi-
    geostrophic flow. Acta Meteorologica Sinica, 45: 257-266.

Liu, Q.Y. and Qin, Z.H., 1989. Instability of barotropic coastal trapped waves. In S.T. Murty et al. (Editor), Storm Surge: Observation and Modelloing, China Ocean Press, Beijing, pp. 101-113.

QIN, Z.H. and LIU, Q.Y., 1990. Barotropic instability of non-linear waves in the tropical atmosphere and ocean. Acta Oceanolgica Sinica (unpubl.), 12 pp.

Pedlosky, J., 1979. Geophysical Fluid Dynamics, Springer-Verlag, 624 pp.

Sun, X.B., et al, 1989. The variation of the Kuroshio during the period 1986-1987. (unpubl.), 10 pp.

White, W.B. and McCreary, J.P., 1976. On the formation of the Kuroshio meander and its relationship to the large-scale ocean circulation. Deep-Sea Res., 23: 37-47.

Zhang, G.Y., 1981. The geometric theory and bifurcation problem of ordinary differential equations. Beijing University Press, Beijing, pp. 160-206.

LABORATORY EXPERIMENTS OF PERIODICALLY FORCED HOMOGENEOUS FLOW IN A ROTATING CYLINDRICAL CONTAINER

JUNGYUL NA and BONGHO KIM
Department of earth and marine sciences, Hanyang University,
Kyunggi-Do, Ansan, KOREA

ABSTRACT
    In order to explore the response of the ocean to an oscillatory wind forcing two separate experiments are carried out. In the first experiment the container having only a sloping planar bottom is forced with the forcing frequency which is tuned to the resonant frequency of the lower inviscid Rossby wave mode. In this case so-called resonant flow pattern with westward propagating vortical cells occupying the whole basin is observed and the overall flow patterns with respect to the various forcing frequencies are very close to the one observed by Beardsley(1975). In the second experiment the bottom includes a ridge topography on the sloping planar bottom, and this additional feature of the bottom topography changes the flow pattern such that asymmetric flows are dominant for the same range of the forcing frequency. Abrupt increase in the flow velocity occurs near the resonant frequency, however the flow pattern remains asymmetric.

1 INTRODUCTION

    Certain gross features of the general circulation of the oceans can be modelled in laboratory experiments, as has been shown by Pedlosky and Greenspan(1967), Beardsley(1975), Kuo and Veronis(1971), Hart (1975), Krishinamurti and Na(1978) and others.

    In particular, Kuo and Veronis(1971) showed how the flow generated by sources and sinks in a rotating pie-shaped basin of homogeneous fluid with a free surface and a sloped bottom is analogous to wind-driven flows in the ocean. It was also shown that the source-sink analogy to the effect of a wind-stress at the surface of the ocean is basically same as the sliced-cylinder model of the wind-driven circulation (Beardsley, 1969). When the wind-stress has a horizontal variation, the Ekman layer sucks fluid up from, or pumps fluid down into, the main body of the ocean. Hence, in a laboratory model the effect of a wind-stress can be simulated by a differential rotation of the upper-lid of the sliced-cylinder or by a suitable source-sink flow. For the case of an oscillatory wind-stress Beardsley(1975) employed a purely sinusoidal forcing mechanism of the driving lid in the sliced-cylinder model. In his model the interior flow is driven by the time-dependent Ekman-layer suction produced by the periodic relative angular velocity of the upper lid. Accordingly using the same analogy employed by Kuo and Veronis(1971), the periodic forcing of

64

the source-sink flow can be easily simulated by changing the duration of the source flow and the sink flow via alternating the direction of the flow into or out from the basin.

Previous laboratory studies on the oceanic response to an idealized atmospheric forcing such as sinusoidal wind fields and standing disturbances(Beardsley 1975, Ibbetson and Phillips 1967, Holton 1971) have demonstrated some of the major features of forced quasi-geostrophic motion. In particular, if the forcing frequency is sufficiently small the interior motion is quasi-geostrophic with the horizontal velocities being independent of depth. When the forcing frequency is tuned to the natural frequency of one of the lower inviscid topographic Rossby wave mode, Beardsley(1975) observed a significant resonant magnification in horizontal velocity field.

Consider now a rapidly rotating right cylinder with a sloping bottom filled with homogeneous fluid. The free surface of the fluid forms a paraboloidal shape, however due to the planar sloping bottom the height of the water column decreases toward one direction, i.e., the shallowest point. This arrangement of basin geometry provides the physical analogy between topographic vortex stretching and the β -effect. If we introduce an arbitrary bottom-topography such as ridge on the sloping bottom, the interior quasi-geostrophic flow that is driven by the source-sink flow may be blocked by the ridge and would go around rather than over it. When the periodic source-sink flow is introduced gyres of positive or negative vorticity will be generated and they will move toward the west because of the β-effect. Eventually a concentration of vorticity tends to build up near the western boundary. Moreover, if the ridge is located within the western boundary layer, where a strong northward(sink) or southward boundary(source) flow may exist the ridge could block the boundary layer flow such that over the ridge positive or negative vorticity may be produced. Hence there will be interactions between the Rossby modes that are concentrated near the western boundary and the topographically produced vortices.

Therefore, the purpose of this experiment is to find the characteristic response of interior flow to the periodic forcing with presence of the bottom topography. In order to control the external parameters we begin with the linearized vorticity equation in a cylindrical geometry.

2 DETERMINATION OF EXTERNAL PARAMETERS

To examine all the imposed effects on the flow pattern we should have at least the governing vorticity equation for the interior flow in the cylindrical container, and based on this equation the magnitude of the external parameter can be determined.

In a rotating cylindrical coordinate(r, θ,z)(Fig. 1) the vertical velocities at

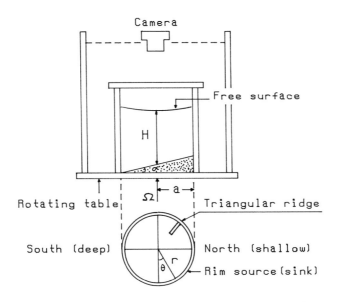

Fig. 1. The general configuration and geometry of the source-sink flow in a
rotating container with a sloping bottom.

the free surface and at the sloping planar bottom can be easily obtained by the
linearized kinematic boundary conditions and so-called the Ekman layer compatibility
condition. Following Kuo & Veronis(1971) and Hart(1972) the dimensionless linearized
vorticity equation can be written as

$$R\frac{\partial}{\partial t}(\nabla^2 P) = -\frac{\partial P}{\partial \theta} + \frac{2\dot{\zeta}}{F} - \frac{E^{1/2}}{F}\nabla^2 P - \left(\frac{\partial P}{\partial r}\sin\theta - \frac{1}{r}\frac{\partial P}{\partial \theta}\cos\theta\right)\frac{\tan\alpha}{2F} \qquad (1)$$

where $E\left(=\dfrac{\nu}{2\Omega H^2}\right)$ is the Ekman number, $R\left(=\dfrac{U}{2\Omega L}\right)$, the Rossby number, and $F\left(=\dfrac{\Omega^2 L}{g}\right)$,
the Froude number. And the other symbols are as following; $\tan\alpha$ is the slope of
the planar bottom, H and L are the mean depth and the radius of the cylinder,
respectively, and $\dot{\zeta}\left(=\dfrac{\partial\zeta}{\partial t}\right)$ is the time rate of change of the free surface
elevation due to source-sink flow and also ν is the kinematic viscosity of the
working fluid and Ω is the constant angular velocity of the basin.

When the sloping bottom contains some general topography, $h=h(r,\theta)$, an
additional term of the form

$$\varepsilon J(h,P) \equiv \varepsilon\left(\frac{1}{r}\frac{\partial h}{\partial r}\frac{\partial P}{\partial \theta} - \frac{1}{r}\frac{\partial h}{\partial \theta}\frac{\partial P}{\partial r}\right) \qquad (2)$$

in which ε is defined as $\dfrac{h}{H} \ll 1$, must be included in equation(1). Therefore
in this case the nonlinearity of the flow will be solely dependent upon the relative

magnitude of ε compared to other parameters such as E, R and F as well as $\dot\zeta$ and tanα. Since relative motion of the fluid is induced by the upwelling(downwelling) of the free surface due to the source(sink), the Rossby number that includes the characteristic velocity must be given by

$$R = \frac{\dot\zeta}{2\Omega L E^{1/2}} \tag{3}$$

for which the characteristic velocity is determined by the free surface kinematic boundary condition. For the periodic source-sink flow two different time scales should be considered. They are the period of source-sink flux(T) and the spin-up time(τ). If we put $\dot\zeta \sim \dfrac{H}{T}$ and $\tau \sim \Omega^{-1} E^{-1/2}$ the Rossby number may be expressed as

$$R \sim \frac{\tau}{T} .$$

Therefore a small Rossby number means that the period of forcing is relatively longer than the spin-up time. In order to have time-dependent vorticity equation we must have the time scale of forcing to be smaller than or comparable to the spin-up time scale.

In the vorticity equation of(1), the first and the last terms on the right hand side represent vortex stretching as columns move over the paraboloidal free surface as well as the sloping planar bottom, and the second and the third terms represent the effects of the change of the free surface elevation and the bottom friction on the vortex stretching, respectively.

In order to linearize the problem, it is necessary to keep $\dot\zeta$ equal to or less than $\dot\zeta = O(E^{1/2})$. Also it is necessary to keep the magnitudes of the other parameters as

$$O(\dot\zeta) = O(E^{1/2}) = O(F) = O(\tan\alpha).$$

The parameter ε, for the effect of the general topography, in the equation (2) needs to be adjusted according to the magnitude of the other terms.

3 EXPERIMENTAL APPARATUS AND METHOD

A plexiglas container of inner radius (a=) 9.3cm and height 20cm was placed on a horizontal turntable with its vertical axis parallel to the rotating axis(Fig. 2).

The rotating rate of the table was changed by a frequency-controlled A.C. motor. To eliminate the influence of fluctuations of the room temperature on the working fluid the container was placed within a square tank filled with a homogeneous fluid (Fig.3). The container was covered with a clear plastic to avoid any influence of

Fig. 2. The experiental apparatus showing a cylindrical container on the rotating turntable and a VCR for flow observations.

air-torque on the free surface. The bottom of the container had a sloping plane with an angle α giving $\tan \alpha = 0.1$ that simulated the β-effect of the oceanic models. A triangular ridge(4.5cm long, 0.7cm wide and 0.4cm high) was placed at some position on the sloping bottom and the position was determined from the preliminary tests that gave the general flow pattern and the width of the western boundary layer.

A triple-tube syringe pump of 150 ml capacity is used for the injection or the withdrawal of fluid into or from the container. The pump mechanism consists of a threaded drive block travelling along a rotating screw shaft which is connected to the shaft of a D.C. motor. The rotating speed of the motor is controlled by a variable D.C. power supply. The polarity of the D.C. power supply was changed to reverse the direction of rotation of the motor for the periodic source(injection)-sink (withdrawal). The motor speed and its polarity were controlled by a timing device that was built for this experiment. The external fluid which is injected or withdrawn by the pump flows through the narrow gap between the vertical wall of the container and the circular edge of the sloping bottom. The size of the gap is about the one tenth of one millimeter which is much thinner than the thickness of the sidewall Ekman layer of $O(10^{-1}cm)$. Several tests of the uniformity of "rim-source" and "rim-sink" fluid were performed and the results showed that the strength of the source or the sink was uniform along the rim.

Fig. 3. The cylindrical container with a triangular ridge at the bottom is placed
within the square-shaped tank and a triple-tube syringe pump is installed
next to it.

For the visualization of flow a pH-indicator technique described by Baker(1966)
was used.   A series of electrode grids of 0.009cm thick stainless-steel-wire were
stretched horizontally across the container at various locations.   The wire was
painted at regular interval to identify the movement of the flow.

A 35mm camera or a VCR was mounted on a stationary frame vertically above the
container with a stopwatch placed next to the container.   In this way the
photographs were analyzed to measure the flow velocities.

4 EXPERIMENTAL RESULTS AND DISCUSSION

Before the experiments of the periodic source-sink flow uniform-source or
uniform-sink driven flows were observed to determine the site of the triangular
ridge.   With a uniform source the interior motion consists of a uniform northward
moving flow and a fast southward moving boundary flow exists in the western
boundary. For the case of uniform sink, the direction of the flows are just
reversed(Fig.4).   It has been shown by Beardsley(1969) that the scale thickness of
this geostrophic western boundary is $E^{1/2}/\tan\alpha$, and for the present experiment, if
the $F=O(E^{1/2})$, the effect of total depth change is mainly due to the sloping bottom
and the scale thickness becomes $E^{1/2}/\tan\alpha$.   We set $\tan\alpha=O(10^{-1}) > E^{1/2}$ for all
experiments and this gives the layer thickness of about 3cm.   Therefore, the
horizontal length of the triangular ridge(4.5cm) which will be placed at the bottom
is large enough to block the flow in western boundary layer as well as the interior.
It should be also noted that the horizontal length of the ridge is much smaller than

the corresponding Rossby radius of deformation that is about 37cm for the present experiment. The frequencies of the source-sink flow were selected based on the spin-up time and the lowest eigenfrequency of the free topographic Rossby modes that was given by Beardsley(1975) as σ_{01}=atanα/k_{01} where k_{01} is the horizontal wave number satisfying $J_0(k_{01})$=0 and J_0 is the zeroth order Bessel function. For the present experiment σ_{01}=0.041 and when this equals the ratio of

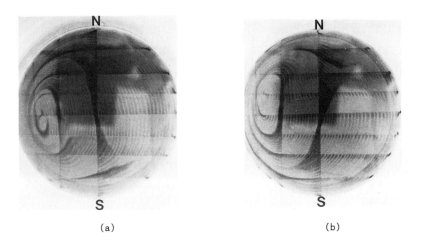

        (a)   (b)

Fig. 4. The flow patterns of (a)uniform source and (b)sink Rossby number R = 3.79 X 10^{-3}.

the forcing frequency of the source-sink flow($2\pi/\omega$) to the angular frequency of the rotating table($2\pi/\Omega$), i.e., σ_{01}=ω/Ω, the meridional velocities at the western boundary and at the center of the container become equal(Bearsley,1975) due to the zonal symmetry of the flow by exciting a quasi-inviscid mode resonantly. Since the spin-up time scale is about 120 seconds for the present experiment, we have selected 60 seconds as lower limit of the corresponding forcing period and it was increased up to the value of 400 seconds which is much longer than the spin-up time. The Rossby number and the Ekman number as well as the Froude number were almost fixed as to give R$\sim$O(10^{-3}),

E$\sim$O(10^{-4}), F$\sim$O(10^{-2}). The change in magnitude of $\dot{\zeta}$ give rise to changes in Rossby number, however, there was no order of magnitude change due to the smallness of $\dot{\zeta}$ which is about O(10^{-2}). Therefore the parameter that we changed most was the forcing frequency of source-sink flow. The velocity measurements were done by using the VCR data from which several pictures of equal time-interval were digitized and processed by the video image processing software. Following the method employed by Beardsley(1975) the maximum north-south velocities at two positions were measured:

one at the center and the other point within the western boundary layer. The velocity difference at these points were interpreted as guideline whether the flow is symmetric or not. Symmetric means that the differences are being minimum or being nearly equal to zero.

Fig. 5 shows the results of the observed velocities at two locations for various forcing frequencies. Our estimation of $\sigma = 0.041$ for the resonance to occur gives excellent agreement with the observations. At this frequency the north-south velocity at the center become maximum to reduce the velocity difference between two points. However it is interesting to note that the velocity difference became zero at $\sigma = 0.0429$ that is higher than the resonant frequency. This discrepancy may be due to the presence of the paraboloidal free surface that gives additional change of

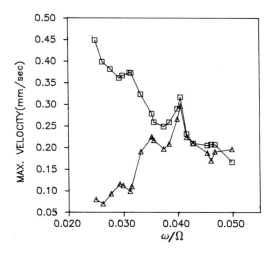

Fig. 5. Maximum north-south velocities observed at two positions versus the forcing frequencies. △ ,at the center; □ ,at the western boundary layer. The peak corresponds to $\omega / \Omega = 0.0429$.

the .depth of fluid column compared with the sliced cylinder model used by Beardsly(1975).

Fig. 6 is the flow pattern of the periodically driven source-sink flow. The dye lines clearly indicate the presence of the Rossby wave that propagates toward the west. The forcing frequency was closely tuned to the eigenfrequency such that $\sigma = 0.0406$. When we follow the dye lines carefully a single cyclonic vortical cell that occupied whole basin could be drawn. When the forcing frequency was small the east-west asymmetry in flow pattern was pronounced(Fig. 5) by increasing the velocity difference between two points.

Fig. 7 shows the concentration of the dye lines toward the western wall for lower forcing frequency, thus asymmetry of flow pattern was increased. When the triangular ridge was mounted at the bottom near the "north-west" corner of the basin(Fig. 8) the flow patterns, in general, were changed appreciably. In Fig. 8, the forcing frequency was the same as the one in Fig. 6. The asymmetry of flow pattern is clearly increased compared to Fig. 6. Since the magnitude of $\varepsilon = O(10^{-2})$, non-linearity of the flow could not be expected and any remarkable blocking effect was not seen. With the triangular ridge placed on the bottom, various forcing

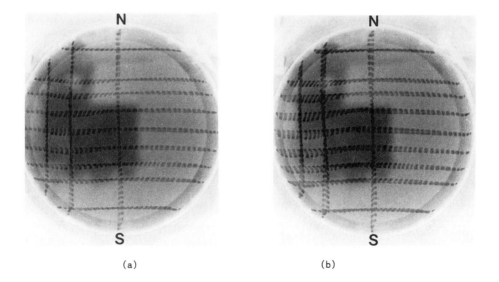

(a)   (b)

Fig. 6. The flow patterns of periodically driven source-sink flow. The time
difference between (a) and (b) is 15 seconds. The wavy dye lines from the
horizontal wires move toward the west, thus showing the westward
propagating Rossby wave. R = 3.79 X 10^{-3}, ω/Ω=0.0406.

frequencies were applied and the results of flow observations are shown in Fig. 9. The velocities at the two locations never tend to be the same. In other words strong asymmetric flow patterns were always existed when the ridge was present in the western boundary layer. The velocity differences between two locations were

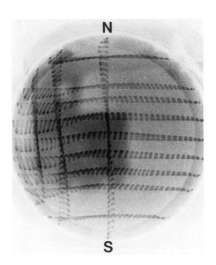

Fig. 7. When the forcing frequency is lowered(ω/Ω = 0.0317) asymmetry of flow is increased by western concentration of dye lines.

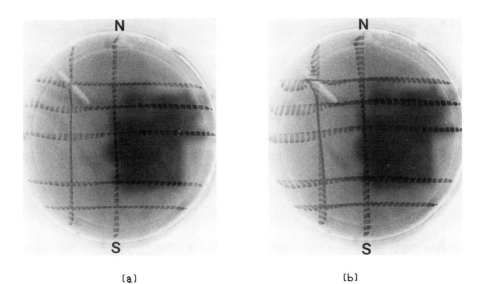

(a)   (b)

Fig. 8. The flow patterns of periodically driven source-sink flow with the triangular ridge mounted at the bottom. R and ω/Ω are the same as in Fig. 6. Asymmetry of flow pattern is quite evident.

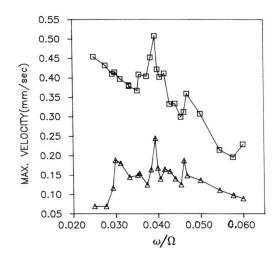

Fig. 9. Maximum north-south velocities observed at two positions. When the triangular ridge was placed at the bottom the maximum west velocities are increased by 70% compared to the ones in Fig. 5.

frequency is small very slow interior flow of Sverdrup-type and almost steady also increased near the resonant frequency, and the velocities at the center were decreased about 30% compared with the case of without-topography, while the velocities at the western boundary were dramatically increased by almost 70%. It is interesting to note that abrupt increase in velocity at two locations occurred near the resonant frequency, but the velocity difference remained still large.

For both cases of experiment, with and without the topography, when the forcing western boundary layer flow were observed. In particular, for the forcing periods which were much larger than the spin-up time, time-dependent flow patterns were barely observable only in the western boundary layer.

5 CONCLUSIONS

We have examined the response of the periodically forced homogeneous source-sink flow in a rotating cylinder with the β-effect. When the fluid in the container of with sloping bottom is periodically forced, so-called resonant flow pattern is observed with westward propagating vortical cells or the Rossby wave, which occupied whole basin. The resonance frequency is found to be very close to that of the inviscid topographic Rossby wave modes shown by Beardsly(1975). When the container includes a topography or a triangular ridge at the north-west corner of the bottom, asymmetric flow patterns are dominant with large differences in north-south

74

velocities at two fixed positions. At the resonant frequency, abrupt increase in
flow velocity occurs but the velocity difference is still large, which means that an
inclusion of even a very weak topographical influence ($\varepsilon \lll 1$), east-west asymmetry
of the flow pattern changes dramatically over the same forcing frequency range.

ACKNOWLEDGEMENTS

This research was supported in part by the Basic Science Research Institute
Program, Ministry of Education, Korea, 1989.

7 REFERENCES

Baker, D. J., 1966. A technique for the precise measurements of small fluid
    velocities. J. Fluid Mech., 26,573-575.
Beardsley, R. C., 1969. A laboratory model of the wind-driven ocean circulation.
    J. Fluid Mech., 38,255-272.
    _____, 1975. The 'sliced cylinder' laboratory model of the wind-driven
    ocean circulation. Part 2.Oscillatory forcing and Rossby wave resonance. J. Fluid
    Mech., 69,41-64.
Hart, J. E., 1972. A laboratory study of baroclinic instability. Geophy. Fluid
    Dynamics, 3,181-209.
Holton, J., 1971. An experimental study of forced barotropic Rossby waves. J.
    Geophys. Fluid Dyn., 2,323-335.
Ibbetson, A. and N. Phillips, 1967. Some laboratory experiments on Rossby waves in
    a rotating annulus. Tellus, 19,81-87.
Kuo, H. H. and G. Veronis, 1971. The source-sink flow in a rotating system and its
    oceanic analogy. J. Fluid Mech., 45. Part 3.441-464.
Krishnamurti, R. and J. Na, 1978. Experiments in ocean circulation Modelling.
    Geophy. Astrophys. Fluid Dyn., 11,13-21.
Pedlosky, J. and H. P. Greenspan, 1967. A simple laboratory model for the oceanic
    circulation. J. Fluid Mech., 27,291-304.

REMOTE SENSING FOR MODELLING OF VARIATION IN PRIMARY PRODUCTION FIELD

AKIRA HARASHIMA
National Institute for Environmental Studies, Japan Environment Agency,
Tsukuba, Ibaraki 305, Japan

ABSTRACT
    The author tried to construct a model to assess the time and space varia-
tion of the ocean primary production by using satellite products as boundary
conditions. The rate of phytoplankton's photosynthesis is basically governed
by supplies of light and nutrients, indications of which were required to be
computed systematically. The insolation was computed from the cloud coverage
data taken by Geostationary Meteorological Satellite. The wind stress, which
governs the upwelling and mixing to induce the supply of nutrients from the
deeper level, was tentatively computed from monthly statistical data, as they
are to be replaced by the microwave scatterometer data in the future.   Comparing the distribution of computed insolation and curl of (windstress /
Coriolis parameter) with the phytoplankton biomass map computed from CZCS
data, qualitative correlations were found between the northward advance of the
spring blooms and the variation of time—space variation in accumulated insolation, and between the area of high production and the wind stress curl.

INTRODUCTION

    It is required to construct some strategies to assess the global environ-
mental changes originated both from anthropogenic impact to the earth system
and from the background variations inherent in the system. Primary production
in the ocean is one of the processes which are controlling the earth's environmental system.   This process and other environmental factors are dynami-
cally linked to each other. A system dynamical model, e.g. an ocean general
circulation model containing several biological and chemical submodels, is one
of the best approaches to analysis of the linkage of those processes in long
term trends.

    Here, it is not always required to solve the oceanic general circulation
primarily depending on whether the system is governed by the internal dynamics
or by the boundary conditions.   In the latter case, data resources of time
and space coverage in question are more crucial rather than the preciseness of
the model structure.   Furthermore, the verification of the model also
requires observational data.   Use of satellite data fits these purposes because recent development of sensor technologies allows us to measure many
kinds of quantities globally. The significance of the remotely sensed data to
ocean modelling has been discussed (e.g., Nihoul, 1984).   Many research
works have started to use remotely sensed information to verify results of

numerical models by comparing predicted and observed 2-dimensional distributions.

The rate of primary production in the ocean is determined by several factors. Recently, attempts are made to express this rate as a function of variables simplified as possible, namely two factors ; solar energy input and concentrations of photosynthetic pigments, chlorophylla(Platt et al., 1983, Eppley et al., 1985). Here, abundance of chlorophylla is not an external parameter and primarily governed by nutrient supply from the lower layer caused by upwelling or vertical mixing due to wind system. Therefore, two variables, light and chlorophylla concentration, can be replaced by two geophysical factors; insolation and wind, as external forcing parameters indirectly determining the time and space variation of primary production.

One of the striking characteristics of primary production is spring blooms of phytoplankton. It occurs in the season when insolation is increasing day by day and nutrients in the euphotic zone are still not used up. Area of spring blooms propagates to higher latitudes in accordance with the seasonal shift of the sun as shown by several literatures. Thus, these two geophysical factors are quite important to assessing the biological processes.

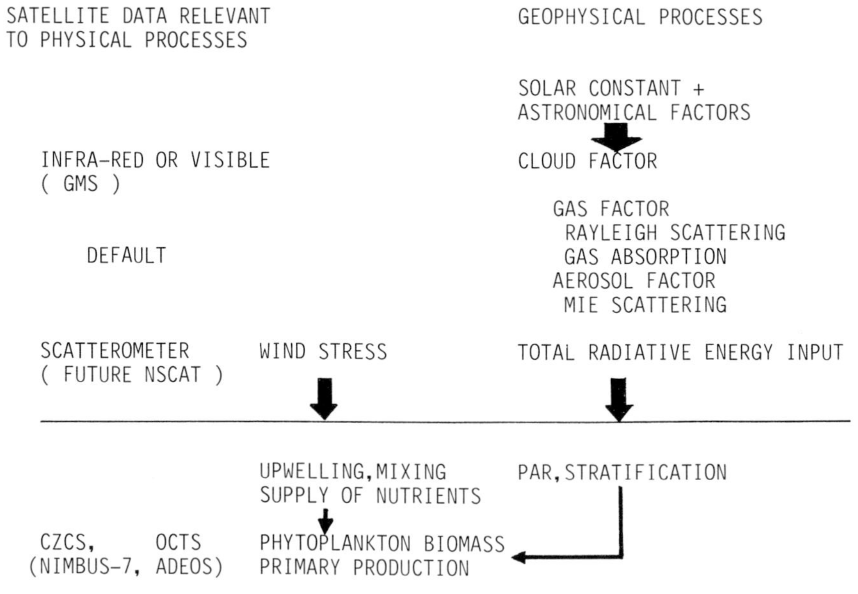

SATELLITE DATA RELEVANT
TO PHYSICAL PROCESSES

GEOPHYSICAL PROCESSES

SOLAR CONSTANT +
ASTRONOMICAL FACTORS

INFRA-RED OR VISIBLE
(GMS)

CLOUD FACTOR

GAS FACTOR
RAYLEIGH SCATTERING
DEFAULT   GAS ABSORPTION
AEROSOL FACTOR
MIE SCATTERING

SCATTEROMETER   WIND STRESS   TOTAL RADIATIVE ENERGY INPUT
(FUTURE NSCAT)

UPWELLING,MIXING   PAR,STRATIFICATION
SUPPLY OF NUTRIENTS

CZCS,   OCTS   PHYTOPLANKTON BIOMASS
(NIMBUS-7, ADEOS)   PRIMARY PRODUCTION

Fig.1 Schematic diagram of physical forcing processes related to the primary production and available satellite data for retrieving information of each process.

As is schematized in Fig.1, several kinds of satellite data contribute to attain distribution of these two geophysical factors. If precise algorithm to obtain the physical values from these satellite data can be established, we are able to construct a simulation model that substantially depends on the data as well as on the dynamical equations to explain the time and space variation of the primary production.

At present, stage of our research is still immature and only the indexes of these quantities can be calculated based on simplified equations.   Here, "index" means the numerical value that bears information of time and space variations of a quantity instead of expressing its precise numerical value. Still it is worthwhile to clarify the overall relation between these indexes and time and space variation of phytoplankton biomass.   This is the main ob-jective of the present study.

Recent advance of the ocean color remote sensing, such as Coastal Zone Color Scanner on Nimbus-7(Esaias et al., 1986, Feldman et al., 1989) enables us to check validity of the model by comparing model results with the remotely sensed distribution of photosynthetic pigments.   In addition, we have developed a data base system compiled from the in situ ship observation of the biological and chemical components to supplement and calibrate the satellite ocean color data(Harashima et al., 1990).

Importance of JECSS area comprising the marginal seas, the continental shelf and the open ocean, is in the fact that both anthropogenic and back-ground variation take place in these ocean system.   The method described in this paper will fit the analysis of such an area, although the present analysis is tentatively concerned with the whole Pacific Ocean.

2 MODELLING OF INSOLATION FIELD

The rate of solar energy input above the atmosphere, which is also called extra-terrestrial radiation, is calculated from the solar constant and astronomical factors (Budyko, 1956).   In addition to these factors, cloud coverage contributes substantially to the actual insolation at the earth's surface.

Part of the radiation that phytoplankton can use called PAR (photosyn-thetically available radiation) is within the spectral range of 300 - 720nm. Several models have been proposed to calculate the spectral characteristics by considering the absorption and scattering due to gases and aerosols (Birds, 1984). However, we did not deal with the detail of the spectral characteris-tics because the present study aims at construction of a first stage of the system. The insolation process is crucial to the primary production through the creation of thermal stratification rather than solely through the con-tribution to PAR. Concerning the initiation of spring bloom, arguments have been made as follows. The creation of stable stratification makes the phytoplanktons stay in the euphotic layer, otherwise their light utilization is limited by the forced convection to the deeper level(reviewed by Parsons et al., 1984).

Index of bulk solar radiation at the sea surface was calculated in this study. Cloud albedo is the primary factor to govern bulk radiation. Data from the Geostationary Meteorological Satellite is suitable for grasping the cloud process.

The infra-red image of GMS-satellite data, which have been processed by Meteorological Satellite Center of Japan into 5-day mean, 1° x 1° gridded, cloud coverage value X_c(in %), were used in this analysis. The insolation index, I_s^*, was calculated following a simplified empirical formula proposed by Budyko(1956),

$$I_s^* = (1 - 0.68(X_c/100)) I_{et}.$$

Here, I_{et} is the extra-terrestrial radiation that varies with location and time. I_{et} is obtained from the solar constant and the astronomical factors. It should be noted that the definition of traditional cloud coverage is not identical to X_c sampled by satellite. Therefore, Budyko's formula cannot apply directly. However, I_s^* contains information of time and space variabilities of cloud coverage and fit the present purpose of deducing an index.

Figure 2 shows the distribution of I_s^* over the western Pacific in early April of 1979. Subtropical belts are characterized by relatively intense insolation. Level of the insolation is relatively low in the equatorial area and in the belts of atmospheric low of the mid latitude.

Figure 3 shows isopleths of the insolation index accumulated from January 1st, 1979 along meridians of 150°E, 180°E and the one in assumed cloud-free hemisphere. Along the cloud-free meridian, each isopleth advances from the equator to the higher latitude from spring to summer. This pattern is modified by the relatively high cloud coverage in the tropical and the sub-arctic region, where atmospheric lows frequently appear. Advance of the isopleth to the high latitude is slower with presence of clouds.

3 MODELLING OF UPWELLING FIELD

The monthly mean, 2° x 2° gridded wind stress data compiled by Hellerman and Rosenstein (1983) was used to construct the wind field. Satellite data will be used to obtain the wind fields, if the microwave scatterometer data coverage is sufficient in space and time.

The well-known relation between the wind stress τ and the divergence of

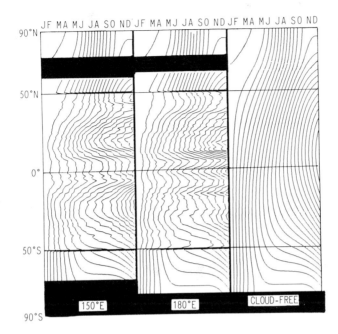

Fig.3 Isopleths of the insolation index integrated from January 1st, 1979 along the meridians. Contour interval is $5 \times 10^4 [cal/cm^2]$. Contour label is omitted because the absolute value is not important.

flow in the Ekman drift current gives the upward velocity w_e at the bottom of the Ekman layer as

$$w_e = curl_z (\mathbf{\tau}/\underline{f})$$ (1)

except at the equator and its vicinity where $\underline{f}$ is very small. Although the absolute value of w_e is insignificant, it would work as an index of wind-induced upwelling.

Figure 4 shows the distribution of w_e over the Pacific Ocean in March. One of the remarkable features is upwelling in the equatorial zone. The cause of the equatorial upwelling is accounted for as follows. In the north(south) of the equator, the direction of water movement is north-west(south-west) due to the sign of $\underline{f}$. This elongation of water element is compensated by the upward water motion. This zone overlaps with the area of relatively abundant chlorophyll as shown by the CZCS images(Esaias et al., 1986 ; Feldman et al., 1989), which suggest the supply of nutrients from the lower layer.

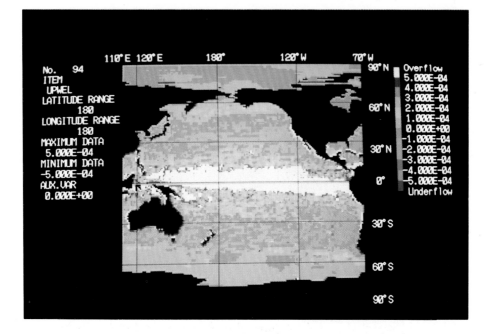

Fig.2 Insolation index of early April, 1979, computed from infrared radiation data of GMS-satellite. This data set covers an area of 90°E to 170° W, 50° S to 50° N. The value was level-sliced and colored [purple, ..., red] so as to indicate the scale from the low to the high.

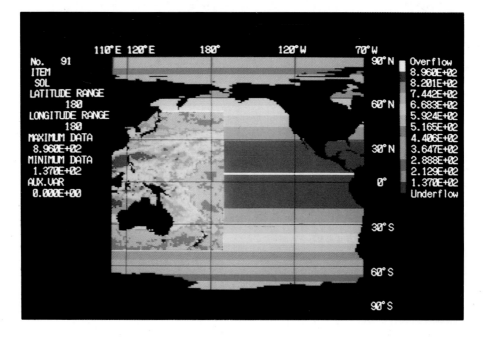

Fig.4 Distribution of the index of the wind-induced upwelling : curl$_z$(wind stress/f) in April, computed from the monthly compiled data sets by Hellerman et al., 1983.

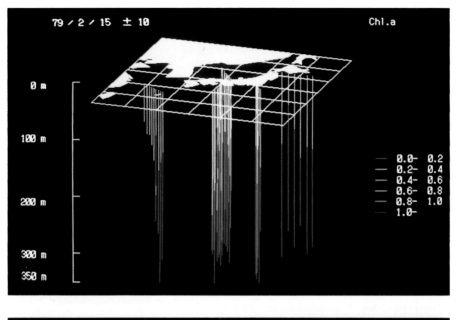

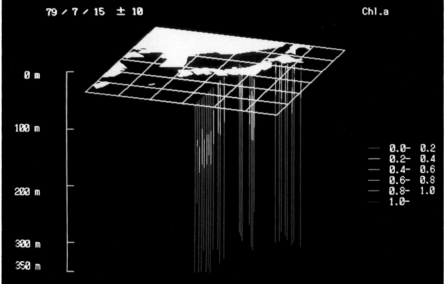

Fig.5 Three-dimensional distribution of chlorophyll-a from observations by
Japan Meteorological Agency performed within the period February 5 to 25th
(top), July 5 to 25th, 1979 (bottom) (reproduced from Harashima et al.,
1990)

4 SEA TRUTH DATA BASE OF CHLOROPHYLLa AND NUTRIENTS

As was described in the Introduction, CZCS image is a strong tool to grasp the distribution of phytoplankton biomass. However, CZCS does not yield qualitative information on the vertical distribution nor quantitative informa- tion of the total chlorophyll abundance in a water column. Particularly, there is a critical problem that CZCS-image reflects both chlorophylla and phaeopigments. The latter does not contribute to the primary production and should be regarded as an index of detritus. Discrimination of the two pig- ments is an important issue.

Therefore, it is preferable to calibrate satellite data by actual observa- tions. A data base system was constructed to search and display existing in situ data that include the chlorophylla and nutrients systematically (Harashima et al., 1990). Figure 5 shows three-dimensional distributions of chlorophylla in February and June. In February, chlorophylla is abundant in the eutrophicated Seto Inland Sea and in the East China Sea. It is not very abundant in the Oyashio region (east of Hokkaido) in February in spite of rather rich nutrients. In July, the Oyashio region is characterized by high concentration of chlorophylla. It is basically due to the fact that the in- solation is insufficient for phytoplankton growth in the northern seas in February, and it reaches the level for blooming up to July. There are several possibilities how the insolation controls the phytoplankton growth; lack of PAR, limitation of water temperature, and the unfavorable physical conditions such as the photoinhibition. It is not fully clarified by the materials that we have which is the primary limiting factor. Regardless of the precise test of hypothesis, area of blooming propagates to high latitudes as the solar radiation increases as shown by Figure 5.

5 DISCUSSION

Features appeared in the CZCS images are picked up to evaluate the results of sections 2 to 4.
 i) In the subtropical gyre, the chlorophyll concentration is low. The sub- arctic ocean is characterized by the abundant chlorophyll.
ii) There is a band of relatively high chlorophyll level at and near the equator due to the equatorial upwelling.
iii) The eastern periphery of the ocean is characterized by the high chlorophyll concentration due to the coastal upwelling.
iv) The continental shelf and inner bays are characterized by high chlorophyll concentration.
 v) The northward advance of blooming is observed in the subarctic ocean from spring to summer. In the western Pacific Ocean, this tendency is not as

remarkable as in the Atlantic Ocean.

The distribution of the insolation index (Fig.2) does not seem to directly correlate with i). The maximum insolation appears on the subtropical seas, where chlorophyll concentration is rather low. This is due to the fact that the semi-permanent stratification of sea water maintained by the strong insolation inhibits the water exchange between the euphotic layer and the deeper, nutrient-rich water. This contrasts to the process creating seasonal thermocline which causes spring blooms.

The minimum isolation area of the subarctic band overlaps with the area of relatively high productivity. Therefore, it seems that the reduction of insolation due to the cloud coverage does not limit the primary production in terms of PAR.

The variation of accumulated insolation (Fig.3) does not correlate with the time variation of chlorophyll distribution shown in the CZCS images since the northward advance of the blooming area is basically not very remarkable in the western Pacific. However, such a tendency is partly seen in the in situ observation data base (Fig.5). Thus, the northward advance of blooming in the western Pacific may be confirmed by considering three dimensional structure of chlorophyll.

The wind stress distribution(Fig.4) qualitatively accounts for ii) and iii). Furthermore, following possibility is suggested concerning the fact that the minimum insolation area of the subarctic band , which is identical to the area of atmospheric low, correlates with the high chlorophyll area, although no strict order estimation is done. The cyclonic wind causes the surface water divergence as shown by (1), which leads to the upwelling of nutrients. Apart from the ocean surface divergence, relatively strong wind under the atmospheric low causes the vertical mixing, which also helps the supply of nutrients from the deeper layer.

Based on the qualitative results, we can conclude that developing the present method to a time-dependent model by assembling the submodels will be useful to account for the time and space variation of primary production. At present, accumulation of the satellite data is not sufficient to make a complete set of time and space coverage, which limits the establishment and validation of the model. In the near future, the method introduced in this report will be an indispensable tool to analyse the linkage between the factors of ocean environmental change.

ACKNOWLEDGMENTS

The author would like to express his sincere thanks to Professor K. Takano and C. Arakawa of the University of Tsukuba, and H. Sasaki of the Meteorological Satellite Center, Japan Meteorological Agency for providing the data for present analysis.   Thanks are extended to Japan Oceanographic Data Center. He also thanks Y. Kikuchi of the University of Tsukuba and H. Yamagami of University of Library and Information Science for helping the computer works.

7 REFERENCES

Birds, R.E., 1984. A simple solar spectral model for direct-normal and diffuse horizontal irradiance, Sol. Energ., 32: 461-471.

Budyko, M.I., 1971. The heat balance of the earth surface, Gidrometeorologi-cheskoe izdatel'stvo, Leningrad, 255pp.

Eppley, R.W., Stewart, E., Abbott, M.R. and Heyman, U., 1985. Estimating ocean primary production from satellite chlorophyll, Introduction to regional differences and statistics for the Southern California Bight, J. Plankton Res., 7: 57-70.

Esaias, W.E., Feldman, G.C., McClain C.R., and Elrod, J.A., 1986. Monthly satellite derived phytoplankton pigment distribution for the North Atlantic Ocean Basin, EOS, 67: 835-836.

Feldman, G., Kuring, N., Ng, C., Esaias, W., Mclain, C., Elrod, J., Maynard, N., Endres, D., Evans, R., Brown, J., Walsh, S., Carle, M., and Podesta, G., 1989. Ocean color: Availability of the Global Data Set, EOS, 70: 634-641.

Harashima, A. and Kikuchi, Y., 1990. Biogeophysical remote sensing: A ground truth data base and graphics system for the Northwestern Pacific Ocean, EOS, 71: 314-315.

Hellerman, S. and Rosenstein, S., 1983. Normal monthly wind stress over the world ocean with error estimates, J. Phys. Oceanogr., 15: 1405-1413.

Japan Meteorological Agency, The results of marine meteorological and oceano-graphical observations (Semiannually published), Tokyo.

Nihoul, J.C.J., 1984. Contribution of remote sensing to modelling, In: J.C.J. Nihoul (Editor), Elsevier Oceanogr. Ser., 38: 25-36.

Parsons, T.R., Takahashi, M., and Hargrave, B., 1984. Biological Oceanographic Processes, Pergamon Press, Toronto, 330pp.

Platt, T. and Herman, A.W., 1983. Remote sensing of phytoplankton in the sea : surface-layer chlorophyll as an estimate of water-column chlorophyll and primary production, Int. J. Remote Sensing, 4: 343-351.

REFLECTION OF THE OCEANIC FRONTS ON THE SATELLITE RADAR IMAGES

L.M. MITNIK and V.B. LOBANOV
Pacific Oceanological Institute, Far Eastern Branch, USSR Academy of Sciences,
7 Radio St., Vladivostok 690032 (USSR)

ABSTRACT

The ocean surface sensing has been accomplished by Side Looking Radar (SLR) from a number of Cosmos and Ocean series satellites. SLR imagery of the ocean is a map of the radar reflectivity of 2-4 cm wavelength ocean waves. The reflectivity (brightness of radar image) depends on wind characteristics and also on hydrological inhomogeneities associated with the oceanic fronts, currents and eddies. This paper presents an analysis of the radar images of the Kuroshio confluence zone obtained in spring 1987 and 1988. Comparisons with concurrent NOAA infrared images, surface weather maps and ship data showed a strong correlation between the radar cross section σ^o and the sea surface temperature (SST) when the wind speed was in range from 2-3 to 8-10m/s A number of possible mechanisms by which an SLR detects the oceanic phenomena have been discussed: variation in σ^o may reflects a change in SST (due to dependence of the kinematic viscosity on water temperature), surface films concentration, air-sea temperature difference, surface current shear through their modulation of the 2-4 cm waves. This ambiguity may be solved by radar sensing of the ocean at different wavelengths.

1. INTRODUCTION

Since September 1983, the radar survey of the ocean has been carried out with the help of the incoherent Side Looking Radar (SLR) from several oceanographic satellites (Cosmos-1500, Cosmos-1766, Ocean and others). The main characteristics of SLR (Mitnik and Victorov, 1990) are:

Swath width		460 km
Resolution { flight direction		2.1-2.8 km
normal to flight direction		0.8-3 km
Wavelength		3.1 cm
Pulse length		3 μs
Polarization on radiation and reception		vertical
Beamwidth of the antenna at -3 dB level		
in azimutal (H) plane		0.2^o
in elevation (E) plane		42°

The SLR is switched on for 7 - 10 min. After the preliminary processing on board the satellite,the data are transmitted to the Earth by radio (Fig.1). The radar image represents two-dimentional distribution of the radar reflectivity with spatial resolution depending on the device and transmitting lines. The image brightness variations reflect the level changes of radar

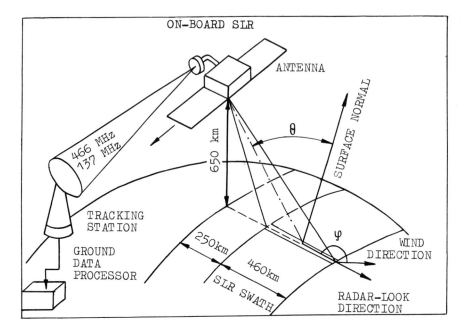

Fig.1. Scheme the ocean surface observation by satellite Side Looking Radar.

signals scattered by the sea surface roughness heterogeneities. The normalized scattering cross-section σ° (dB) is a quantitative measure of those changes.

It is known that the radar signal scattering for angles of incidence θ away from nadir has a resonance nature and occurs primarily at sea waves of wavelength $\Lambda = \lambda/(2\sin\theta)$ where λ is radar wavelength. The main factor influencing the roughness characteristics is the surface wind. Increase of the wind speed W results in increase of σ° (radar image brightness). Figure 2 is the radar image showing a polar low above the Sea of Japan. The cyclone centre position 1 is determined rather exactly - especially if the radar image is analyzed together with infrared one (Fig.2(b)), where a comma-shaped cloud system of the cyclone looks brilliantly white. The cyclonic curved belts of weak winds 2 and 3 link up in the centre region. The cloudy spirals, brighter than background, consist of cumuli elements, and fit them on IR image. Wind speed to the south from the centre (region 4) increases sharply. Ship data give wind speed of 15-20 m/s there. Strong winds are observed west of Honshu regions. The areas of Tokyo-Yokohama 6 and Nagoya 7 are characterized by high level backscattered signals.

Macro- and mesoscale structures in the surface wind field and their relations with the cloud field are investigated in detail (Mitnik and Victorov, 1990).Besides wind, the spectral density of capillary-gravity waves depends on physical properties of waters. Because of this, the heterogeneities

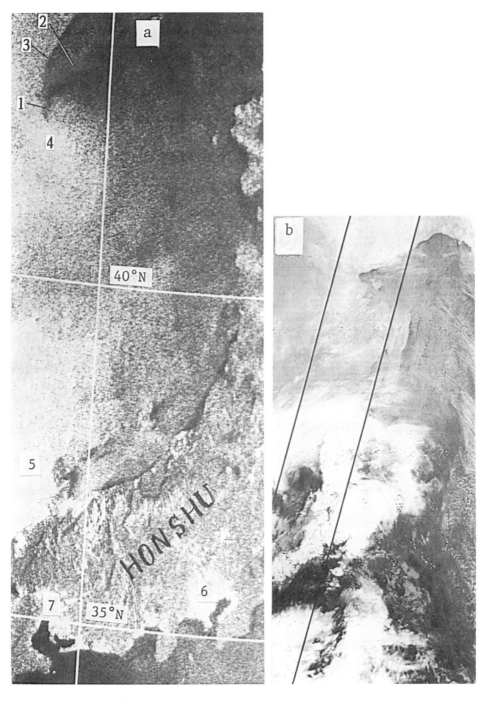

Fig.2. Satellite images of the Sea of Japan on February 2, 1988.
(a) Cosmos-1766 SLR image at 0317 GMT showing sea roughness distribution.
(b) NOAA-9 IR image of cyclone cloudiness.

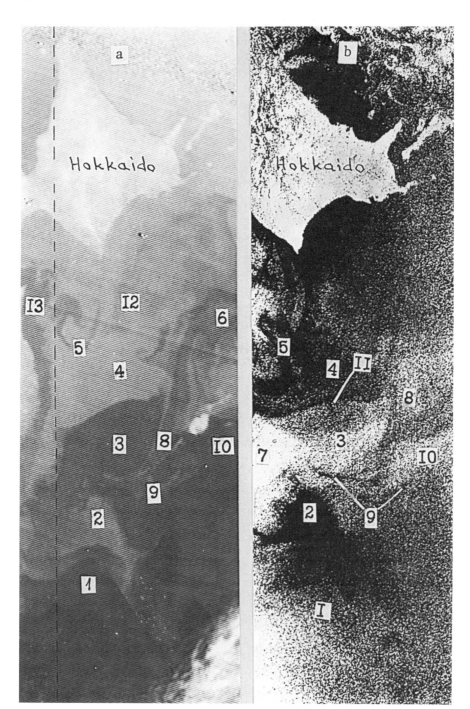

Fig. 3. Satellite images of an area east of Japan on April, 15, 1987.
(a) NOAA-9 IR image at 1747 GMT. (b) Cosmos-1766 SLR image at 1605 GMT.

of the waters structure and dynamics would be reflected on the radar images.

The principal aims of this paper are to analyze those radar images where the oceanographic phenomena are fixed, and to discuss the mechanisms defining their remote indication. Areas adjacent to Japan and the Kuril Islands, where the warm Kuroshio and the cold Oyashio Currents having water masses of very different properties come in close contact and interaction, were chosen as the regions of investigation.

2. SATELLITE IMAGES ANALYSIS

Hydrological fronts patterns in the Kuroshio confluence zone are prominent in the SST field as revealed on satellite IR images. The similar patterns were detected on radar images obtained during 1987 and 1988. Let us perform mutual analysis of the radar and IR images obtained in spring. Spring is the most favourable for this purpose because the region east of Japan is characterized by strong thermal contrast at the surface and minimum cloudiness.

2.1. Spring 1987

Figure 3 depicts synchronous radar and IR images covering the confluence zone east of Japan. A surface weather map obtained two hours after satellite observations (Fig. 4) shows a low gradient baric field between a high pressure area above the Sea of Japan and a cyclone to the southeast of Japan. The wind speed W was 2-10 m/s. Wind is strengthened up to 12-15 m/s by the atmospheric front in the southern part of the images.

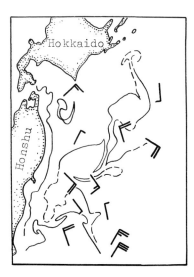

Fig. 4. Surface weather map at 1800 GMT on April 15, 1987.

Interaction of the Kuroshio and Oyashio waters exhibits an abrupt contrast of SST on the boundary reaching up to 5-8°C per 1-3 resolution element (about 4-10 km) and well marked on the IR image (Fig. 3(a)). Thermal heterogeneities outlines are traced in the brightness (roughness) field of radar image (Fig. 3(b)). The swath of space radar intersected the eastern portion of the Kuroshio anticyclonic meander 1 (t = 19-20°C, dark tone on the IR image), comparatively cold 2 and warm 3 areas of the frontal zone (t = 2-4 and 9-10°C, light gray and dark tone on the radar image, respectively), cold subarctic waters 4 (t = 2-4°C, light gray tone on the IR image) and warm areas of eddy-shaped form 5 and 6 (t = 8-10°C, dark gray tone on the IR image) in the subarctic zone.

The correlation of the brightness fields on the radar and IR images is best marked in areas 3 and 5. The former shows an anticyclonic eddy A3 having a warm core of 90-100 miles diameter and of water temperature change on the boundaries 6-8°C. The eddy boundary is distinct on the radar image by the brightness (roughness) gradients corresponding to the thermal front on the IR image. Only the western part of the eddy shows no radar contrast due to wind increase (bright area 7 where W > 10m/s). The radar image tone inside the anticyclone 3 did not remain constant. Brightness is lower in the northern (warmer) part of the eddy. Brightness is enhanced in the southern part where the vortex movement centre was likely located and the narrow stream of Oyashio waters was observed. However, as a whole, on the radar image warm eddy waters have a light tone and the cold subarctic ones a dark one.

Correlation between the SST and the backscattered radar signal level σ° is visible at periphery of A3. Thus, the narrow stream (about 10 miles) of Oyashio waters 8, involved in the vortex movement and its branches 9 appear darker on the radar image than the eddy. The warm area 10 east of these branches is as light as the eddy. The dark streak 11 denoting the minimum roughness zone adjoins the eddy from the north. The rather warm zone 12 with light tone extends farther to the north. The coldest Oyashio waters are located here. On the IR image the zone 12 stands out indistinctly, which may be explained by the weak thermal contrast.

The spiral vortex formation 5 developed by waters of the Tsugaru Current 13 is traced north of the anticyclonic eddy 3. The width of the warm spiral changes from about 5 miles in its tail to 25-30 miles in the rotational centre. The central part of the spiral 5 with t = 9-10°C is characterized by higher values of σ° on the radar image. Water temperature and radar image brightness are decreased in the narrow portion of the spiral. The boundaries of the thermal and radar contrasts correlate well.

The Oyashio cold water zone with a low σ° level stands out east of the Tsugaru Strait. The warm area 6 formed by the anticyclonic eddy of the

Kuroshio is not visible on the radar image, because it is located at the far edge of the swath where only the significant changes of σ^o can be registered (Mitnik and Victorov, 1990).

In spring 1987 the anticyclone A3 and other heterogeneities in the SST field were registered on several radar images. During the radar survey on April 1 (Fig. 5(a)) a baric field with low gradients was located above eddy 3. Wind speed measured almost simultaneously with satellite observations was 2-3 m/s in the vicinity of the Honshu, and south of the eddy at 0600 Z. The weak wind zones 1 are reflected by a dark gray tone. A warm front north of Hokkaido is seen in the surface weather map. A strong (15 m/s) westerly was noted in the frontal rear and resulted in σ^o increase (area 2). On the IR image obtained by NOAA-10 satellite 3.5 hours after radar one the larger part of the eddy was shielded by clouds, which makes it difficult to get detailed correlation.The position and form of the contrast zone 5 coincide with the bank of cumuli clouds and are distinguished well on the IR image. Alternate light and dark belts about 16-18 km apart are noted in almost all areas of the anticyclone to the southwest of the boundary 5. Variations in sea surface wind caused by lee waves (atmospheric internal gravity waves) resulting from interaction of the eastward air flow over mountains on Honshu are responsible for such modulation of the backscattered radar signal level (Mitnik and Victorov, 1990). These waves are expressed in the cloud field also. Quasiperiodic brightness variations are visible in the strong wind area 6 where the western part of the anticyclonic eddy is located (this eddy designated by 6 in Fig.3(a)). The eddy can be recognized on the IR image. However, its indication on the radar image is practically impossible due to strong wind, in spite of the fact that it is situated at the near-edge of the radar swath.

During radar measurements on May 2 (Fig. 5(b)) wind speed did not exceed 7 m/s in the greater part of the area under investigation, as in the previous two cases. The nearest IR image obtained two days before radar one permits to interpret the brigtness variations. In the anticyclone area 3 the sea surface is rougher than in surrounding colder waters, except for area 1 to the west, where the brightness increase is associated with 'the wind action. Compared to image on April 1 the eddy A3 was shifted by about 100 km to the north. This estimation is supported by IR images analysis.

The arced line 2 fits the boundary of waters with t = 3-5°C (area 4) and 10-12°C (area 5). The higher radar contrast features the eastern boundary 6 of the first meander of the Kuroshio. The Kuroshio waters 7 have temperatures of 19-20°C. It is 2-4°C higher than in adjacent waters 8 and 9. The brightness heterogeneities in area 7 where the SST variations from IR data are small can be attributed to variations in the wind speed and direction.

In spring 1987 eight radar images east of Japan were obtained. The eddy A3

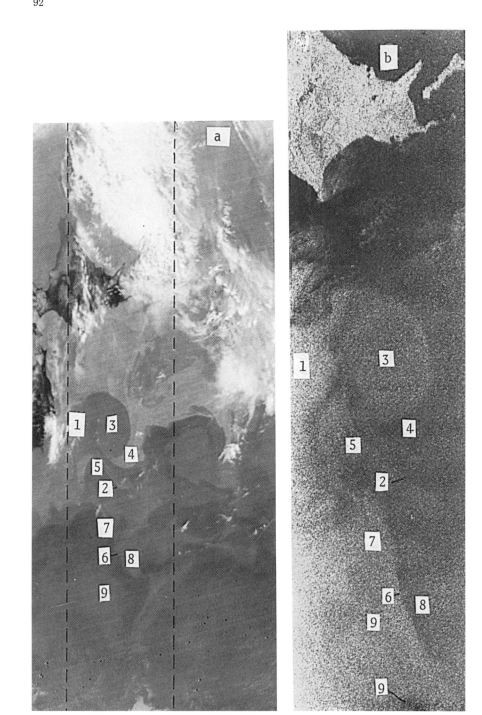

Fig. 5. Cosmos-1766 SLR images of an area east of Japan. (a) At 0605 GMT on April 1, 1987. (b) At 1350 GMT on May 2, 1987.

area is traced with different degree of contrast on six images. On two images dated April 12 and 24 the eddy is not revealed because of strong wind (10-15 m/s according to the synoptic maps).

2.2. Spring 1988

In April-May 1988 fifteen radar images of areas east of Japan were obtained. Most of them showed definite elements of the frontal zone. However, those elements were absent on two images. In one case (April 11), the influence of SST field was not visible, because of the atmospheric front passage with very strong winds. In the other case (May 21), no temperature contrasts was not revealed in the roughness field, because of calm weather with winds ranging from 0 to 3-5 m/s. In May hydrometeorological measurements were carried out from a weather ship "Ocean" to estimate the SST, air temperature and wind speed variations in the areas observed on radar images.

On the radar images in 1988 the oceanic fronts were observed not only in the region of the first Kuroshio meander, but also to the east of it. Two anticyclonic eddies of the second branch of the Kuroshio ($147\text{-}151^\circ$) A5 and A6 stand out sharply on IR image in April 28 (Fig. 6). Their temperature was about 13-14 and $15\text{-}16^\circ C$, respectively. The SST contrast on the eddy boundaries was about $4^\circ C$. Both eddies had a high radar contrast against the frontal zone waters. In the vicinity of eddies the thermal front structure coincides well with σ° field variations.

Because of clouds, the radar and IR images (Fig. 6) can not make clear whether the southern boundary of the eddy A6 in the SST field fits the contrast zone 1 position. The cloudiness continued to cover this area in the succeeding days. However, the distinctive outlines of eddy 1 southern boundary and Kuroshio meander stepwise boundary 2 were observed clearly again on the radar image on May 1. Over three last days the eddies' configurations were kept, while their contrast to the background substantially decreased. Apparently, the hydrological characteristics of the eddies could not change greatly for such a short period, as shown by IR images. A possible reason of contrast decrease is the W weaking down to 0-3 m/s by formation of a vast low-gradient baric area by May 1 over the region considered. In contrast, wind speeds of 2-7m/s were observed on April 28.

The highest radar contrasts on radar images were associated with intrusion of the Kuroshio waters to the Oyashio area. In the middle of April the Kuroshio waters ($18\text{-}19^\circ C$) occupied the eddy A4 area and formed meander stretched to the north (Fig. 6(a)). Then a warm streamer of $16\text{-}17^\circ C$ was separated from the eddy A4, approached the eddy A3 boundary and began to enter into its area curving in clockwise direction (Fig. 7).

Waters of the warm streamer 1 and the eddy A4 2 were characterized by a very high level backscattered radar signals. The lowest values of σ° were

94

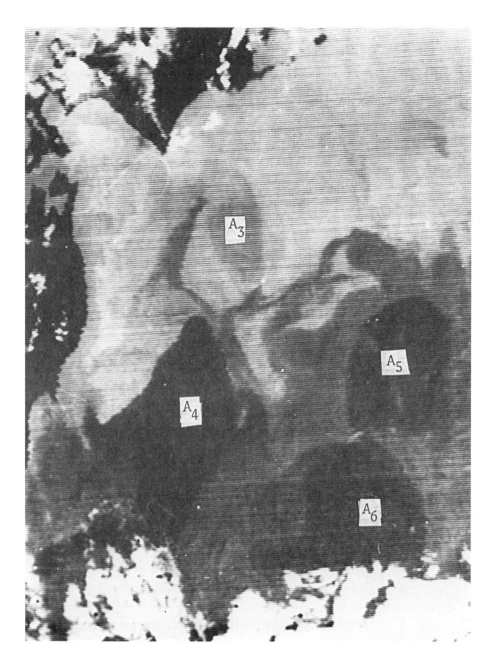

Fig. 6. Satellite images of the Kuroshio frontal zone on April 28, 1988 .
(a) NOAA-9 IR image at 0553 GMT.

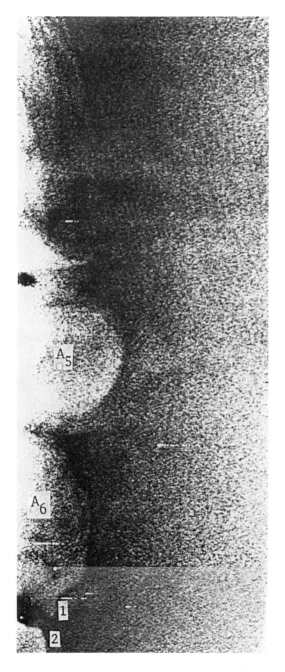

(b) Cosmos-1766 SLR image at 1525 GMT.

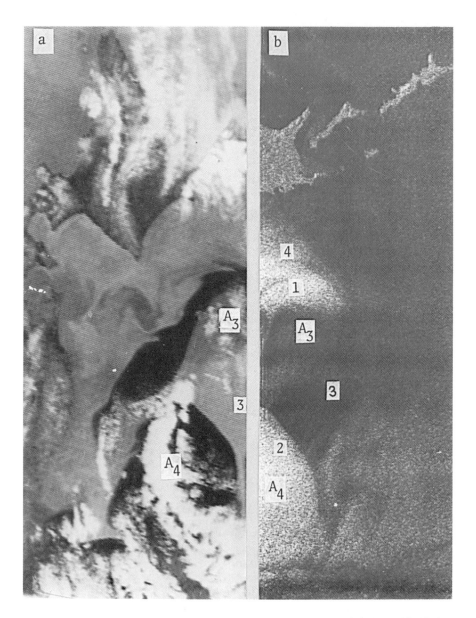

Fig. 7. Satellite images of the Kuroshio first meander. (a) NOAA-10 IR image at 2227 GMT on May 8. (b) Cosmos-1766 SLR image at 1435 GMT on May 7, 1988.

found in the Oyashio waters 3 going around the eddies A3 and A4 from the east. The northern boundary of the warm streamer was not remarkable because it coincided with the wind action area 4 while the eddy A4 boundary was in high contrast with the surrounding waters.

The horizontal distribution of SST across the eddy A4 eastern boundary on May 8 is shown in Fig. 8. The SST changes at the eddy boundaries are characterized by an abrupt jump by about 12°C over a distance of only 2-3 km. That area corresponds to a zone of sharp brightness contrast on the radar image. A sharp front on the edge of warm streamer occupied eddy A3 remained in the succeeding days and was visually observed from the weather ship" Ocean" on May 17-22. During 5 days of low windy weather differences in the surface state between lighter Oyashio and darker waters of warm streamer were marked well from the ship within a distance of 1-3 miles.

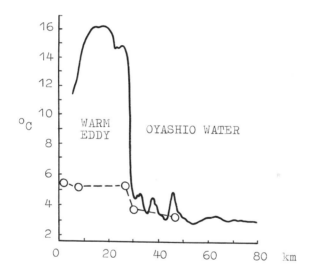

Fig. 8. Sea surface temperature (solid line) and air temperature (dashed line) distributions across the eastern edge of warm eddy A4 obtained by weather ship "Ocean" at 10-14 GMT on May 8.

Roughness differences of the Kuroshio and Oyashio surface waters are confirmed by visible images in sun glitter area. One of such images obtained on May 2 is given together with synchronous IR image in Fig. 9. The region of the first Kuroshio meander with eddies A3 and A4 is in sun glitter area. Higher brightness of cold waters 1 on the visible image is explained by the fact that the sea surface is smoother. Agreement between SST (IR image) and roughness (visible image) fields is high enough: in the brightness field the SST fine structure in warm water intrusion 2 and areas surrounding the frontal zone are outlined.

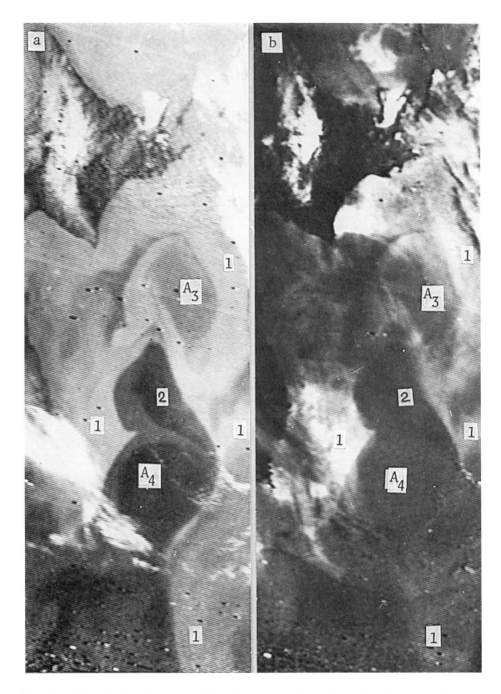

Fig. 9. NOAA-10 IR (a) and visible (b) images of the Kuroshio first meander at 2259 GMT on May 2, 1988.

3. DISCUSSION

Analysis of radar and IR images shows that the spectrum of ripples is correlated with the SST. The following problems arise. By which mechanisms is this relation accomplished and which hydrometeorological and other processes work for the indication of temperature inhomogeneities by radar technique?

The influence of the SST variations on spectral density of capillary-gravity waves may depend on change of the sea water characteristics and stratification of the atmosphere in the air-sea boundary layer caused by these variations. Two parameters, the surface tension coefficient α and a kinematic viscosity ν, which are functions of temperature, govern the spectrum of short surface waves of scales of centimeters. The dependence of σ^o on the SST obtained from processing of scatterometer data by Seasat satellite was explained by decrease of α with temperature increase (Woiceshyn et al., 1986). However, the change of α did not exceed 1.6% at a temperature difference of 6-8°C, which brings about no marked change in the spectrum for waves of $\Lambda=2$-4cm.

The dependence of the kinematic viscosity on water temperature is much stronger: ν = 1.84 cSt at t = 0°C, ν = 1.36 cSt at t = 10°C and ν = 1.06 cSt at t = 20°C (at a salinity of 35 ppt). As a result, the increase of water temperature leads to decreasing the minimum wind speed $W_o(\Lambda)$ at which waves of wavelength Λ begin to form. W_o values for edges of the radar swath are different since wavelength Λ, for which resonance scattering condition holds true, depends on incidence angle θ. At the near edge where θ = 21° and Λ = 4.1 cm, W_o is 2.7 m/s at t = 0°C, 2.4 m/s at t = 10°C and 2.1 m/s at t = 20°C. At the far edge (θ = 46°, Λ = 2.1 cm) W_o decreases from 3.8 m/s to 2.9 m/s with variation of t from 0°C to 20°C (Donelan and Pierson, 1987). For example, the dark tone of areas 2 (Fig.3(b)) and 1 (Fig.5(a)) possibly results from $W < W_o$. If the wind speed fluctuates, then the SST variation must affect the value of σ^o for $W > 4$-5 m/s also. Actually, if there an abrupt change in local gusts, patches of enhanced intense of ripples arise. The ripple amplitudes will depend on the SST because the damping decrement is given by $\gamma(t)=8\pi^2\nu(t)/\Lambda^2$ and the wave amplitude is damped with time τ in proportion to $\exp(-\gamma\tau)$. At τ = 10 s, the wave amplitude with wavelength Λ = 3 cm (resonance scattering condition for this wave is fulfilled at θ = 30°) at t=10°C will be about 1.5 times larger than that at t =0°C.

The above estimation refers to the pure sea water without surfactant film. However, in the Kuroshio confluence zone under consideration, water masses are different not only in temperature, but also in chemical and biological properties. In particular, the plankton concentration may be one order higher in cold Oyashio waters than in warm frontal waters and in the Kuroshio waters (Kun, 1969). The difference in plankton concentration is also revealed by remote sensing, as shown by satellite images obtained by multichannel scanning device from Meteor (Ginsburg and Fedorov, 1986) and CZCS from Nimbus-7 (Thomas and Emery, 1988).

The surface-active films are a product of the vital functions and decomposition of plankters. These films damp effectivly short capillary-gravity waves (Ermakov, 1987; Ermakov et al., 1987). Damping mechanism of films is not thoroughly understood. In the open ocean, three possible processes may come into play: change of energy inflow from wind, change of nonlinear interactions between waves, and direct influence of the surface film - damping. The damping of waves may be associated with decrease of surface tension of water. Then, when the surfactant concentration exceeds a value G_{min}, the elasticity of film $P = -G\partial\alpha(G)/\partial G$ and surface wave damping decrement increases by almost one order of magnitude within a narrow transition area of width $0.1G_{min}$, and then decrease slowly (Monin and Krasitskii, 1985; Ermakov, 1987). The viscosity increase in waters with films is an additional factor which increases γ (Carlson,1987).

There are a number of indirect mechanisms increasing σ^o of warmer waters. The main one is formation of unstable stratification of the atmosphere above them which increases the wind stress and therefore the surface roughness increase. For example, Fig. 8 shows SST and air temperature profiles across the eddy A4 boundary.

The horizontal current shear is another possible factor affecting the level of scattered radar signals. The water speed in synoptic eddies, streamers and the Kuroshio major branch is 1-2m/s. The abrupt drop of the speed down to 0.1-0.3 m/s occurs several miles apart from the flow boundary. Similar inhomogeneities of currents at wind speeds of 2 to 7 m/s cause a striking (two or three orders of magnitude) change of spectral intensity of the capillary- . gravity waves due to nonlinear effects (Van Gastel, 1987). Wind speed variations with respect to moving water surface should be also taken into account. Such cross-stream variations may reach 20-50% at wind speed of 4-6m/s.

Such an effect is observed in an eddy, where wind and current directions are the same on one side of the eddy and the opposite on the other side. Probably, inhomogeneities of the brightness field in the anticyclone A3 region (Fig. 3) are associated with this.

4. CONCLUDING REMARKS

Similarity of radar reflectivity and temperature fields can be used to study the oceanic frontal zones. Radar information is particularly useful in the regions covered by clouds. The observation of the oceanographic phenomena on radar images is, however, limited by the rather narrow range of the surface wind speed variability (from 2 to 8-10 m/s). To explane the relationship between σ^o and the SST, one should take into account the dependence of water viscosity on the temperature, the close connection between surfactant concentration and SST field, and the variation in the stability of the air flow above the sea surface caused by changes in SST. The determination of the relative contribution of each of the above mechanisms in radar contrast formation is a complicated problem and requires special satellite and subsatellite measurements. The effects of water temperature, film concentration, current speed variation and other parameters on roughness spectrum change with wave length Λ (Donelan and Pierson, 1987; Ermakov et al., 1987). Therefore, radar sensing of the sea surface at different wavelengths λ is of importance to solving this problem and to estimating potential power of radar techniques in the detection and study of oceanographic phenomena.

ACKNOWLEDGEMENTS

We wish to thank N.V. Bulatov (Pacific Institute of Fishery and Oceanography) for helpful discussion and reviewers for their comments and suggestions.

REFERENCES

Carlson, D.J., 1987. Viscosity of sea-surface slicks. Nature, 329: 823-825.
Donelan, M.A. and Pierson, W.J., 1987. Radar scattering and equilibrium ranges in wind-generated waves with application to scatterometry. J. Geophys. Res., 92: 4971-5029.
Ermakov, S.A., 1987. Film slicks on sea surface. In: A.V. Gaponov-Grekov and S.A. Khristianovich (Editors), Methods of Hydrophysical Research. Waves and Vortices. Inst. Pricl. Phys., Gorky, pp. 259-277 (in Russian).
Ermakov, S.A., Zuikova, A.M. and Salashin, S.G., 1987. Transformation of short wind wave spectra in film slicks. Izv. Acad. Sci. USSR, Atmos. Oceanic Phys., Engl. Transl. 23: 707-715.
Ginsburg, A.I. and Fedorov, K.N., 1986. Near-surface water circulation in the subarctic frontal zone from satellite data. Issled. Zemli iz Cosmosa, No. 1: 8-13 (in Russian).
Kun, M.S., 1969. Seasonal changes in mesoplankton constitution and distribution in the Kuroshio (observations of 1965-1966). Izv. TINRO, 68: 93-110 (in Russian).
Mitnik, L.M. and Victorov, S.V. (Editors), 1990. Radar Sensing of the Earth Surface from Space. Gidrometeoizdat, Leningrad, 200 pp. (in Russian).
Monin, A.S. and Krasitskii, V.P., 1985. Phenomena on the Ocean Surface. Gidrometeoizdat, Leningrad, 375 pp. (in Russian).
Thomas, A.C. and Emery, W.J., 1988. Relationships between near-surface plankton concentrations, hydrography, and satellite-measured sea surface temperature. J. Geophys. Res., 93: 15733-15748.
Van Gastel, K., 1987. Imaging by X band radar of subsurface features: nonlinear phenomenon. J. Gephys. Res., 92: 11857-11865.

SST STRUCTURE OF THE POLAR FRONT IN THE JAPAN SEA

Y. ISODA[1], S. SAITOH[2] and M. MIHARA[3]

[1]Department of Ocean Engineering, Ehime University, Bunkyo 3, Matsuyama 790 (Japan)

[2]Research Institute, Japan Weather Association, Kaiji Center Bldg.,5,4-chome, Kojimachi, Chiyoda-ku, Tokyo 102 (Japan)

[3]ECOH; Enviromental Consultant for Ocean and Human, Co., Ltd., Minamisenju, 59-7, 1-chome, Arakawa-ku, Tokyo 116 (Japan)

ABSTRACT

Horizontal structure and seasonal variability of the Polar front in the Japan Sea are examined by analyzing NOAA 9 infrared images (IR) and hydrographic data in 1987. The Polar front oriented in the west-east direction spreads from the Korean coast to the Tsugaru Strait. Its northern end corresponds to the location of a synoptic warm eddy trapped over the Yamato Rise. Besides, the Polar front has a different structure between western and eastern parts of the Yamato Rise. The western Polar front spreads over the Korean coast from 39°N to 42°N. Its structure is not so distinct that it disappears in an IR image in summer by the development of stratification due to a surface heating. On the other hand, the eastern front has a relatively remarkable SST (Sea Surface Temperature) gradient in all seasons.

INTRODUCTION

The Tsushima Current system is a major hydrographic feature in the Japan Sea. Warm and saline water of this current enters through the Tsushima Strait and flows northward above a homogeneous water of low temperature ($0° \sim 1°C$) and low salinity ($34.0 \sim 34.1$ psu) which is called the Proper Water in the Japan Sea. Then, most of the warm water flow out through the Tsugaru Strait. It is well known that this warm water can not spread to the northern end of the Japan Sea due to the formation of a thermal front in the central region. This thermal front has been called "Polar front" and is one of the important physical phenomenon for understanding the horizontal circulation patterns in the Japan Sea.

However, little is known about the observational characteristics of the Polar front in the Japan Sea. This lack in our knowledge mainly results from the limitation in time and space resolution of shipboard observation due to the large horizontal scale of the Polar front, about 1000 km. Recently, some numerical experiments, which were done by basically inflow-outflow model, were carried out in order to supplement the knowledge about dynamical structure of the

104

Tsushima Current system (for example, Yoon, 1982a,b; Kawabe,1982a and Sekine,1986). However, the Polar front was not reproduced in these models and its formation mechanism could not be clarified. Thus, we can not yet say that the fundamental structure of the Polar front in the Japan Sea has been understood fairly well. The Polar front should be interpreted as the boundary between subarctic and subtropical waters, i.e. the Proper Water in the Japan Sea and Tsushima Current water. A further work will be important to name a more appropriate terminology for this front. But we shall use "Polar front" through this paper for convenience.

Satellite infrared (IR) images are useful to examining the horizontal SST structure of the Polar front. Previous studies (Toba et al.,1984; Isoda and Saitoh,1988 and Sugimura et al.,1989) indicate a usefulness of IR images for investigating the variability of the Tsushima Current. After a search of all images published in the JMA-NOAA data catalog(1987), it was found that 105 partial cloud free images are available in the Japan Sea for 1987. In the present study, we discuss the horizontal structure and seasonal variation of the Polar front on the basis of the results from analysis of IR images and hydrographic data in the Japan Sea for 1987. Although the present study is a descriptive one, it will give useful information about future investigation on the genesis of the Polar front.

PROCESSING OF THE SATELLITE IR IMAGES

The data utilized in the present analysis consist of 105 IR images collected by NOAA 9 AVHRR from January to December, 1987 as shown in Table 1. Our study area is shown by the thick line in Fig. 1. We chose the AVHRR channel 4 data for observing SST. These selected IR images are partially affected by clouds. Isoda and Saitoh(1988) developed a composite method removing the cloud data by

TABLE 1

Temporal distribution of NOAA 9 AVHRR images selected in this study.

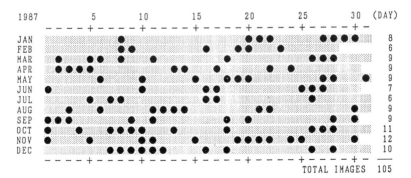

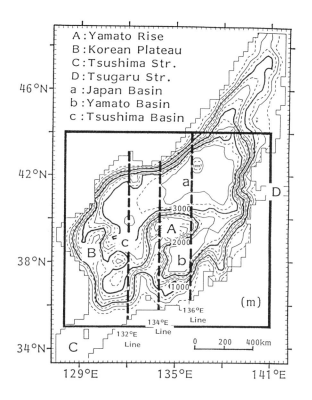

Fig. 1. The bottom topography in the Japan Sea. Thick outlined region is the area covered by the NOAA 9 AVHRR imagery utilized in this study.

divided into appropriate grids according to the spatial scale for the phenomena of concern. Although information of small scale eddies with short life time is crashed by this method, the synoptic SST structure can be represented by selecting an appropriate grid size and composite period.

The major hydrographic features in the southern interior region of the Japan Sea are meandering paths of the Tsushima Current or synoptic scale eddies, of which scales are a few hundred km (e.g. Kawabe,1982b). Figure 2 shows the trajectories of warm and cold eddies in 1987. Their positions show considerable temporal variations; for example, the northeastward shift of a warm eddy lying to the west of the Sado Island and the counter-clockwise rotation of a warm eddy between the Noto Peninsula and the Oki Islands. However, their moving distance in one month is shorter than several tens of kilometers. This may indicate that the synoptic eddy structure can be also represented by the monthly composite

(a) Warm eddy

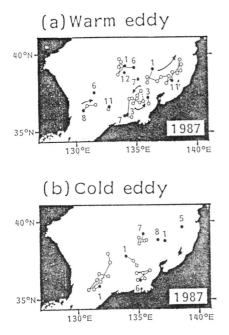

(b) Cold eddy

Fig. 2. The trajectories of (a)warm and (b)cold eddies in 1987(after the intantaneous ocean map of JMSA; 1987). Number indicates the month.

Fig. 3. Temporal change in the spatial mean SST (open circle) and spatial standard deviation (thick line) of the composite IR images in 1987.

images. Therefore, we chose 0.25°mesh (approximately 27×20km) as the spatially averaged grid size and one month as the composite period. Then, we made up 12 monthly composite IR images in 1987 with a method by Isoda and Saitoh(1988). The right hand side numbers in Table 1 denote the composite numbers in each month.

MONTHLY COMPOSITE IR IMAGES

Figure 3 shows the spatial mean and standard deviation of SST in each monthly composite IR image. The seasonal variation of spatial mean SST is characterized by the minimum (about 5°C) in January through April, and the maximum (about 17°C) in July through September. Such a seasonal variation of SST may be affected by both the heat exchange through the sea surface and the volume transport of warm waters through the Tsushima Strait. Although the volume transport increases in summer, spatial standard deviation of SST tends to become minimum. This may be because a strong stratification is developed at shallower than about 50 m depth due to a surface heating (Minami et al.,1987). On the other hand, oceanic structure in winter may easily appear in an SST due to the vertical mixing of a surface cooling. Figure 4 shows the composite IR images for even months in 1987. Warm waters extend over the southern part of the Japan Sea in all seasons. These warm waters seem to converge at the south of the Tsugaru Strait (the location of Tsugaru Strait is denoted by the black arrow in Fig. 4) and afterward to flow northward along the Japanese coast. In the interior region of the Japan Sea, a strong SST gradient region, i.e. a thermal front, runs in the east-west direction with significant meandering, particularly from winter to spring. Note that this front is not connected to the southern entrance of warm water and in direct contact with the Korean coast. We hereafter call this thermal front in the central region of the Japan Sea, the Polar front.

Meanders of the Polar front from winter to spring correspond well to the bottom topography around the Yamato Rise (Figs. 1 and 4). That is, the existence of a synoptic eddy just above the Yamato Rise can be inferred from such a meandering. Moreover, the sharpness of the Polar front is different between the eastern and western parts of the Yamato Rise. That is, the western front structure is weaker than the eastern one. The western Polar front spreads over the Korean coast and is divided into two parts around 39°N and 42°N. The Polar front around 39°N tends to become sharp only near the Korean coast. Then, this western Polar front cannot be traced in an IR image in the heating season from June to August and another thermal front is formed from the Tsushima Strait to the Yamato Rise in August. The SST structure of the Polar front in the western region begins to appear again in October, when a sea surface cooling begins. On the other hand, the eastern Polar front from the Yamato Rise to the Tsugaru Strait has a relatively strong SST gradient. Especially, the Polar front around the Tsugaru Strait is confined to the Japanese coastal region and can be detected in all seasons as maintaining the sharp thermal front. These results may indicate that the Polar front is influenced by a synoptic scale structure above the Yamato Rise and has a different SST structure between its western and eastern parts. Hence, we shall investigate the relationship between the Polar front and the bottom topography around the Yamato Rise. Figure 5 shows SST distributions in each monthly IR images and the section of bottom topography along 132°E, 134°E and 136°E (thick broken lines in Fig. 1). Seasonal variations of SST common to the three lines are as follows. The increase of SST from March to August is uniform everywhere and the Polar front loose its clearness. On the other hand, the decrease of SST from September to Febrauary makes the Polar front clear again. Such SST variations show that SST is mainly influenced by the heat exchange through the sea surface. That is, oceanic structure will be masked or weakened by the development of a surface stratification in summer. Therefore, we cannot judge that the western Polar front does not exist in summer from only SST data. The sharpness of the Polar front along 134°E is stronger than those

108

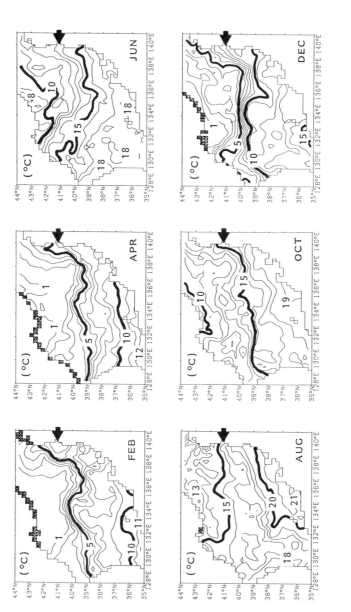

Fig. 4. The composite IR images for even months in 1987. ⊠ denotes the grid lacked by the removal of cloud data. The black arrow shows the location of the Tsugaru Strait.

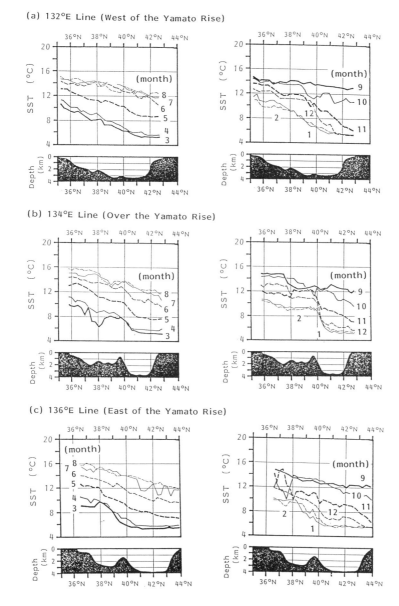

Fig. 5. SST profiles in each composite IR image (upper) and the section of bottom topography (lower) along 132°E(a), 134°E(b) and 136°E(c) lines shown in Fig. 1.

along other lines. Besides, its location Ts also stable in time above the northern end of the Yamato Rise. Thus, it is suggested that a strong oceanic structure trapped above the Yamato Rise exists in all seasons.

HORIZONTAL DISTRIBUTION OF TEMPERATURE AT 100 M DEPTH

Japanese hydrographic data are quantitatively insufficient to cover the whole Japan Sea, and we investigate the oceanic conditions only around the Yamato Rise. Figure 6 shows the horizontal distributions of temperature at 100 m depth in 1987 after the "ocean instantaneous map" compiled by JMSA(1987). The 2000 m isobath is indicated by a thick broken line.

The cold water region occupies both sides' basins of the Yamato Rise, i.e. Yamato Basin and Tsushima Basin. The warm water region is stably confined above the Yamato Rise in all seasons. A warm eddy is formed above the western part of the Yamato Rise, although its horizontal temperature gradient is not always large. This warm eddy location is consistent with one inferred from the SST meandering of the Polar front in an IR image. Thus, it is confirmed that the spatial pattern of warm and cold water regions well corresponds to the bottom topography. This relationship is also found in other reports based on the saline water distributions (Tanioka ,1962 and Ogawa, 1974). Moreover, Kolpack (1982) suggested that the position of the Polar front in the Japan Sea is controlled by the bottom topography. These reports support the present view that the synoptic eddies in the Japan Sea are affected by the bottom topography, especially over the Yamato Rise.

CONCLUSION AND DISCUSSION

The Polar front oriented in the east-west direction is formed in the central region of the Japan Sea. Some fundamental properties of this Polar front are understood through the present study using the monthly composite IR images in 1987. We point out the following three points which are considered to be important subjects.

(1) The northern end of the Polar front corresponds to the location of a synoptic warm eddy trapped over the western part of the Yamato Rise. This warm eddy stably exists in all seasons. Therefore, the Polar front depicts significant meandering around the Yamato Rise.

(2) The sharpness of the Polar front is different between in the eastern part and western part of the Yamato Rise. The eastern front is stronger and more stable than the western one in all seasons, particularly around the exit of the Tsugaru Strait.

(3) The eastern end of the Polar front connects to the Tsugaru Strait or along the Japanese coast. Such a feature shows that warm waters of Tsushima Current flow out through the Tsugaru Strait or Soya Strait. On the other hand, its western end is in direct contact with the Korean coast form 39°N to 42°N. It is known that the East Korean Warm Current flows northward from the Tsushima Strait along the Korean Peninsula. Nevertheless, a significant northward flow can not be inferred from the composite IR images. In a cloud free image off the Korean

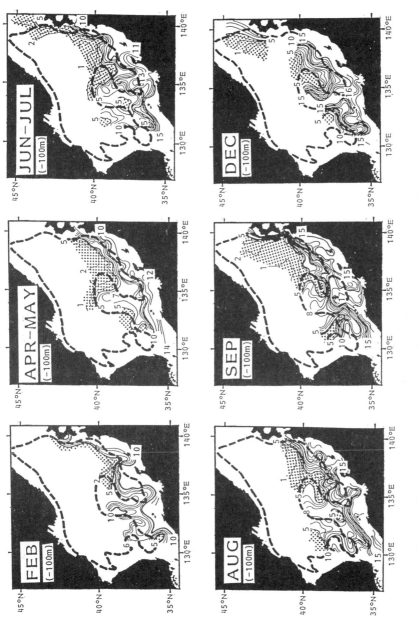

Fig. 6. Horizontal distributions of temperature at a depth of 100 m in 1987 (after JMSA;1987). The 2000 m isobath is indicated by a thick broken line.

coast (not shown), some synoptic eddies with the fork-like splitting of warm water are often observed; however, it seems that the behaviors of eddies are complicated.

These fact suggest that a synoptic warm eddy over the Yamato Rise plays an important role in the formation of the Polar front. Therefore, more detailed observation, focusing on the dynamic structure of this warm eddy should be undertaken in the feature. In addition, intensive works are needed to investigate the dynamical structure of the East Korean Warm Current.

ACKNOWLEDGMENTS

The authors express their sincere thanks to Dr. T. Yanagi of Ehime University for his discussion and encouragement during this study. The data analysis was carried out on a FACOM M 360 AP of Ehime University.

REFERENCES

Japan Weather Association, 1987. NOAA data catalog.

Japan Maritime Safety Agency, 1987. Ocean instantaneous map.

Isoda,Y. and S.Saitoh, 1988. Variability of the sea surface temperature obtained by the statistical analysis of AVHRR imagery. - A case study of the south Japan Sea -. J.Oceanogr.Soc.Japan, 44, 52-59.

Kawai,H., 1974, Transition of current images in the Japan Sea. In: The Tsushima Warm Current - Ocean structure and fishery. ed.by Fishery Soc. Japan,Koseisha-Kouseikaku, pp.7-26. (in Japanese)

Kawabe,M., 1982a. Branching of the Tsushima Current in the Japan Sea, Part 2. Numerical experiment. J.Oceanogr.Soc.Japan, 38, 183-192.

Kawabe,M., 1982b. Branching of the Tsushima Current in the Japan Sea, Part 1. Data analysis. J.Oceanogr.Soc.Japan, 38, 95-107.

Kolpack,R.L., 1982. Temperature and salinity changes in the Tsushima Current. Lar mer, 20, 199-209.

Minami,H,Y,Hashimoto,Y.Konishi and H.Daimon, 1987. Statistical features of the oceanographic condition in the Japan Sea. Umi to Sora, 4, 163-175. (in Japanese)

Ogawa,Y., 1974. The relation between the high saline water in the Japan Sea and Tsushima Current. Bull.Japan Soc.Fish.Oceanogr., 24, 1-12. (in Japanese)

Sekine,Y., 1986. Wind-driven circulation in the Japan Sea and its influence on the branching of the Tsushima Current. Progress Oceanography, 17, 297-312.

Sugimura,T.,S.Tanaka and Y.Hatakeyama, 1984. Surface temperature and current vectors in the sea of Japan from NOAA-7/AVHRR data. pp.133-147. In: Remote sensing of shelf sea hydrodynamics, ed.by Jacques C.J.Nihoul,Elsevior, 38.

Tanioka K., 1962. A review of sea conditions in the Japan Sea(2) - on the cold, warm and saline water regions -. Umi to Sora, 38, 115-128. (in Japanese)

Toba,Y.,H.Kawamura,F.Yamashita and K.Hanawa, 1984. Structure of horizontal turbulence in the Japan Sea. pp.317-332. In: Ocean Hydrodynamics of Japan and East China Sea. ed.by T.Ichiye,Elsevier, 39.

Yoon,J.H., 1982a. Numerical experiment on the circulation in the Japan Sea. Part 1. Formation of the East Korean warm Current. J.Oceanogr.Soc.Japan, 38, 43-51.

Yoon,J,H., 1982b. Numerical experiment on the circulation in the Japan Sea. Part 2. Formation of the nearshore branch of the Tsushima Current. J. Oceanogr.Soc.Japan, 38, 119-124.

A NUMERICAL EXPERIMENT ON THE SEASONAL VARIATION OF THE OCEANIC CIRCULATION IN
THE JAPAN SEA

YOSHIHIKO SEKINE
Institute of Oceanography, Faculty of Bioresources, Mie University,
1515 Kamihamachou, Tsu Mie 514 (Japan)

ABSTRACT
    The seasonal variation of the general circulation in the Japan Sea is
studied numerically with special reference to the branching of the Tsushima
Current. A two-layer model with bottom and coastal topograpgy of the Japan Sea
is used. The seasonal change in the inflow corresponding to the Tsushima
Current and two outflows corresponding to the Tsugaru and Soya Currents are
given as open boundary conditions. The wind-driven circulation dominates in
winter, whereas the circulation in summer is due to the in-and outflow . In
winter, a strong anti-cyclonic circulation is formed in the deep Japan Basin
by the intensified wind stress with negative curl. In summer, the intensified
inflow flows along the Japanese Coast and generates an anti-cyclonic
circulation in the Japan Basin. The circulation in spring and autumn is
transitional from the winter to summer regime and from the summer to winter
regime, respectively. On the whole, the Tsushima current has three dominant
branches in the Japan Sea, the coastal branch along the Japanese main Islans,
the offshore central branch and the western boundary branch, of which
remarkable seasonal variations depend on the variations of the in- and outflow
and wind stress.

INTRODUCTION

    A major feature of the current system in the Japan Sea is an inflow of the

Tsushima Current through the Tshushima Strait. The Tsushima Current splits

into three branches (Suda et al.,1933; Uda,1934,36); one branch flows along

the Japanese coast, the second branch flows in a central region and the third

branch runs along the Korean Coast (Fig. 1). However, Moriyasu (1972a,b)

considered this flow pattern as a large meander path rather than the branching

into the three barnches, because the Tsushima Current shows a clear large

meander at the southern Japan Sea. Because of the interannual and seasonal

variation of the inflow, the details of the Tsushima Current in the Japan Sea

have not been clarified yet.

    So far, some theoretical studies on the Tsushima Current were carried out

(Table 1). The flow along the Japanese Coast is due to the topographic guiding

effect along the geostrophic contour (f/h), where f is the Coriolis parameter

and h is the water depth (Yoon,1982c), which is hereafter referred to as the

topographic branch (TPB). However, the direct current measurements by

Matsuyama et al. (1986) showed the baroclinic structure of the TPB, which is

not always controlled by the barotropic flow character along f/h. The second

dominant flow is along the Korean coast, and is identified with the Korea Warm

114

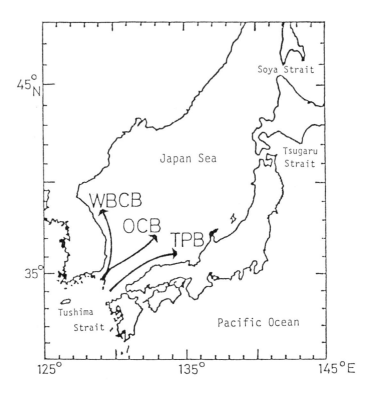

Fig. 1. Schematic representation of three representative branches of the Tsushima Current flowing into the Japan Sea. Letters TPB, OCP and WBCB mean the topographic branch along the Japanese Coast, the offshore central branch and the western boundary current branch corresponding to the Korean Warm Current, respectively.

TABLE 1

Model characteristics of the Japan Sea so far developed.

Author	Density stratification	Coastal topography	Bottom topography	Wind	Heat*	Inflow
Yoon (1982 a)	Yes	Rectangular	Flat	Constant	Constant	Stationary
Yoon (1982 b)	Yes	Realistic	Flat	Seasonal variation	Seasonal variation	Stationary
Yoon (1982 c)	No	Rectangular	Simplified	No	No	Stationary
Kawabe (1982b)	Yes	Rectangular	Simplified	No	No	Variable
Sekine (1986)	Yes	Realistic	Realistic	Seasonal variation	No	Stationary
Present study	Yes	Realistic	Realistic	No	No	Seasonal variation
Present study	Yes	Realistic	Realistic	Seasonal variation	No	Seasonal variation

* Constant, Seasonal variation and No mean stationary thermal condition, seasonally varying thermal boundary condition and no consideration on the thermal condition, respectively.

Current. It is formed by the planetary β effect (Yoon,1982a,b), referred to as the western boundary current branch (WBCB). The last dominant flow in an offshore central region (OCB) has not been clarified well by the historical observation. Kawabe(1982b) showed that this branch is caused by the propagation of the lowest two modes of upper shelf waves which are generated by the increase in the inflow of the Tsushima Current in summer. On the other hand, Sekine(1986) pointed out that because the wind stress is very weak in summer, this branch is formed by the westward shift of the kinetic energy of the TPB carried by planetary Rossby waves.

Kawabe(1982a) showed from historical data analysis that the TPB along the Japanese coast exists at least from spring to summer, the OCB in summer only and the WBCB throughout the year. Sekine (1986) demonstrated by a numerical model that the observed branching of the Tsushima Current except for summer season is well simulated by the seasonally varying wind stress; the WBCB predominates in winter under the strong wind stress, whereas the TPB is intensified in spring to summer because of the positive wind stress curl over the Japan Sea, which intensifies cyclonic circulations such as the TPB, but weakens anti-cyclonic circulations such as the WBCB. However, Sekine (1986) failed to simulate the WBCB in summer, which is observed throughout the year. This may be due to a constant in-and outflow given at the boundary (see, Table 1).

Figure 2 displays the observed current velocity at the Tsushima Strait by Inoue et al.(1985). A weak inflow with small vertical shear is seen in winter, but the inflow is intensified in summer and the vertical shear becomes prominent in autumn. Sekine(1989) pointed out that because the surface mixed layer penetrates to the bottom in the Tsushima Strait from January to April, the inflow is blocked by the sill topography of the strait, which is suggested by the barotropic (no density stratification) flow character along the isopleth of depth (e.g., Pedlosky, 1979). In contrast to this, density stratification develops in late summer to autumn and induces a strong inflow

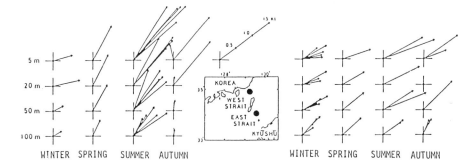

Fig. 2. Seasonal change in the current velocity at two stations in the Tsushima Current (after: Inoue et al.,1985). Locations of the two stations are shown by two solid marks in the map.

with large vertical shear.

Although a constant inflow is given in the model of Sekine (1986), there is a possibility that the seasonal variation in the inflow of the Tsushima Current has an important role in the circulation of the Japan Sea with reference to its branching. For this reason, the effects of seasonal change in the in- and outflow together with the effects of seasonal change in wind stress are examined in the present study (Table 1). The numerical model will be described in the next section. Results are presented in section 3. Summary and discussion will be given in section 4.

2 NUMERICAL MODEL

Figure 3 shows the schematic representation of the model ocean. Coastal and bottom topography are modelled in a simplified form. A two-layer model is used. The basic equations by hydrostatic balance, ß-plane, rigid lid and Boussinesq approximations are the same as in Sekine (1986). The coefficient of horizontal eddy viscosity is 10^7 cm^{2}s^{-1} and the reduced gravity 2.77cms^{-2} (for details of the basic equations, see Sekine, 1986). The annual cycle of the wind stress is obtained by linear interpolation of the monthly mean wind stress data by Japan Meteorological Agency (1972) and Kutsuwada and Sakurai(1982), which is displayed in Fig. 4. In winter, a strong northwesterly from the Siberian High prevails, while the wind stress is very weak in summer. The annual change over the whole domain is shown in Fig. 4(b). The average of wind stress curl is negative in winter, but positive in April to July.

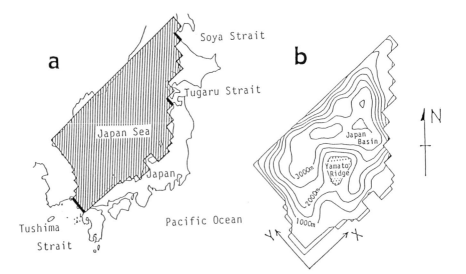

Fig. 3. (a) Domain of the model ocean (hatched region), (b) bottom topography of the model shown by isobaths (m). Contour interval is 500m and shallow area is stippled. Three thick lines on the coastal boundary show the region where in-and outflow are imposed.

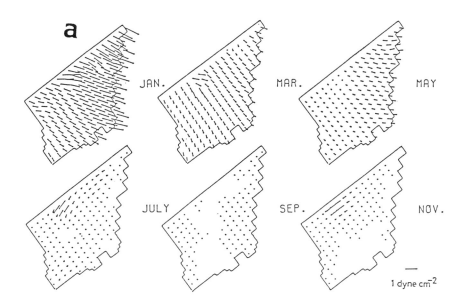

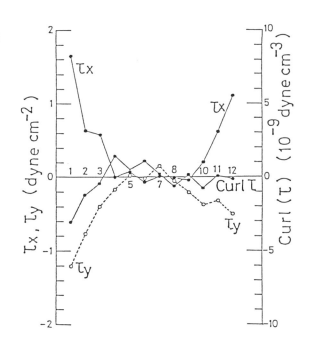

Fig. 4. (a) Monthly mean wind stress data for 1958 – 75 (after Japan Meteorological Agency,1972; Kustuwada and Sakurai,1982). No vectors less than 0.05 dyne cm⁻² are plotted. (b) Mean x-compenent of monthly mean wind stress (τx), mean y-component (τy) and wind stress curl over the whole model domain (for the directions x-and y coordinate ,see Fig.3).

The inflow velocity in the upper and lower layers is assumed to be $U_0 \sin(\pi y/L)$, where U_0 is the maximum inflow velocity and L is the width of the inflow. Similar velocity profiles are assumed at the two outflow boundaries. The annual change in the inflow is shown in Fig. 5 in terms of U_0 for both layers. The velocity and total volume transport represent the observed features shown in Fig. 2. It is also assumed that in all seasons, 65% (35%) of the inflow volume flow out through the Tsugaru (Soya) Strait.

In order to see the individual influences of the in- and outflow and wind stress on the current system in the Japan Sea, numerical experiments are performed in three cases. The first case is driven by the in- and outflow and no wind stress is imposed. The second case is driven by both the in- and outflow and wind stress. The third case is the same as the second case except for the inflow condition. Since the seasonal variation in the inflow is more remarkable in the western part of the Tsushima Strait (Kawabe, 1982a), the seasonal change in the inflow is confined to the western two thirds of the boundary , while the stationary inflow with the minimum velocity of January is given at the eastern one third. For these three cases, the time integration of the basic equations starts by the wind stress on 1 January. The initial flow is given along the southern boundary from the inflow region to the two outflow boundaries.

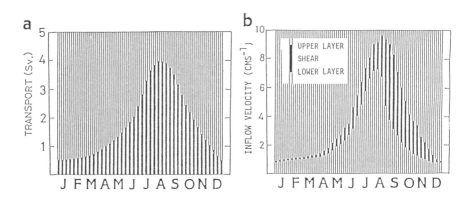

Fig. 5. Seasonal change in the inflow given in the model.(a) Total volume transport, (b) maximum velocity (U_0) of the inflow with the sinusoidal distribution in the horizontal direction across the inflow boundary: u = $U_0 \sin(\pi y/L)$, where L is the width of the inflow.

3 RESULTS

A time integration was carried out for 10 years for the three cases. As was shown by Sekine (1986), periodic annual changes in kinetic energy and flow patterns are clearly seen in the numerical solution. Therefore, the solution for the last one year is examined in each case.

The annual velocity field in the first case is displayed in Fig. 6. Except in late spring, the upper layer inflow has a tendency to flow along the northern boundary, which corresponds to the WBCB. However, in summer, the upper layer inflow flowing along the southern coast also dominates, which corresponds to the TPB. From late summer to autumn, a flow is formed in the offshore central region by the westward shift of the kinetic energy of the intensified TPB. Although this flow may be different from the OCB denoted by Kawabe(1982a,b), it is possibly formed in an area west of the TPB by this process. In late autumn, it is carried further westward and absorbed in the WBCB. From autumn to winter, most of the upper layer inflow feed the WBCB and the TPB disappears in the upper layer in November.

Because of the predominance of bottom intensifeid mode (Rhines, 1970; Suginohara, 1982), the velocity field in the lower layer has a strong tendency to flow along the geostrophic contour, which is approximated by the isobath in a local area where change in the Coriolis parameter is relatively small. Because there exists a relatively wide continental slope in offshore of the Japanese Coast, an inflow in the lower layer is forced to flow along the southern boundary in all the season. From summer to autumn, a prominent anti-cyclonic circulation is formed in the northern Japan Basin with a deep and relatively flat bottom (see, Fig. 3b). This circulation includes the northern boundary current and the westward flow along an isobath north off Yamato Ridge. In particular, the latter westward flow along the bottom slope may be connected with a sharp front observed here. Although the main generation process of the observed front may be due to other processes, an important role of the continental slope off Yamato Ridge is suggested from this experiment.

The annual variation in total volume transport function in the upper and lower layer is shown in Fig. 7. To the south of the outflow corresponding to the Tsugaru Strait, most of the TPB turn westward and flow along an isobath, which generates an anti-cyclonic circulation in the northern half of the basin. This anti-cyclonic circulation is more remarkable from late spring to summer, but decays in autumn. It is clear that this circulation is formed by the increase in inflow transport. At the inflow region, the WBCB is weak in spring, but intensified gradually in late summer to autumn. This is caused by the development of the ocean response to the strong inflow and the kinetic energy shifts to the western boundary region with lapse of time. If the stationary in- and outflow is imposed, the WBCB is not clear in summer (Sekine, 1986). Therefore, the formation of the WBCB in summer is conclusively due to the strong inflow transport that gives the WBCB the kinetic energy.

The seasonal change of the velocity field in the second case driven by both the in-and outflow and wind stress is characterized by the formation of a strong anti-cyclonic circulation in winter (Fig. 8). This circulation is generared by the strong wind stress in winter with the negative curl (see, Fig. 4). However, as the wind stress is weakened in spring, the center of the

120

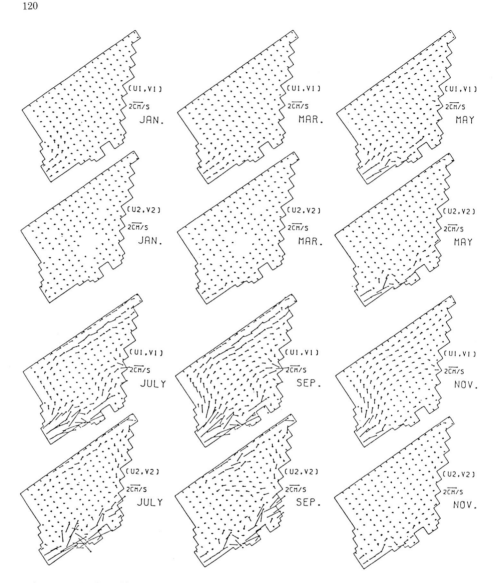

Fig. 6 Velocity field of the upper (U1,V1) and lower (U2,V2) layer in the case of no wind stress. Vecotors are plotted at every other grid point. No vectors smaller than 0.05 cms⁻¹ are plotted.

anti-cyclonic circulation moves westward and decays in late spring. The TPB and offshore central flow are generated in summer. Because the wind stress is weak in the latter half of the year, the velocity field is similar to that in the first case in this period, which is displayed more clearly by stick diagrams of the velocity vectors (Fig. 9). It is thus concluded that the wind-driven circulation is dominant in winter and the circulation is mainly due to the in-and outflow in summer. Furthermore, the process of the oceanic responce to these two events is achieved : the circulation in spring is

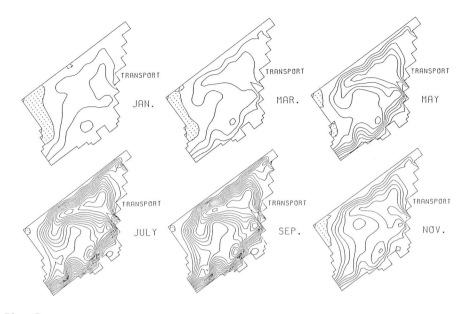

Fig. 7. Seasonal change in the calculated volume transport function. Contour interval is 0.2 Sv ($1Sv = 10^{12}cm^3s^{-1}$) and regions with negative transport function (cyclonic circulation) are stippled.

characterized by spin-down of the anti-circulation circulation formed by strong wind stress in winter. On the other hand, the main process in autumn is a westward shift of the kinetic energy given by the enhanced TPB in summer. The difference in these flow patterns from those of Sekine (1986) with a constant inflow of 2 Sv is small in winter but significant in summer to autumn when an inflow larger than 2 Sv is specified in the present study.

The volume transport function in the second case and the net wind-driven circulation which is obtained by subtracting the transport function in the first case from that in the second case, are displayed in Fig. 10. In the former half year, the seasonal change in the total volume transport is characterized by the formation of the anti-cyclonic circulation in the northern basin in winter and its gradual spin-down in the following period. From winter to spring, a northeastward flow in the TPB contacts a southwestward counter flow in the southern part of the anti-cyclonic ciruclation, which attenuates the TPB. The process depends on the relative intensity of the in- and outflow and the anti-cyclonic circulation driven by the wind stress. Because the simulated volume transport of the anti-cyclonic circulation exceeds 4 Sv in the present model, larger than an observed inflow (2 Sv at most) in winter (Inoue et al., 1985), the TPB becomes weak . This results is common to the case with the larger inflow of 2 Sv (Sekine, 1986). Hence, the TPB is expected to be weakened in winter. In spring, cyclonic eddies are formed in the contact region. In the northern region, the formation of the anti-circulation is slow in this case in comparison with the first

122

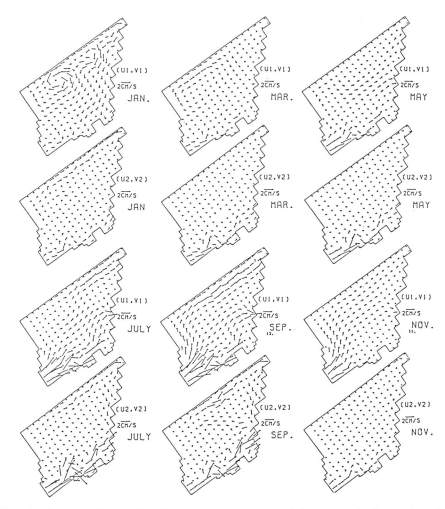

Fig. 8. Same as Fig. 6 but for the second case with seasonal change in wind stress.

case, which is clearly seen in the transport function in May. The anti-cyclonic circulation is weakened by the positive wind stress curl in late spring to early summer, which generates the cyclonic circulation in the Japan Basin (Fig. 10b). In late summer to autumn, because the wind stress is very weak, no remarkable difference is detected between the two cases.

Figure 11 shows the results in the third case with the enhanced seasonal variation of inflow at the western part of the boundary. The difference between the second and third case is relatively remarkable from summer to late autumn and in mid winter (Fig. 11b). In summer, the flow in the central region is more intensified than in the second case. However, this intensified flow is shifted westward more quickly and absorbed in the WBCB in late autumn. The upper layer velocity field in November indicates that the WBCB in autumn is

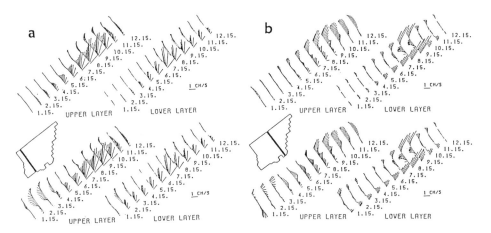

Fig. 9. Stick diagram of the upper and lower mean velocities across (a) near the inflow boundary, (b) central region. Case with no wind stress (upper panel) and case with seasonal change in wind stress(lower panel). The thick lines show the sections where the mean velocities are calculated.

caused by the westward shift of the kinetic energy of the TPB formed by the intensified inflow in summer. The difference between the two cases in winter is caused by the combined effect of continuing westward shift of the kinetic energy of the TPB and the strong wind stress in winter. Figure 12 implies that the westward shift of the kinetic energy is accomplished in January. After this, the difference is relatively small until the increase in the inflow begins to be intensified.

Two cyclonic circulations are combined near the inflow and outflow regions, which is clearly shown in the volume transport function (Fig. 13b). One cyclonic circulation is caused by intensified shear of the western part of the inflow, while the other is caused by the enhanced flow in the central region. On the whole, there exists no significant difference in the total flow pattern between the second and third cases. However, the seasonal variation of the inflow confined into the western two thirds of the boundary yields fast oceanic response to the increased inflow in summer, and the central offshore flow and the WBCB are formed faster than in the second case.

4 SUMMARY AND DISCUSSION

The seasonal variation of the circulation in the Japan Sea has been examined numerically. A two-layer model is used and the observed in- and outflow is imposed through the boundaries. The main results are summarized as follows:
(1) In winter, a strong anti-cycyclonic circulation is formed in the northern Japan Basin by the intensified wind stress with negative curl. Because the inflow through the Tsushima Strait is weak, the TPB is suppressed by a counter current generated by the anti-cyclonic circulation. These results agree with those of Sekine (1986) with a stationary inflow of 2Sv.

124

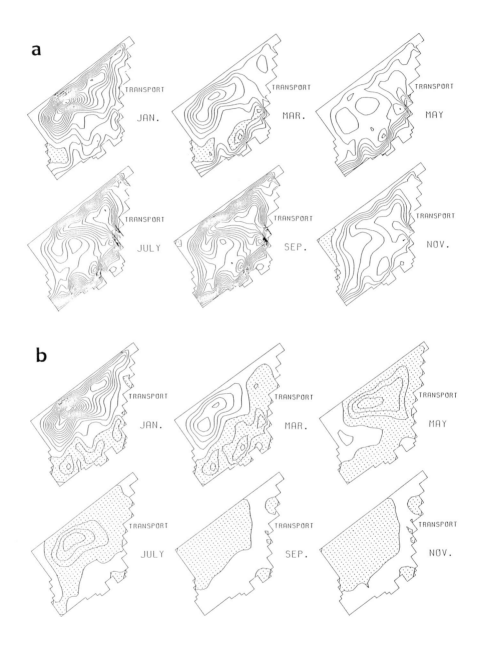

Fig. 10. (a) Same as Fig. 7 but for the second case with seasonal change in wind stress. (b) Transport of the wind-driven circulation obtained by subtracting the transport function of the first case (Fig. 7) from that of the second case shown in (a).

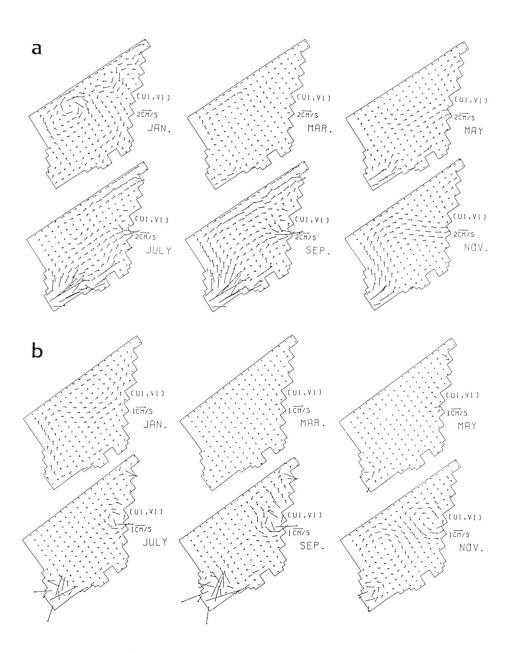

Fig. 11. (a) Seasonal change in the upper layer velocity in the third case with the seasonal change confined to the western part of the inflow. (b) Difference of the velocity vectors. (third case) minus (second case).

126

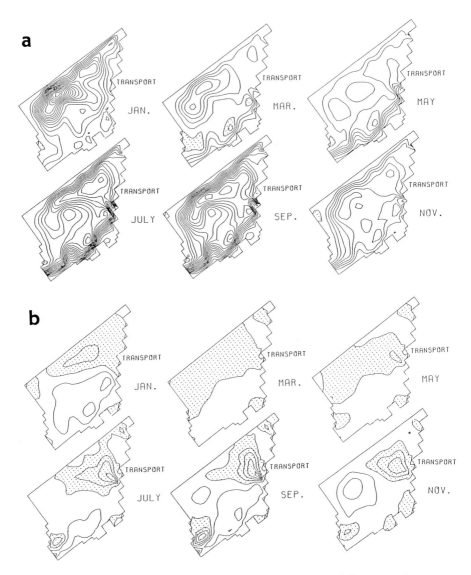

Fig. 12. (a) Same as Fig. 7 but for the third case.(b) Difference of transport function . |third case| minus |second case|.

(2) In spring to summer, the increased inflow flows along the Japanese Coast and forms the TPB. But most of the increased inflow does not flow out directly through the Tsugaru Strait, but forms an anti-cyclonic circulation. In late summer, the WBCB is formed gradually by the westward shift of the kinetic energy of the TPB. These results are not simulated in Sekine(1986), which suggests that the intensified inflow in summer has an important role in the circulation in the Japan Sea.

(3) In autumn, the flow pattern is chracterized by the process toward

accomplishing the oceanic response to the increased inflow in summer. In early autumn, formed are the central flow to the west of the TPB in the upper layer and a westward current along the northwestern bottom slope of the Yamato Ridge. In late autumun, the TPB and offshore central flow are absorbed in the WBCB. This flow pattern is maintained until the strong anti-cyclonic circulation is formed by strong wind stress in January. (4) On the whole, the total flow pattern in the Japan Sea is influenced by the seasonal change in wind stress and in- and outflow. The Tsushima Current has three possibile branches in the southern Japan Sea; the first branch along the Japanese Coast formed by the topographic guiding effect of the continental slope, the second branch to the west of the first branch formed by westward shift of the kinetic energy of the first branch, and the third branch along the Korean Coast formed by the planetary β effect (westward intensification).

(5) The seasonal variation in the current system of the Japan Sea is characterized by the wind-driven circulation in winter and by the enhanced increased in-and outflow in summer. The spring and autumn flow patterns are of transitional character from winter to summer regime and from summer to winter regime, respectively.

A next problem is to investigate the role of thermohaline circulation. However, modeling of the thermohaline ciculation is very difficult because of its sensitive dependence on the value of the coefficient of vertical eddy diffusivity (e.g., F.Bryan, 1987); if a larger (smaller) value of the coefficient is assumed, the thermohaline (wind-driven) ciculation controlls the current system. Another problem is that the frontal structure of the current system can not be properly handled with the layer model. The circulation in Japan Sea will be more realistically simulated by use of level models.

ACKOWLEDGMENTS

The author would like to express his sincere thanks to Professor K. Takano of Tsukuba University for his critical reading of the Manuscipt. Thanks are extended to Dr. M. Matsuyama of Tokyo University of Fisheries, Dr. J.H. Yoon and Dr. M. Kawabe of University of Tokyo for their valuable comment and discussions.

REFERENCES

Bryan.F. 1987. Parameter sensitivity of primitive equation ocean general circulation models. J. Phys. Oceanogr., 17: 970-985.
Inoue,N., Miita, T. and Tawara, S. 1985. Tsushima Strait II Physics. In H. Kunishi et al. (Editors), Coastal oceanography of Japanese Islands, Tokai University Press, 914-933 (in Japanese).
Japan Meteorological Agency 1972. Marine meteorological study of the Japan Sea. Technical Report of the Japan Meteorological Agency, 80: 1-161.
Kawabe, M. 1982a. Branching of the Tsushima Current in the Japan Sea. Part I Data analysis. J. Oceanogr. Soc. Japan, 38: 95-107.
Kawabe, M. 1982b. Branching of the Tsushima Current in the Japan Sea. Part II

128

Numerical experiment. J. Oceanogr. Soc. Japan, 38: 183-192.

Kutsuwada, K. and Sakurai, K. 1982. Climatological maps of wind stress field over the north Pacific Ocean. Oceanogr. Mag., 32: 25-46.

Matsuyama, M., Nazumi, T. and Takahata, T. 1986. Some characteristic velocity fields near the Tajima Coast. Bull. Coast. Oceanogr., 23: 129-138 (in Japanese).

Moriyasu, S. 1972a. The Tsushima Current. In, H. Stommel and K. Yoshida (Editors), Kuroshio - Its physical aspects. Univ. Tokyo Press, Tokyo, 353-369.

Moriyasu, S. 1972b. Hydrography of the Japan Sea. Marine Sciences (Kaiyo Kagaku, 4, 171-177 (in Japanese with English abstract).

Pedlosky, J. 1979. Geophysical fluid dynamics, Springer Verlag, 624 pp.

Rhines, P.B. 1970. Edge-bottom and Rossby wave in a rotaing stratified fluid. Geophis. Fluid. Dyn., 1: 273-302.

Sekine, Y. 1986. Wind-driven circulation in the Japan Sea and its influence on the branching of the Tsushima Current. Prog. Oceanogr., 17: 297-313.

Sekine, Y. 1989. On the seasonal change in the in-and outflow of the Japan Sea. Prog. Oceanogr., 21: 269-279.

Suda, K. and Hidaka, H. 1932. The results of the oceanographic observation on board R.M.S. Synmpu Maru in the southern part of the Japan Sea in the summer of 1929, Part I. J. Oceanogr. Imper. Mar. Obs., 3: 291-375 (in Japanese).

Suginihara, N. 1981. Quasi-geostrophic waves in a stratified ocean with bottom topography. J. Phys. Oceanogr., 11: 107-115.

Uda, M. 1934. The results of simultaneous oceanographic investigations in the Japan Sea and its adjacent waters in May and June,1932. J. Imp. Fish. Exp. St., 5: 57-190 (in Japanese with English abstract).

Uda, M. 1936. Results of simultaneous oceanographic investigations in the Japan Sea and its adjacent waters during October and November, 1933. J. Imp. Fish. Exp. St., 7: 91-151 (in Japanese with English abstract).

Yoon, J. H. 1981a. Numerical experiment on the circulation in the Japan Sea Part I Formation of east Korean warm current. J. Oceanogr. Soc. Japan, 38: 43-51.

Yoon, J. H. 1981b. Numerical experiment on the circulation in the Japan Sea Part II Influence of seasonal variation in atmospheric condition on the Tsushima Current. J. Oceanogr. Soc. Japan, 38: 125-130.

Yoon, J. H. 1981c. Numerical experiment on the circulation in the Japan Sea Part III Mechanism of nearshore branch of the Tsushima Current J. Oceanogr. Soc. Japan, 38: 125-130.

ON THE INTERMEDIATE WATER IN THE SOUTHWESTERN EAST SEA (SEA OF JAPAN)

C.H. KIM[1], H.-J. LIE[1] and K.-S. CHU[2]
[1]Korea Ocean Research & Development Institute, P.O.Box 29, Ansan, Seoul 425-600, Korea
[2]Hydrographic Office, P.O.Box 56, Incheon 400-600, Korea

ABSTRACT

The spatial and temporal variability of the East Sea Intermediate Water (ESIW) in the East Sea is examined by incorporating the CTD data, taken over three different surveys in August 1986, December 1987 and May 1988, with the historical hydrographic data obtained bimonthly. The ESIW is clearly identified by the salinity minimum and the dissolved oxygen maximum off the mid-east coast of Korea and in Korea Strait in August 1986. A core of the salinity minimum water appeared at the coastal region in December 1987, offshore in May 1988, and at both region in August 1986. In the historical bottle data location of salinity minimum core is not so clear as CTD data. Temporal variation in the properties of the ESIW suggests that this water appears interannually.

1. INTRODUCTION

Water masses in the East Sea (Sea of Japan) have been discussed by various authors (Akagawa, 1954 ; Kajiura et al., 1958 ; Gong and Park, 1969 ; Miyajaki and Abe, 1960). Moriyasu (1972) has classified four water masses according to Kajiura et al. (1958) ; Surface and Intermediate Waters, the Proper Water, and the fourth water mass.

Along the east coast of Korea a cold water exists forming a strong front with a warm water carried by the Tsushima Warm Current flowing north. The cold water is known to show year-to-year variation in the width and strength (Gong and Son, 1982 ; Gong and Lie, 1984). Kim and Kim (1983) have observed that the cold water of 2-5 ℃ is characterized by high concentration of dissolved oxygen(> 6.5 ml/l) and low salinity(< 34.00 ‰). They have suggested that this water is brought to the southwestern East Sea by the North Korean Cold Current, which has been believed to flow southward along the northern Korean coast (Uda, 1934). This cold water is a distinct water mass, which can be identified by the salinity minimum and the dissolved oxygen maximum in the vertical sections (Kim and Chung, 1984). The cold water mass off Korea is also found to have the same characteristics as those of the fourth water mass that has been observed at the polar front in the middle of the East Sea (Kajiura et al., 1958 ; Moriyasu, 1972). This fact suggests that the cold water is widely distributed in the East Sea. For that reason Kim and Chung (1984) have proposed to call the cold water as the East Sea Intermediate Water (ESIW). Compared with the Tsushima Warm Current Water (TWCW) and the Proper Water, the ESIW takes a small portion of the total water volume in the East Sea (Yasui et al., 1967). However, it may play an important role in the circulation of the East Sea. Unfortunately, little is known about the spatial and temporal variability of the ESIW ; Kim and Kim (1983) concerned mainly on the existence and origin of the ESIW ; Kim and Chung (1984) focused on the characteristics of the ESIW.

In this paper we have examined the spatial and temporal variations of the ESIW by analyzing

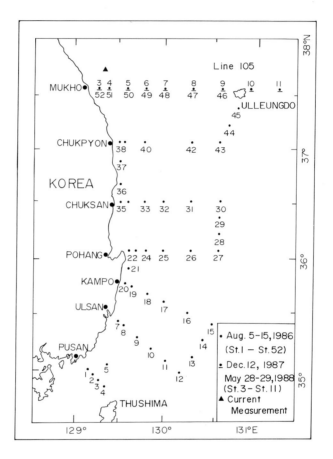

Fig. 1. Locations of hydrographic stations occupied in August 1986, December 1987 and May 1988. The stations between Mukho and Ulleung Island are the same with line 105 which is occupied bimonthly by the Fisheries Research and Development Agency of Korea.

both CTD data, observed by the Korea Ocean Research & Development Institute (KORDI), and historical hydrographic data collected by the Fisheries Research and Development Agency (FRDA) of Korea.

2 SPATIAL DISTRIBUTION OF THE INTERMEDIATE WATER

2.1 August 1986

CTD casts were made over three different periods in the southwestern East Sea : August 1986, December 1987 and May 1988 (Fig. 1). Vertical profiles of temperature, salinity and dissolved oxygen along the 36° 30'N line (stations 35-30 in Fig. 1) show that isotherms are almost parallel to the sea surface and a strong seasonal thermocline is formed at the depth of 10-30 m (Fig. 2a). The upper surface water has high temperature (16-25 ℃) and low salinity (32.40-34.00 ‰). This water is originated from the Tsushima Warm Current Surface Water (TWCSW) (Lim and Chang, 1969). The Tsushima Warm Current Middle Water (TWCMW) below the upper surface water appears at about 50 m with two separated cores of high salinity (> 34.40 ‰) and low concentration of dissolved oxygen (< 5.6 ml/l).

A permanent thermocline is found at 70-120 m underneath the TWCMW. Temperature below 150 m is lower than 2 ℃, so that the Proper Water seems to occupy the whole water column under the TWCMW. However, the salinity profile (Fig. 2b) distinguishes a salinity minimum layer (< 34.05 ‰) from the Proper Water (< 2 ℃, 34.08-34.09 ‰). The water of low salinity corresponds to the ESIW mentioned by Kim and Chung (1984). The location of the salinity minimum layer of the ESIW coincides with that of the dissolved oxygen maximum layer. The thickness of the salinity-minimum and dissolved oxygen-maximum layer (< 34.05 ‰, > 7.5 ml/l) is about 100-130 m in the vertical section. Two cores of the salinity minimum and dissolved oxygen maximum appear : one in the coastal region and the other in the offshore region. The offshore core is more distinct in the salinity and dissolved oxygen character.

The salinity minimum layer of the ESIW is also observed clearly in other sections. Fig. 3a shows the depth where the lowest salinity is observed in the salinity minimum layer and Fig. 3b is the spatial distribution of the lowest salinity at the depths corresponding to Fig. 3a. A core of salinity minimum, less than 33.98 ‰, appears at 130 m near Mukho and at 200 m depth west of Ulleung Island. In the south of 37° 00'N a salinity minimum core of 34.00-34.02 ‰, which is slightly higher than at northern section, appears at depths of 130-140 m near the coast. Cores of salinity minimum also exist at the stations east of 130° 30' E : at 120 m depth of station 15, 155 m depth of station 27 and 160 m depth of station 30. The ESIW extends to coastal region along the east coast of Korea and to the northern end of Korea Strait. As the bathymetry becomes shallow toward Korea Strait, the ESIW appears near the bottom. Salinity minimum layer does not appear outside the boundary shown as a broken line in Fig. 3. In the southwestern area of Ulleung Island the TWCMW occupies from 20 m almost to the maximum sampling depth of 250 m. Therefore, no salinity minimum layer is found within the sampling depth in the area.

2.2 December 1987 and May 1988

It is worthwhile to examine the spatial distribution of water masses in other seasons. Fig.

132

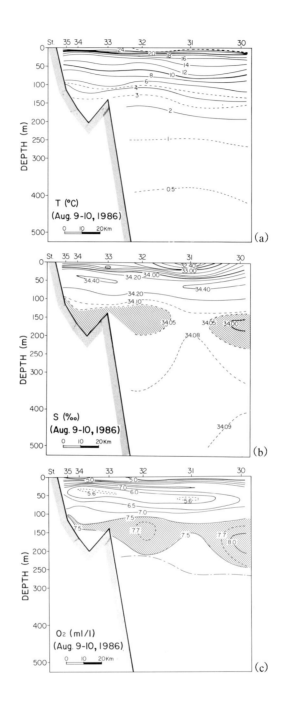

Fig. 2. Vertical sections of (a) temperature, (b) salinity and (c) concentration of dissolved oxygen along 36° 30′ N in August 9-10, 1986. Layers of salinity minimum and dissolved oxygen maximum are shaded. Dissolved oxygen was not measured below the dot and dashed line.

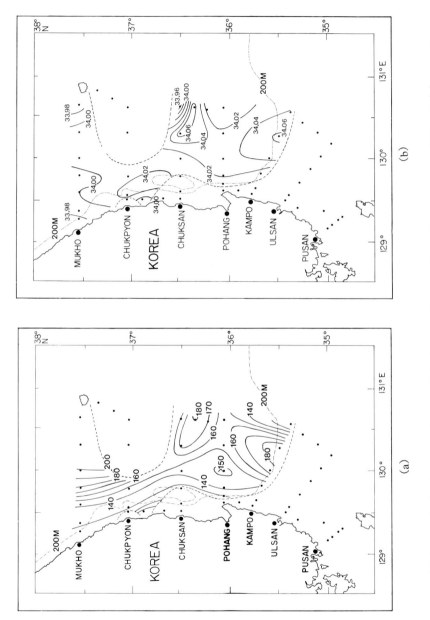

Fig. 3. (a) Depth distribution of the lowest salinity in the salinity minimum layer in August 1986 and (b) its spatial distribution at the depths corresponding to (a). Isobath of 200 m is indicated.

4 is the vertical sections of temperature and salinity along the 37° 33′N (line 105 in Fig. 1) surveyed in December 1987 and May 1988. The TWCSW composed of well-mixed water (14-16 ℃, 33.90-34.00 ‰) is lying from the surface to a depth of 100 m (Fig. 4a and 4b). Around the permanent thermocline there exists the saline TWCMW (8-14 ℃, 34.20-34.40 ‰). Though the dissolved oxygen was not measured, the ESIW (0.7-4 ℃, 33.96-34.02 ‰) can be readily identified by the existence of salinity minimum layer which is found between the TWCMW and the Proper Water of homogeneous salinity (34.02-34.03 ‰). The salinity minimum layer deepens from the coast to offshore.

In May 1988 the ESIW(1-4 ℃, < 34.02 ‰) exists below 200 m at the east of station 6 and goes upto a depth of 50 m in the coastal area (Fig. 4c and 4d). Salinity of the ESIW is lower in December 1987 and May 1988 than in August 1986 by about 0.03 ‰. The core of salinity minimum (1-2 ℃, < 33.96 ‰) locates at 90-130 m depth near the coast in December 1987, while it locates at about 250-330 m at the offshore stations 8-9 in May 1988.

2.3 Historical hydrographic data

Bimonthly section of salinity along 105 line is obtained for 1982-1987 to examine temporal variation of the ESIW (Fig. 5). The data were obtained at the standard depths (0 m, 10 m, 20 m, 30 m, 50 m, 75 m, 100 m, 125 m, 150 m, 200 m, 250 m, 300 m, 400 m, 500m) by bottle casts except in December 1987. The ESIW is depicted by the isohaline of 34.05 ‰ beneath the permanent thermocline except for the months showing abnormally low-salinity such as June and December 1983 when the isohaline of 33.90 ‰ is chosen. The ESIW does not show the vertical structure of salinity minimum layer clearly because of coarse sampling interval. For example, the salinity section for December 1987 in Fig. 5f is plotted using CTD data at the standard depth, while Fig. 4b at 1 meter depth interval using the same data.

The TWCSW shows the seasonal variation (Moriyasu, 1972). That is, the TWCSW of low salinity appears in the surface layer in summer and autumn, while in winter and spring the TWCSW disappears and saline TWCMW occupies from surface to mid-depth. On the other hand the ESIW does not seem to show seasonal change in the vertical distribution of salinity. There is no evidence of its presence in June and October 1982 and February 1985-January 1986. The core of salinity minimum appears only in the nearshore area in August of 1982 and 1983, February and October 1984, March 1986 and December 1987. Meanwhile, the two cores of salinity minimum appear both near the coast and offshore in June and December 1983, January 1985, June-December 1986 and August and October 1987.

3 TEMPORAL VARIATION OF THE INTERMEDIATE WATER

In order to see the variations in properties of the ESIW at stations 3-11 along the line 105 time series of T-S diagram are constructed for 1967-1987. In Fig. 6 salinity range is 33.80-34.30 ‰ and range of temperature is 0-4 ℃ which includes the temperature of the ESIW and the Proper Water. Reference line indicates the salinity of 34.05 ‰ and the lowest salinity in each month is connected. As discussed above core of salinity minimum in the ESIW is less than 34.05 ‰ in general. So water masses below the reference line in Fig. 6 may indicate the appearance of the ESIW.

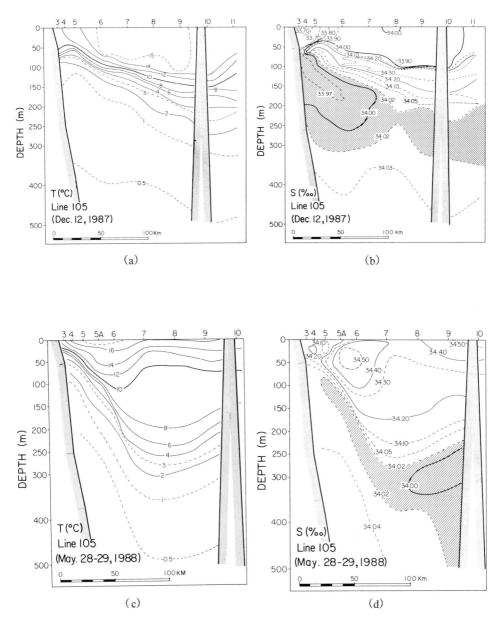

Fig. 4. Vertical sections of (a, c) temperature and (b, d) salinity along 37° 33′N (105 line) in December 12, 1987 and May 28-29 1988.

136

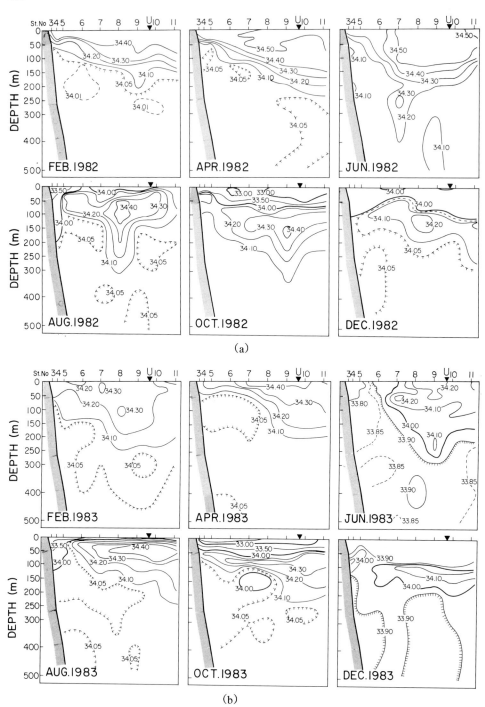

(a)

(b)

Fig. 5. Time sequences of vertical section of salinity (a-f) along 105 line for 1982-1987. The location of Ulleung Island is indicated by a symbol ▼.

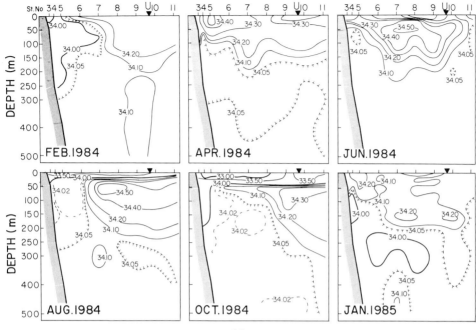

(c)

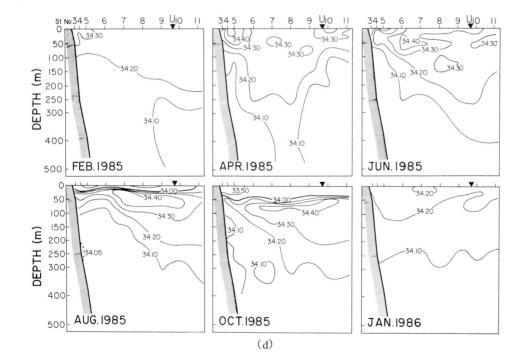

(d)

Fig. 5. Continued.

138

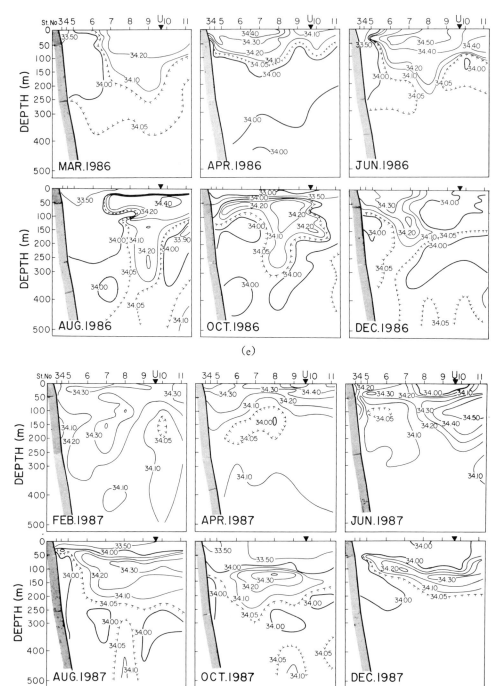

(e)

(f)

Fig. 5. Continued.

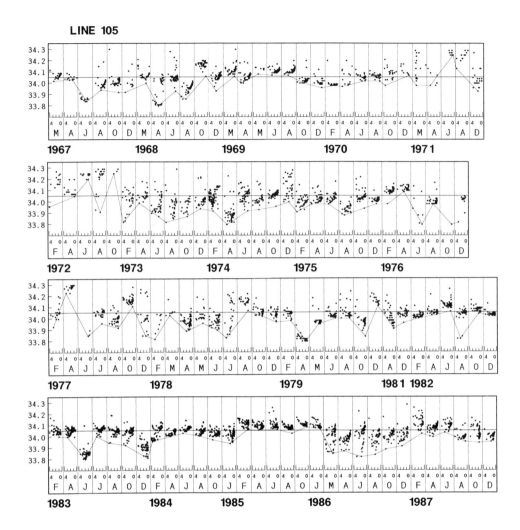

Fig. 6. Time sequences of temperature-salinity relationship for stations along 105 line. Range of temperature and salinity is 0-4 ℃ and 33.80-34.00 ‰ .

The ESIW does not seem to be present in this section for a relatively long period of February 1985-January 1986. However, it appears persistently for the periods of March 1967-August 1968, October 1969-October 1970, February 1973-December 1975, December 1977-October 1979, August 1982-January 1985, and March 1986-December 1987. Change of salinity in salinity minimum core is less than 0.10 ‰ in October 1969-October 1970, February 1984-January 1985 and March-December 1986, while it is as much as 0.20 ‰ from March to April 1968, February to March 1978 and June to July 1983.

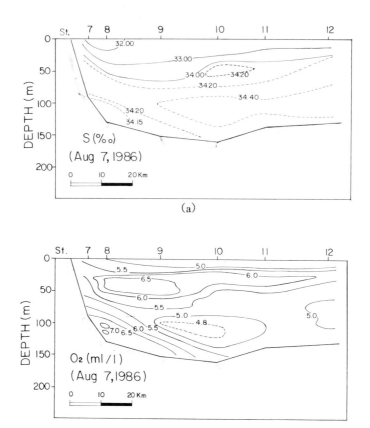

Fig. 7. Vertical sections of (a) salinity and (b) concentration of dissolved oxygen in August 7, 1986 at stations 7-12 shown in Fig. 1

4 DISCUSSION

CTD observations conducted in August 1986 show that the ESIW appears at the stations in the Korea Strait. The typical character of salinity minimum and maximum concentration of dissolved oxygen becomes obscure in the vertical section as the sill depth shoals southward. However, it can be traced to station 8 and station 9 (Fig. 7), where salinity is low and dissolved oxygen content is high near the bottom. The appearance of this water extends the southern limit of ESIW to about 35° 25′ N, much further to the south as compared with the observation of Kim and Chung (1984).

Lie(1984) has reported the persistent southward flow in summer 1980 near the coast off Chukpyon and Lie et al.(1989) have observed a deep southward flow with a mean speed of 3 cm/sec more than 2 months in autumn 1986 at a location shown in Fig. 1. As the measured depths are 5 m (Lie, 1984) and 620 m (Lie et al., 1989), respectively, it is difficult to associate these current measurements with the southward flow of the ESIW. It is necessary to verify the movement of the ESIW directly in the future.

ACKNOWLEDGEMENT

The authors wish to thank crew of Korea Hydrographic Office and Fisheries Research and Development Agency, for their assistance in data collection. We express sincere thanks to Prof. K. Kim of Seould National University for his helpful comments and useful suggestions. Thanks are also due to Prof. Y.H. Seung of Inha University and Dr. M.W. Han of KORDI who carefully read the manuscript.

REFERENCES

Akagawa, M., 1954. On the oceanographical conditions of the north Japan Sea (west off the Tsugaru Straits) in summer. Jour. Oceanogr. Soc. Japan, 10(4) : 189-199.

Gong, Y. and Park, C.G., 1969. On the oceanographical character of the low temperature region in the eastern sea of Korea. Bull. Fish. Res. Dev. Agency, Korea, 4 : 69-91.

Gong, Y. and Lie, H.-J., 1984. Distribution of thermal fronts in the southeast sea of Korea (southern Japan Sea). KORDI Rep. BSPE00055-86-7B, 215pp.

Gong, Y. and Son, S.J., 1982. A study of oceanic thermal fronts in the southwestern Japan Sea. Bull. Fish. Res. Dev. Agency, Korea, 28 : 25−54.

Kajiura, K., Tsuchiya, M. and Hidaka, K., 1958. The analysis of oceanographical condition in the Japan Sea. Rep. Develop. Fisher. Resour. in the Tsushima Warm Current, 1, 158-170.

Kim, C. H. and Kim, K., 1983. Characteristics and origin of the cold water mass along the east coast of Korea. Jour. Oceanol. Soc. Korea, 18(1) : 71-83.

Kim, K. and Chung, J.Y., 1984. On the salinity-minimum layer and dissolved oxygen-maximum layer in the East Sea (Japan Sea). In : Ocean Hydrodynamics of the Japan and East Chi na Sea, T. Ichiye (Editor), Elsevier, Amsterdam, pp. 55-65.

Lie, H.-J., 1984. Coastal current and its variation along the east coast of Korea. In : Ocean Hydrodynamics of the Japan and East China Seas, T.Ichiye(Editor), Elsevier, Amsterdam, pp. 399-408.

Lie, H.-J., Suk, M.S. and Kim, C.H., 1989. Observations of southeastward deep currents off the east coast of Korea. Jour. Oceanol. Soc. Korea, 24(2) : 63-68.

Lim, D.B. and Chang, S.D., 1971. On the cold water mass in the Korea Strait. Jour. Oceanol. Soc. Korea, 4(2) : 71-82.

Miyazaki, M. and Abe, S., 1960. On the water masses in the Tsushima Current area. Jour. Oceanogr. Soc. Japan, 16(2) : 59-68.

Moriyasu, S., 1972. The Tsushima Current. In : Kuroshio, its physical aspects, H. Stommel and K. Yoshida (Editor), Univ. of Tokyo, pp. 353-369.

Uda, M., 1934. Results of simultaneous oceanographic investigations in the Japan Sea and its adjacent waters during May and June, 1932. Jour. Imp. Fish. Exp. Station, 5 : 57-190.

Yasui, M., Yasuoka, T., Tanioka K., and Shiota, O., 1967. Oceanographic studies of the Japan Sea (I). Oceanogr. Mag., 19(2) : 177-192.

MODERN   SEDIMENTATION OFF SAN'IN DISTRICT IN THE   SOUTHERN   JAPAN
SEA

K. IKEHARA
Marine Geology Department, Geological Survey of Japan, Higashi 1-
1-3, Tsukuba, Ibaraki 305 (Japan)

ABSTRACT
        Modern   sedimentation   off San'in district in   the   southern
Japan   Sea   is   discussed on the basis   of   the   distribution   of
surface   sediments, composition of sediments, bedform   morphology
and   seismic   records.   The sediments   are   transported,   by   the
Tsushima Current, eastward on the shelf and southeastward on   the
marginal   terrace   at   the east of the Oki Islands   and   the   Oki
Ridge. The wave and current actions prevent mud deposition on the
shelf.   Muddy   sediments are transported from the land   over   the
shelf   to   the   marginal   terrace and   are   deposited above   the
marginal   terrace   and in the trough through the   slowly   flowing
Japan   Sea   Proper Water.   Low   temperature   and   high   oxygen
concentration   of the Japan Sea Proper Water forms the   oxidizing
bottom   condition   and dissolves the calcium   carbonates   in   the
surface sediments. Therefore, there is close relationship between
modern sedimentation and oceanographic conditions.

1 Introduction

        The Japan Sea is a marginal sea between the Japanese Islands
and   the   Asian   continent.   Geological,   sedimentological   and
paleontological   works have been conducted by many   workers.   For
example,   the genesis of the Japan Sea has been discussed by   many
authors in various fields of geological and geophysical sciences.
Also   on the surface sediments in the Japan Sea, there have   been
many   studies   (e.g. Maritime Safety Agency,   1949;   Sato,   1961;
Iwabuchi,   1968;   Ikehara, 1989) and   the   size   distribution   of
surface sediments have already been known.

        The   oceanography of the Japan Sea has been investigated   by
many workers and the main water masses have been discussed   (e.g.
Yasui   et al., 1967). Most of the Japan Sea water is   homogeneous
and consists of deep and bottom water with   temperatures of 0-1 ℃
and   salinities of about 34.1‰. In the southern Japan Sea,   above

the deep and bottom water mass is a surface water mass which belongs to the Tsushima Warm Current (Asaoka et al., 1985). The main thermocline between these two water masses occurs at the water depth of around 200 m.

These oceanographic conditions are very different from those of the Pacific side of Southwest Japan where a strong ocean current, Kuroshio, is prevailing along the coast. The difference in the oceanographic conditions may be clearly reflected on the modern sedimentation. Few studies, however, have been carried out regarding the relationship between sedimentation and oceanographic conditions such as currents, waves and sea water stratification around the Japanese Islands (e.g. Hoshino, 1952, 1958; Ikehara, 1988a). This paper reports the results of four geological and sedimentological survey cruises off San'in district in the southern Japan Sea. The surveys were carried out during R/V Hakurei-Maru cruises GH86-2(June-July, 1986), GH86-4(September-October, 1986), GH87-2(June-July, 1987) and GH87-4 (September-October, 1987) of the Geological Survey of Japan. At first, I shall describe the characters of surface sediments and bedforms distributed off San'in district. Next, I shall show the relationship between modern sedimentation and oceanographic conditions.

2 Physiography and oceanography

The major morphological features in the study area are the Oki Spar in the west, a marginal terrace, several ridges and troughs and the Yamato Basin in the north (Fig.1). The Oki Spar is a relief which extends in the north-south direction from the Shimane Peninsula through the Oki Islands and further to the north. There are three ridges extending east-northeastward. These are, from the south to the north, the Echizen Bank Chain, the Wakasa Sea Knoll Chain and the Oki Ridge. The Oki Trough lies between the last two ridges. The marginal terrace is a wide terrace 200-400 m deep. The shelf is narrow off Tottori with a width of less than 10 km at the narrowest part and is wide only on the Oki Spar. The shelf edge is about 140-150 m deep. A short but distinct slope forms the boundary between the shelf edge and the marginal terrace (Fig.1-b). The slope between the marginal terrace and the trough is 2-10°.

The southern Japan Sea water is divided into the surface water and the deep and bottom water at the water depth of around

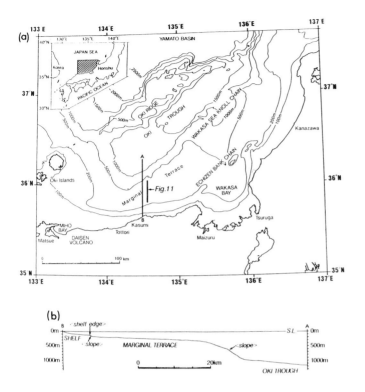

Fig.1 Submarine topography of the study area (a) and topographic cross section between A and B (b).

200 m (Fig.2). According to Asaoka et al. (1985), surface water with high temperature, high salinity and low oxygen concentration belongs to the Tsushima Current. The Tsushima Current inflows through the Tsushima Strait and flows eastward in the region of the present study (e.g. Kawabe, 1986). One of the oceanographic features of the Japan Sea is the occurrence of homogeneous Japan Sea Proper Water (JSPW) having low temperatures (0-1 ℃), rather low salinities (34.0-34.1‰) and high oxygen concentration (5.0 to more than 7.0 ml/l) (Suda, 1932; Nitani, 1972). The origin of JSPW is considered to be off the Siberian coast at the time of ice formation in winter (Nitani, 1972).

3 General remarks on the surface sediments and bedforms

The sediment distribution map (Fig.3) indicates a predominance of muddy sediments. In general, the surface sediments become finer in grain size with increasing water depth. That is, sandy sediments are distributed on the shelf. Silt and clayey silt are present on the marginal terrace. Finer silty clay

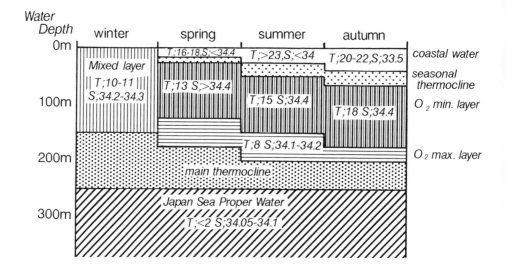

Fig.2 Schematic diagram of oceanographic conditions off Wakasa Bay (modified from Asaoka et al., 1985).

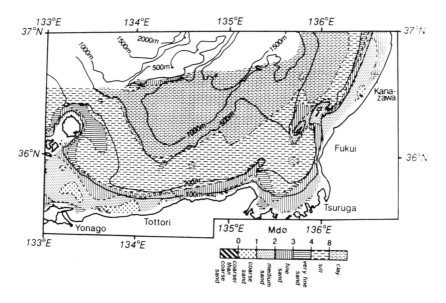

Fig.3 Median diameter distribution off San'in district.

occurs in the trough. Brown clay covers the sea floor below the water depth of 1000 m. Sub-bottom volcanic ash layers of known age in core samples from the marginal terrace and the trough show that the sedimentation rates were 10-25 cm/1000 years during the late Quaternary (Machida and Arai, 1983; Ikehara et al., 1990a).

The percentage of clay fraction in sediments becomes higher with increasing water depth (Fig.4). Two inflection points are recognized in Fig.4. The first occurs at the water depth of 200 m and the clay content increases to 20-30%. Mud content also increases clearly at this depth. The second occurs at 800 m depth. Between the first and the second, clay content increases gently, but beyond the second point, the clay content is more or less stable at 60-70%.

The characteristics of the grain composition of the coarse fraction off Tottori is summarized as follows (Fig.5; Ikehara et al., 1990a). The dominant constituents of the sandy shelf sediments are quartz and rock fragments. Iron-coated quartz grains occur at a water depth of 50-110 m (mid-outer shelf). Heavy mineral compositions of the shelf sediments are classified

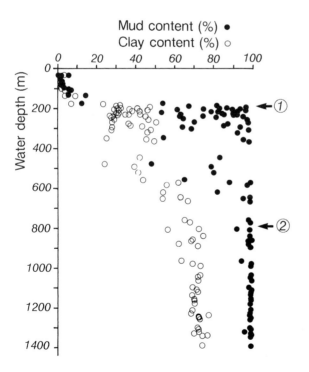

Fig.4 Depth distribution of mud and clay contents offshore Tottori. Arrow with 1 or 2 shows the position of inflection point.

into two groups (Fig.6; Yokota et al., 1990). Group I contains hypersthene, oxyhornblende and brown hornblende and is distributed on the inner shelf. Group II is composed of green hornblende, clinopyroxene and olivine and is distributed on the mid-outer shelf (Yokota et al., 1990). Sand grains of the surface sediments on the marginal terrace are composed of remains of several kinds of planktonic and benthic organisms, pumice, volcanic glass shards and quartz. Those of the trough sediments are composed of quartz, pumice, volcanic glass shards and benthic foraminifers.

Muddy sediments are diatomaceous, which are more clearly marked below a water depth of 700 m because of dissolution of $CaCO_3$. The assemblages of planktonic remains such as calcareous nannoplanktons (Tanaka, 1987), foraminifers (Oda and Ikehara, 1987), diatoms (Tanimura, 1981) and silicoflagellates (Shimonaka et al., 1970) are warm water assemblages living in the territory of the Tsushima Warm Current.

The values of the oxidation-reduction potential (ORP) are

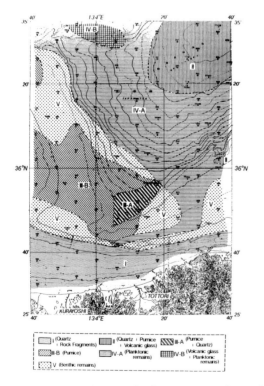

Fig.5 Province of the coarse fraction of the surface sediments offshore Tottori.

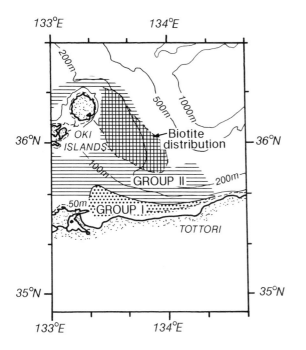

Fig.6 Province of heavy mineral and the biotite distribution of
the surface sediments (modified from Yokota et al., 1990).

positive on the shelf and the trough, and negative on the
marginal terrace (Fig.7; Ikehara et al., 1990a). Values of more
than +50 mV occur where the brown clay is distributed in the
trough.   Therefore, it seems that the shelf and trough are now
under oxidizing conditions but the marginal terrace is under
reducing ones.

     Bedforms are observed only on the sandy shelf in the Oki
Strait.   In the Oki Strait, three kinds of bedforms, sandwaves,
megaripples and ripple marks, are recognized by the bathymetric
survey using a 12kHz echosounder, sea floor mapping using a side-
scan sonar and sea bottom photography (Ikehara et al., 1987;
Ikehara and Katayama, 1987). The morphology of bedforms is
briefly described below. Sandwaves have 50-490 m wavelengths,
1.6-7.4 m waveheights and 13-104.8 wavelength-to-waveheight
ratios.   Vertical profiles of sandwaves show asymmetrical forms
with steeper eastern slopes. Plan patterns of sandwaves show
straight or gently undulating and rarely branched.   The

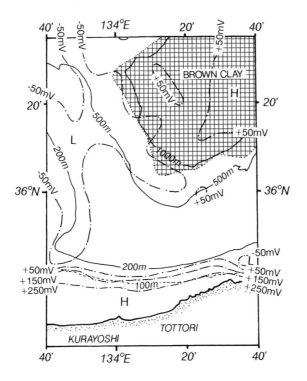

Fig.7 Distribution of the oxidation-reduction potential (ORP) and brown clay (hatched area).

wavelengths of megaripples are 8-20 m, but their waveheights are unknown because of their low relief. Plan patterns of megaripples are straight or weakly undulating and occasionally branched. Ripple marks are of straight form in plan patterns, rounded crest and 10-20 cm wavelengths.

Geographical distributions of these bedforms are shown in Fig.8. Sandwaves are found at water depths of 70-100 m in the eastern part of the Oki Strait. Megaripples are distributed on the eastern side of the sandwave fields. Ripple marks are found only at one station in the Strait.

4 Discussions

4.1 Shelf sediments - relationship between sedimentation and wave and current conditions

Coastal sediments in zones shallower than 50-60 m depth off Tottori are well- to moderately well-sorted sandy sediments

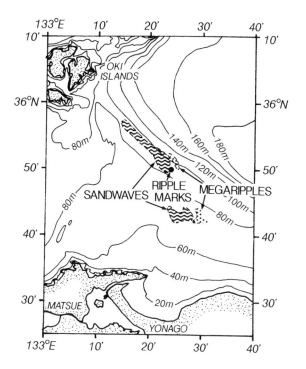

Fig.8 Bedform distribution in the Oki Strait.

(medium to coarse sand) or moderately to poorly sorted gravelly sand with little amount of mud and contain hypersthene, oxyhornblende and brown hornblende (heavy mineral Group I). Off the City of Tottori, well-sorted medium sand with little mud amount occurs. In shallow water, muddy sediments can be distributed where the amount of the suspended load supply is greater than that by the reworking by wave action (Saito, 1988). Therefore, present wave actions are strong enough to remove the muddy sediments from the bottom, which is confirmed by low mud content in the coastal sediments. From offshore wave data measured at Tottori (Sugawara et al., 1986) and an equation for calculating the critical depth for sand to move in the wave direction (Sato et al., 1963), it is found that medium sand is unable to be moved by the storm wave if the depth is greater than 40-50 m. This supports that present wave actions are able to agitate the sandy bottom sediments distributed in the coastal area. Depths of 40-50 m are almost equal the depths of offshore

mudline   in   the Sendai Bay (50-60 m, Saito, 1989a,   b)   and   the
Akita Bay (50-70 m, Saito, 1989c). Gravelly or rocky bottoms   are
recognized   in the coastal area off Daisen Volcano   and   offshore
area   between the City of Tottori and Kasumi. It is thought   that
these are residual deposits due to sea bottom erosion by   present
wave action, because the depths of gravelly or rocky bottoms   are
almost   equal   to   the   depth of   wave-base   in   Iceland   (50   m,
Sunamura,   1990)   and to the critical water   depth   for   bedrock
abrasion which is the maximum water depth beyond that no abrasion
can   extend   at a site off the Niijima Island   (40   m,   Sunamura,
1986),   and because sandy sediments around the gravelly or   rocky
bottoms have been affected by storm waves as discussed above. The
sand   grains   are derived from volcanic rocks exposed   on   nearby
land   based on heavy mineral compositions (Yokota et   al.,   1990)
and   may   be   transported   by   waves   or   wave-induced   currents.
Therefore,   the coastal sediments are presently being formed   and
are distributed where the sea bottom has been influenced by storm
waves.

The shelf sediments on the mid-outer shelf are moderately to
poorly-sorted   sandy sediments (medium to fine sand) with   little
amount   of   mud and contain green hornblende,   clinopyroxene   and
olivine (heavy mineral Group II). It is thought that they are old
(relic) sediments because the quartz grains coated by iron oxides
occur and finer modern sediments occur closer to the shore   (e.g.
in   the   Miho Bay). Heavy mineral content of these   sediments   is
similar to that near the Oki Strait (Fig.6; Yokota et al., 1990).
The   geographical   distribution   of grain   size   of   the   surface
sediments   (Fig.3) and the vertical profiles of sandwaves at   the
Oki Strait as mentioned earlier (Ikehara et al., 1987)   indicates
eastward movement of the sediments.

The   sediment   transportation direction inferred in this   way
is   concordant   with the direction of the current in   the   Strait
(Japan Oceanographic Data Center, 1979) which is the first branch
of   the Tsushima Current (Kawabe, 1986). Therefore, the   sediment
transportation and bedform formation are certainly accounted   for
by a current related to the Tsushima Current. The present current
velocity   observed   in the Oki Strait (Minami et al.,   1984)   is,
however,   only   about   0.6-0.8   knots   and   is   insufficient   to
transport   the sand grains. Ripple marks are observed only   at   a
limited area in the Oki Strait (Fig.8). Its occurrence in the Oki
Strait   is   quite   different   from   that   where   the   sediment

transportation is active under present hydraulic conditions in European and Japanese shelves (Belderson et al., 1982; Ikehara, 1988a). That is, the bedform arrangement in the Oki Strait lacks in rippled sand sheet. The occurrence of ripple marks in the Oki Strait suggests that the currents are not strong enough to form the ripple marks. They are not strong enough, either, to form larger bedforms such as sandwaves and megaripples under normal flow condition because the velocity necessary to form larger bedforms is greater than that of smaller ripple marks (Simons et al., 1965). Then, it is inferred that the sediment transport rate is small under present hydraulic conditions. On the other hand, little amount of mud fraction in the shelf sediments show that the muddy sediments can not be deposited; the amount of mud removed by wave and current actions is greater than that of mud supply.

4.2 Deposition of muddy sediments – relationship between sedimentation and water stratification

Muddy sediments are distributed on the marginal terrace and trough (Fig.3). As mentioned above, muddy sediments from land can not be deposited on the shelf by wave and current actions.

The Japan Sea Proper Water (JSPW) exists below water depths of around 200 m (Fig.2) and forms a quiet condition. The percentage of clay fraction in sediments increases at this boundary (the first inflection point in Fig.4). Therefore, muddy sediments transported to this zone begin settling down, whereas

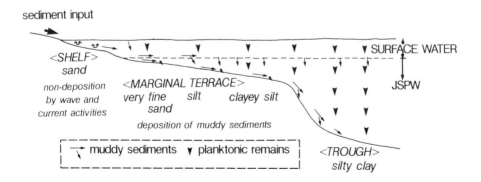

Fig.9 Schematic representation of mud deposition off San'in district.

finer clay particles are transported farther from land. As a result, the sediment particles become finer with increasing water depth (Fig.9).

Planktonic remains (foraminifers, diatoms, calcareous nannoplanktons, etc.) which were living in the surface water (the Tsushima Warm Current water) settle down to the bottom. Therefore, terrigenous muddy sediments and planktonic remains constitute the hemipelagic mud distributed on the marginal terrace and the trough (Fig.9).

The calcium carbonate compensation depth (CCD) in the Japan Sea (about 2000 m) is shallower than that in the Pacific (Ichikura and Ujiié, 1976). This estimate is based on the preservation of calcareous fossils in the surface sediments. In the study area, calcareous planktonic remains such as calcareous nannoplanktons and planktonic foraminifers are absent or very poorly preserved where the water depth is deeper than 700-1000 m (Tanaka, 1987; Oda and Ikehara, 1987). Also, benthic foraminifers change their assemblages from calcareous assemblage on the marginal terrace to the agglutinated assemblage on the trough (Nomura and Ikehara, 1987). Brown clay layer distributed in the trough below 1000 m is considered to be the oxidizing layer from its colour and high ORP (Fig.7). The bottom layer of JSPW located deeper than 2000 m has highest oxygen concentration in the JSPW (Gamo et al., 1986). Although the water depth of brown clay is shallower than that of the bottom layer, the low temperature and high oxygen bottom layer of JSPW makes the oxidizing layer and dissolves calcium carbonates.

4.3 Lower limit of the influence of the current

As the Tsushima Current flows in the surface water, it is believed that the current can not affect the bottom sediments below about 200 m depth. The sandy sediments, however, are distributed on the marginal terrace to the east of the Oki Islands (Fig.3), which indicate the southeastward transportation of the sediments. The transportation in the same direction is found in the distribution of biotite east of the Oki Islands (Fig.6; Yokota et al., 1990), in the distribution of the turbid bottom water zone on the marginal terrace observed by sea bottom photography and in the distribution of the volcanic glass shards from the western part of the Oki Ridge to the Oki Trough in the composition of sand grains (Fig.5; Ikehara et al., 1990a). Since

Japan Oceanographic Data Center (1979) shows that the surface current (the Tsushima Current) flows southeastward in this area, the transportation is probably due to the current related to the Tsushima Current.

On the western end of the Oki Ridge (st.196 in Fig.5), water depth: 557 m, the volcanic ash layer named the Aira-Tn Ash (AT) was deposited during the last glacial age (22000 years B.P., Machida and Arai, 1983) just beneath the sea bottom (18cm from the sea bottom, Ikehara et al., 1990a). This fact shows the sea bottom have been eroded since the inflowing of the Tsushima Current to the Japan Sea at 10000 or 8000 years B.P. (Oba, 1987).

4.4 Sediment accumulation on the marginal terrace and in the trough

Muddy sediments are deposited in the sedimentary basins which have various shapes and sizes on the marginal terrace and the troughs. These basins have been formed by local tectonic movements. The largest basin on the marginal terrace is located offshore of Kasumi (Fig.10). In general, each sediment layer recognized in seismic records (Fig.11) becomes thicker offshoreward reflecting the northward tilting movements (Yamamoto et al., 1989). Although the sediment layers are deposited from homogeneously suspended water mass and become thicker with

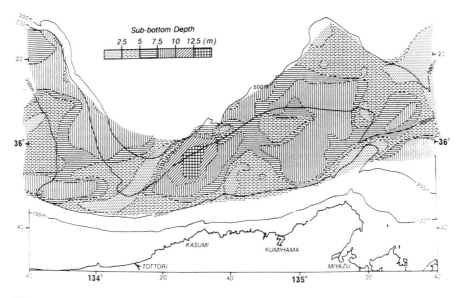

Fig.10 Depth contour of the first reflector on the marginal terrace.

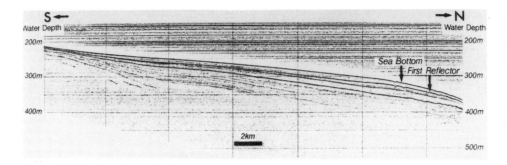

Fig.11 Typical example of the seismic (3.5kHz) record on the marginal terrace. Note the sedimentary layer between the sea bottom and first reflector becomes thicker northward.

increasing water depth (Inouchi, 1988), here, tectonic movements probably exert stronger influence on the offshoreward change in the thickness of a sedimentary layer than sedimentological processes. This is inferred from the fact that the geographical thickness distributions in each sedimentary layer are more concordant with the geological structures such as depth of the acoustic basement rather than the water depth. Also, all of sedimentary units deposited after the Miocene become thicker offshoreward (Yamamoto et al., 1989) and then, the thicknesses of the sedimentary layers may be controlled by tectonic movements after the Miocene (Yamamoto et al., 1989). In modern environments, most of terrigenous sediments, especially coarser than silt particles are trapped in the basins on the marginal terrace and only clay is transported to the deeper trough.

Seismic records obtained from the Oki Trough show a remarkable development of acoustically chaotic layers (Yamamoto et al., 1989; Ikehara et al., 1990a, b), which are formed by slope failures between the marginal terrace and the trough and are submarine debris flow deposits (Nardin et al., 1979; Damuth, 1980; Chough et al., 1985; Ikehara et al., 1990a, b). The triggering mechanism of these slope failures is considered to be the differential tectonic movements between the acoustic basements (ridges) and sedimentary layers (basins) (Yamamoto et al., 1989; Ikehara et al., 1990b). The tectonic movements are closely related to the filling up processes of the Oki Trough.

5 Summary

There are close relationships between modern sedimentation and oceanographic conditions such as wave, current, water stratification and water properties in the southern Japan Sea (Fig.12). That is, the Tsushima Current transports the sediments eastward on the shelf and southeastward on the marginal terrace east of the Oki Islands and on the Oki Ridge. Wave and current actions prevent mud deposition on the shelf. Muddy sediments are deposited under quiet condition of the JSPW. The low temperature and high oxygen concentrated bottom layer of JSPW forms the oxidizing condition and dissolves the calcium carbonates in the surface sediments in the trough. On the other hand, local tectonic movements are forming the sedimentary basins and are closely related to the filling up processes of the trough.

As mentioned first, on the Pacific side of Southwest Japan, a strong ocean current, Kuroshio, is prevailing along the coast and makes a quite different sea condition. Modern sedimentation of the Pacific shelf, slope and basin is highly influenced by the Kuroshio (Okamura et al., 1986; Ikehara, 1988a, b, 1990). The difference of the oceanographic conditions are clearly reflected

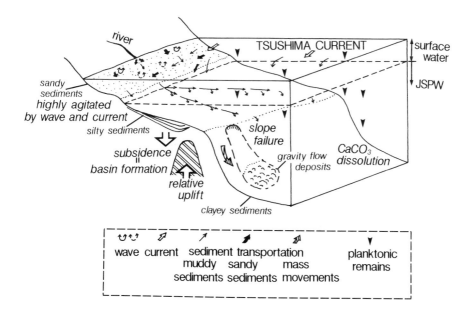

Fig.12 Schematic representation of the modern sedimentation off San'in district.

158

on the modern sedimentation. For example, the distribution of sandy sediments is limited on the shelf and the bedforms are poorly developed in the Japan Sea where the current flows slowly along the shelf, while the sandy sediments are widespread from the shelf to the upper slope and several types of bedforms are found in the Pacific where the strong current is prevailing along the shelf.

Regarding further work, the comparative study of modern sedimentation in different oceanographic environments would provide the most important clues to understanding the marine sedimentology around the Japanese Islands.

Acknowledgements
    I express my hearty thanks to the officers, crew and scientists of the R/V Hakurei-Maru for kind help throughout the surveys. I am grateful to Dr Yoshihiko Shimazaki of Nikko Exploration and Development Co. for his critical reading of an early version of the manuscript. Thanks are due to Dr Masafumi Arita and Mr Yoshiki Saito of the Geological Survey of Japan for their instructive discussions and encouragement. I would like to thank Prof Hiroshi Machida of Tokyo Metropolitan University for identification of volcanic ash and Mrs Hirofumi Yamamoto, Hajime Katayama and Mikio Satoh of the Geological Survey of Japan for their cooperation in the field and laboratory works and fruitful discussions. Acknowledgements are also due to anonymous reviewers for their valuable suggestions. This work was supported by the special research program, "geological mapping on the continental shelf areas in Japan", of the Geological Survey of Japan.

6 REFERENCES

Asaoka, O., Hashimoto, Y. and Katayama, K., 1985. Physical oceanography in the Wakasa Bay. In: Coastal Oceanogr. Res. Commit., Oceanogr. Soc. Japan (editor), Coastal Oceanography of Japanese Islands. Tokai Univ. Press, Tokyo, 958-968 (in Japanese).
Belderson, R.H., Johnson, M.A. and Kenyon, N.H., 1982. Bedforms. In: A.H. Stride (editor), Offshore Tidal Sands, Processes and Deposits. Chapman and Hall, London, 27-57.
Chough, S.K., Jeong, K.S. and Honza, E., 1985. Zoned facies of mass-flow deposits in the Ulleung (Tsushima) Basin, East Sea (Sea of Japan). Mar. Geol., 65: 113-125.
Damuth, J.E., 1980. Use of high-frequency (3.5-12kHz) echograms in the study of near-bottom sedimentation processes in the deep-sea: a review. Mar. Geol., 38: 51-75.
Gamo, T., Nozaki, Y., Sakai, H., Nakai, T. and Tsubota, H., 1986. Spacial and temporal variations of water characteristics in the Japan Sea bottom layer. J. Mar. Res., 44: 781-793.
Hoshino, M., 1952. On the muddy sediments of the continental shelf adjacent to Japan. J. Geol. Soc. Japan, 58: 41-53 (in Japanese with English abstract).
Hoshino, M., 1958. The shelf sediments in the adjacent seas of Japan. Monogr. Assoc. Geol. Collab., 7: 1-41 (in Japanese with English abstract).

Ichikura, M. and Ujiié, H., 1976. Lithology and planktonic foraminifera of the Sea of Japan piston cores. Bull. Natl. Sci. Mus., Tokyo, ser. C, 2: 151-178.

Ikehara, K., 1988a. Ocean current generated sedimentary facies in the Osumi Strait, south of Kyushu, Japan. Prog. Oceanog., 21: 515-524.

Ikehara, K., 1988b. Sedimentological map of Tosa Wan (with explanatory notes). Mar. Geol. Map Ser., 34: Geol. Surv. Japan, 29p. (in Japanese with English abstract).

Ikehara, K., 1989. Some physical properties of shelf to basin deposits off San'in and Hokuriku district, southern part of Japan Sea. Bull. Geol. Surv. Japan, 40: 239-250 (in Japanese with English abstract).

Ikehara, K., 1990. Modern sedimentation of shelf to basin deposits offshore of Southwest Japan - effects of the strong ocean current, Kuroshio to the sedimentation. J. Sed. Soc. Japan, 32: 89-90 (in Japanese with English abstract).

Ikehara, K. and Katayama, H., 1987. Sea bottom photography offshore of San'in district. In: M. Arita, Y. Okuda and T. Moritani (editors), Preliminary Reports of Geological Mapping Program of the Continental Shelf Areas in Japan. Geol. Surv. Japan, 77-80 (in Japanese).

Ikehara,K., Katayama, H. and Satoh, M., 1990a. Sedimentological map offshore of Tottori (with explanatory notes). Mar. Geol. Map Ser., 36: Geol. Surv. Japan, 42p. (in Japanese with English abstract).

Ikehara, K., Kinoshita, Y. and Joshima, M., 1987. Preliminary reports on the bedforms in the Oki Strait. In: M. Arita, Y. Okuda and T. Moritani (editors), Preliminary Reports of Geological Mapping Program of the Continental Shelf Areas in Japan. Geol. Surv. Japan, 63-75 (in Japanese).

Ikehara, K., Satoh, M. and Yamamoto, H., 1990b. Sedimentation in the Oki Trough, southern Japan Sea, as revealed by high resolution seismic records (3.5kHz echograms). J. Geol. Soc. Japan, 96: 37-49 (in Japanese with English abstract).

Inouchi, Y., 1988. Sedimentation models of Lake Biwa. Clastic Sed. (J. Res. Gr. Clastic Sed. Japan), 5: 49-72 (in Japanese with English abstract).

Iwabuchi, Y., 1968. Submarine geology of the southeastern part of the Japan Sea. Contr. Inst. Geol. Palaeont., Tohoku Univ., 66: 1-76 (in Japanese with English abstract).

Japan Oceanographic Data Center, 1979. Marine Environmental Atlas, Currents - adjacent seas of Japan. 71p.

Kawabe, M., 1986. Study on the Kuroshio and the Tsushima Current - variations of the current path. J. Oceanogr. Soc. Japan, 42: 319-331 (in Japanese with English abstract).

Machida, H. and Arai, F., 1983. Extensive ash falls in and around the Sea of Japan from large late-Quaternary eruptions. J. Volc. Geotherm. Res., 18: 151-164.

Maritime Safety Agency, 1949. Bottom sediment chart of the adjacent seas of Japan, sheet 1-4. Chart No.7051- 7054.

Minami, H., Hashimoto, Y., Konishi, Y. and Shuto, K., 1984. On the current pattern in the Oki Channel. Umi to Sora (Sea and Sky), 59: 115-125 (in Japanese with English abstract).

Nardin, T.R., Hein, F.J., Gorsline, D.S. and Edward, B.D., 1979. A review of mass movement processes, sediment and acoustic characteristics, and contrasts in slope and base-of-slope systems versus canyon-fan-basin floor systems. In: L.J. Doyle and O.H. Pilkey (editors), Geology of Continental Slopes. SEPM Special Publication, 27: 61-73.

Nitani, H., 1972. On the deep and bottom waters in the Japan Sea.

   In: D. Shoji (editor), Research in Hydrography and
   Oceanography. Hydr. Dept. Japan, 151-201.
Nomura, R. and Ikehara, K., 1987. Benthic foraminiferal
   assemblage off San'in district. In: M. Arita, Y. Okuda and T.
   Moritani (editors), Preliminary Reports of Geological Mapping
   Program of the Continental Shelf Areas in Japan. Geol. Surv.
   Japan, 165-176 (in Japanese).
Oba, T., 1987. Upper Pleistocene of the Japan Sea floor. Quat.
   Res. (Daiyonki-Kenkyu), 25: 319-321 (in Japanese with English
   abstract).
Oda, M. and Ikehara, K., 1987. Preliminary reports on the
   planktonic foraminiferal assemblage off San'in district. In:
   M. Arita, Y. Okuda and T. Moritani (editors), Preliminary
   Reports of Geological Mapping Program of the Continental Shelf
   Areas in Japan. Geol. Surv. Japan, 162-164 (in Japanese).
Okamura, Y., Tanaka, T. and Nakamura, K., 1986. Diving survey of
   the knolls on the trench slope break off Kochi, Southwest
   Japan. JAMSTECR Deep-sea Res., 173-192 (in Japanese with
   English abstract).
Saito, Y., 1988. Relationships between coastal topography and
   sediments, and wave base. Earth Monthly (Gekkan Chikyu), 10:
   458-466 (in Japanese).
Saito, Y., 1989a. Modern storm deposits in the inner shelf and
   their recurrence intervals, Sendai Bay, Northeast, Japan. In:
   A. Taira and F. Masuda (editors), Sedimentary Facies in the
   Active Plate Margin. Terra Scientific Publishing Company,
   Tokyo, 331-344.
Saito, Y., 1989b. Late Pleistocene coastal sediments, drainage
   patterns and sand ridge systems on the shelf off Sendai,
   Northeast Japan. Mar. Geol., 89: 229-244.
Saito, Y., 1989c. Sediments of the Akita Bay. In: Submarine
   Topography and Sediments in the Akita Bay -Reports of
   Anthropogenic Influence of the Sedimentary Regime of an
   Open Type Bay. Geological Survey of Japan, 14-70 (in
   Japanese).
Sato, S., Ijima, T. and Tanaka, N., 1963. A study of critical
   depth and mode of sand movement using radioactive glass sand.
   Proc. 8th Intern. Conf. Coastal Eng., Mexico City, 304-323.
Sato, T., 1961. The types of grain size distribution in shallow
   sea sediments. J. Geol. Soc. Japan, 67: 58-65 (in Japanese
   with English abstract).
Shimonaka, M., Ogawa, F. and Ichikawa, W., 1970.
   Silicoflagellatae remains in the deep-sea sediments from the
   Sea of Japan. Nihonkai (Japan Sea), 4: 1-14 (in Japanese with
   English abstract).
Simons, D.B., Richardson, E.V. and Nordin, Jr., C.F., 1965.
   Sedimentary structures generated by flow in alluvial channels.
   In: G.V. Middleton (editor), Primary Sedimentary Structures
   and Their Hydrodynamic Interpretation. SEPM Special
   Publication, 12, 34-52.
Suda, K., 1932. On the bottom water in the Japan Sea (preliminary
   report). J. Oceanogr., 4: 221-241 (in Japanese).
Sugawara, K., Kobune, K., Sasaki, H., Hashimoto, N., Kameyama, Y.
   and Narita, A., 1986. Wave statistics with 15-year data in the
   coastal wave observation network. Tech. Note Port and Harb.
   Res. Inst., 554: 1-872 (in Japanese with English abstract).
Sunamura, T., 1986. Coastal cliff erosion in Nii-jima Island,
   Japan: present, past, and future -an application of
   mathematical model. In: V. Gardiner (editor), International
   Geomorphology. John Wiley and Sons, Chichester, Part I, 1199-
   1212.

Sunamura, T., 1990. A dynamical approach to wave-base problems. Trans. Japan. Geomorph. Union, 11: 41-48.

Tanaka, Y., 1987. Calcareous nannofossils assemblage around the Oki Islands. In: M. Arita, Y. Okuda and T. Moritani (editors), Preliminary Reports of Geological Mapping Program of the Continental Shelf Areas in Japan. Geol. Surv. Japan, 177-185 (in Japanese).

Tanimura, Y., 1981. Late Quaternary diatoms of the Sea of Japan. Sci. Rep. Tohoku Univ., 2nd ser.(Geol.), 51: 1-37.

Yamamoto, H., Joshima, M. and Kisimoto, K., 1989. Geological map offshore of Tottori (with explanatory notes). Mar. Geol. Map Ser., 35: Geol. Surv. Japan, 27p. (in Japanese with English abstract).

Yasui, M., Yasuoka, T., Tanioka, K. and Shiota, O., 1967. Oceanographic studies of the Japan Sea (I) -water characteristics. Oceanogr. Mag., 19: 177-192.

Yokota, M., Okada, H., Arita, M., Ikehara, K. and Moritani, T., 1990. Distribution of heavy minerals in the bottom sediments of the southern Sea of Japan, off the Shimane Peninsula, Southwest Japan. Sci. Rep., Dept. Geol., Kyushu Univ., 16: 59-86 (in Japanese with English abstract).

EFFECTS OF WINTER COOLING ON SUBSURFACE HYDROGRAPHIC CONDITIONS OFF KOREAN
COAST IN THE EAST (JAPAN) SEA

Young-Ho Seung and Soo-Yong Nam
Dept. of Oceanography, Inha Univ., Incheon 402-751, Korea

ABSTRACT

Analyses of 24-year hydrographic data suggest that the appearence of
cold/fresh subsurface water off the Korean coast is related to winter cooling.
After the passage of winter and during the following spring, the cold
subsurface water generally extends southward along the coast. The area under
influence is generally limited in both onshore-offshore and north-south
directions. After the winter which is colder than the preceding one, however,
the southward extension of the subsurface cold water, which is greatly
enhanced, takes place for longer period until finally almost the whole area
becomes under influence. It is shown that the observed cold subsurface water is
a consequence of propagating coastal trapped waves. These waves can be
generated near the initial front developed by winter cooling. The concept of
coastal trapped waves are further discussed and partially confirmed by a simple
numerical experiment.

INTRODUCTION

The East (Japan) Sea is very susceptible to the atmospheric cooling in

winter. In fact, Manabe (1957) pointed out that outbursts of cold and dry

continental air from the northwest onto the East Sea in winter involve a large

amount of heat flux from sea to atmosphere. It was shown that this air mass,

while crossing the basin, undergoes a rapid transformation and acquires

maritime characteristics when, or even before, they reach to Japanese coast.

This fact means that the atmospheric cooling in winter is very strong near

continental side and rapidly weakens toward Japanese side. This differential

cooling certainly affects dynamics of the East Sea circulation as it does in

the Labrador Sea (Seung, 1987).

The upper part of the East Sea is generally divided into two sectors: a cold

sector in the north and a warm sector in the south. Between them, a

quasi-permanent thermal front is formed. This front seems the most distinctive

around 100m depth. In the western part of the basin, it is usually observed

around 38°N latitude line (c.f., JODC, 1975). Near the western coast, the front

164

at 100m depth slowly changes its orientation toward the south such that the coastal water is colder than the offshore water. The climatological charts provided by the Fisheries Research and Development Agency (FRDA, 1986) show that this change of orientation begins to develop in February until finally, after spring has passed, the isotherms run nearly in the north-south direction parallel to the coastline. This fact indicates that the cold water found north of the front begins to intrude southward along the coast near the end of winter. The intensity of the cold water intrusion seems to vary from year to year. For example, in spring 1981, abnormally cold/fresh subsurface water was found in large area off the south Korean coast extending far to the south and forming there a strong front against the warm/saline East Korea Warm Current

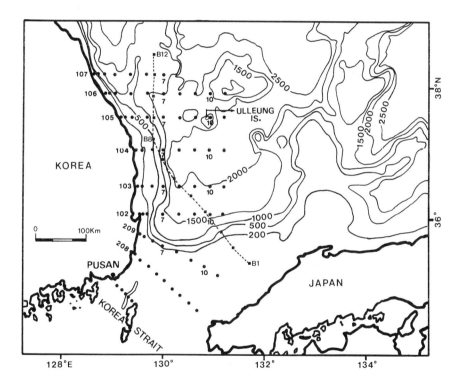

Fig. 1. The area under consideration. Dots are FRDA (1962-1985) observation points: Observation lines are aligned along the east-west direction and numbered from 107 through 208. Each line contains observation stations with numbers increasing offshore. Dotted line is a BT section taken by MMO (1974) in February. Air temperature is measured at Ulleung Island by CMO (1961-1984). Bathymetry is in meters.

Water flowing northward (Hong et al, 1984; Kim and Legeckis, 1986).

In this paper, we investigate the southward intrusion of cold subsurface water in more detail by analyzing 24-year hydrographic data obtained by FRDA (1964-1985) off the Korean coast (see Fig.1 for location). We then try to understand the mechanism how hydrographic conditions, especially those of subsurface layer, change in response to winter cooling in this area.

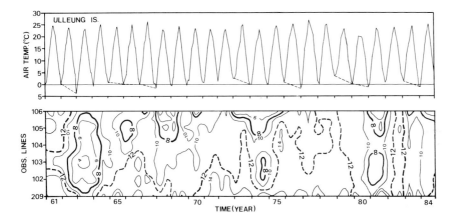

Fig. 2. Variation of air temperature at Ulleung Island and space-time diagram of water temperature (in °C) at 100m depth measured along stations 09 (along 130° 30′ E longitude line).

WINTER CONDITIONS AND SUBSURFACE TEMPERATURES

To obtain a general idea of subsurface temperature variation in space and time, we present in Fig.2 a space-time diagram of temperature at 100m depth. The space here means the north-south line passing through stations 09 (130° 30′ E longitude line; see Fig.1). The subsurface is represented by 100m depth. At this depth, the seasonal and longer-time variabilities trapped at the surface may well present without being disturbed by surface noises of smaller vertical scale. In deeper layers, measurements do not seem to be accurate enough to resolve slight changes of hydrographic condition. Also presented in Fig.2 is a time variation of air temperature at Ulleung Island measured by Central Meteorological Office (CMO, 1961-1984) during the same period. Hydrographic data are normally bi-monthly but those obtained before 1968 are somewhat sparse. In qualitative estimation of long-term variability, however, this fact

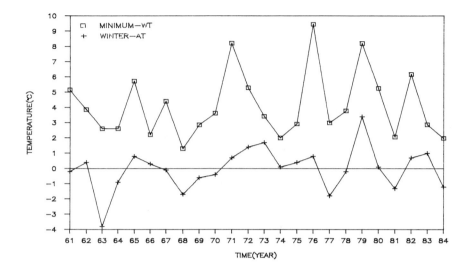

Fig. 3. Variation of air temperature in winter (February) and the yearly minimum water temperature found along stations 09 (longitude $130°30'E$) at 100m depth.

does not seem to be serious.

A comparison between subsurface water and air temperatures (Fig.2) seems to suggest that the distribution of cold subsurface water along this line generally becomes more prominent when winter is colder than the preceding one. A direct relationship between the subsurface water and air temperatures is also seen by a comparison made between air temperatures in winter (February) and the yearly minimum water temperatures among those measured along stations 09 (along the longitude $130°30'E$) at 100m depth (Fig.3). Further inspection of Fig.2 indicate that the minimum water tempereature normally occurs at the northern edge of the line (about $38°N$) except when the colder winter has passed. In the latter case, isolated patches of cold water are frequently observed along this line and the usual north-south gradient of temperature is disturbed. It will be shown in later sections that these processes may be a manifestation of propagating coastal trapped waves which do or do not pass the line under consideration. These waves are baroclinic disturbances generated near the initial front developed by winter cooling. Observations (Fig.2) indicate that the disturbances are so much strong as winter is colder than the preceding one.

Physical meaning of this fact will be discussed in the next section.

SOUTHWARD PROPAGATION OF COOLING EFFECT

Figure 4 and 5 show temporal changes of temperature and salinity distributions on 100m-depth plane in 1974; and Figure 6 shows that of temperature in 1975. In Figure 2, the year 1974 can be considered as a typical year when the distribution of cold subsurface water (or cold patches) is prominent. Likewise, the year 1975 can be considered as a normal year. In these figures, isolines of 8°C or 34.2 ‰ may be considered as a forefront of cold/fresh subsurface water. In February, 1974, the cold subsurface water is initially confined to the north of 37°N which can be considered as the initial front developed by winter cooling. It begins to extend southward along the coastline until, in August, it occupies most of the area considered. Then it

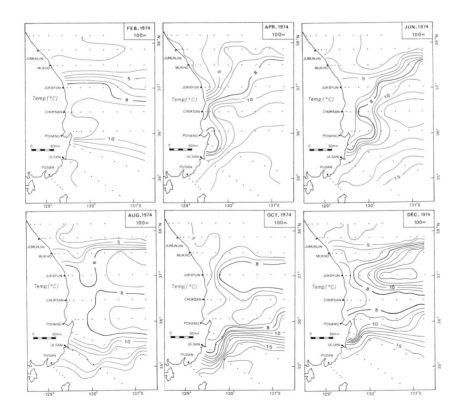

Fig. 4. Change of temperature distributions in the year 1974 on 100m-depth plane.

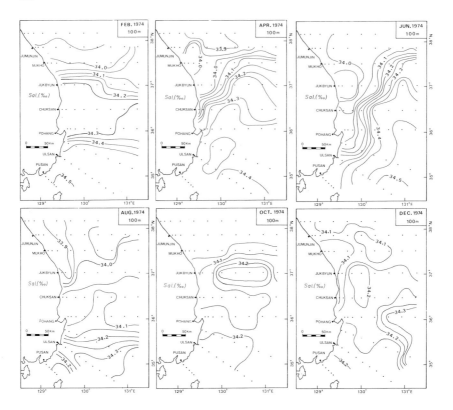

Fig. 5. Change of salinity distributions in the year 1974 on 100m-depth plane.

begins to decay from October probably due to the effects of northward flowing Tsushima Warm Current. The same process is observed in spring, 1975 but for shorter period and thus with much reduced extension. It is quite clear that, in normal condition, the presence of cold subsurface water is generally limited to the coastal area. If conditions are favorable, such as when winter is colder than the preceeding one, it extends further south then turns around along the topographic discontinuity (c.f., Fig.1) until finally it affects most of the area considered. The one-dimensional feature shown in the preceding section thus becomes more clear here.

At this time we have to consider how the initial front develops in response to the winter cooling (represented by air temperature) since it seems to give rise to, and determines the amplitude of, the southward propagating baroclinic

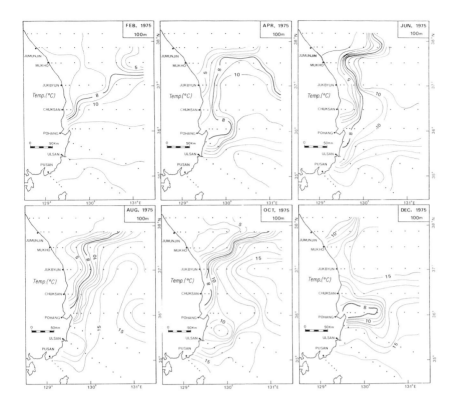

Fig. 6. Change of temperature distributions in the year 1975 on 100m-depth plane.

disturbances. In most part of the study area (Fig.1), the buoyancy removed by the local cooling can be partially supplemented by the Tsushima Warm Current entering the basin through the Korea Strait. In the north of the front, however, horizontal advection of buoyancy by this current is negligible and the surface/subsurface water temperature is mainly controlled by local surface heat flux. The front can therefore develop by cooling. In particular, further stronger (or deeper) front is expected when winter is colder than the preceding one based on the following hypothesis: Assume that a water column is nearly homogeneous at the end of a winter. Buoyancy is stored during the following summer season and begins to be removed again as the next winter approaches. If this winter is the same cold as the preceding one, exactly the same amount of buoyancy as stored during the last summer will be removed. For colder winter, the buoyancy in surface layer becomes less than that in deeper layers and renewal can take place resulting in deep front. For the same cooling, the development of front may be more effective near the coast than in the open sea because the subsurface water is closer to the surface near the coast of the area considered, i.e., the thermal inertia is smaller there.

Some places, such as near station 103-09, are affected by the cold subsurface water only in such winter condition as favorable for large baroclinic disturbances. The years 1974 and 1981, for examples, meet this condition. For these two years, annual cycles of water temperature at each depth are presented at the station 103-09 (Fig.7). For reference, the annual

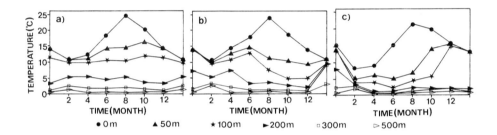

Fig. 7. Annual variation of water temperature at each depth (in m) measured at station 103-09: a), average over the period 1961-1975; b), the year 1974; and c), the year 1981.

cycle averaged over long period (1961-1975) is also shown. Generally, the seasonal cycle of water temperature is similar to that usually observed in temperate zone: highest in summer and lowest in winter at the surface with decreasing amplitude and increasing phase lag as depth increases. However, in both 1974 and 1981, much change is seen. In February, 1974, upper 100m is weakly homogenized probably due to local cooling. But the decrease of temperature in layers 50-100m from June to December cannot certainly be explained by local cooling. In 1981, the same kind of subsurface temperature decrease is observed in February. The water then does not change its temperature until August. Though the sudden decrease of temperature happens at the coldest time of the year, it does not seem due to local cooling because the rate of decrease is larger than that at the surface, as well as because the

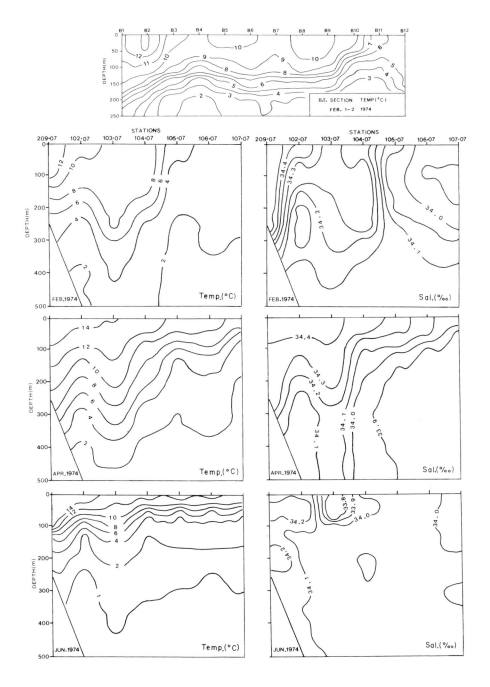

Fig. 8. Vertical section of temperature taken along the BT section of MMO and those of temperature and salinity along 07 stations of FRDA. The latters are taken about 15 days later than the former

temperature change thereafter seems to be decoupled from that at the surface. Compared with that in 1974, the cold subsurface water appears much earlier in 1981. This is probably because winters have become successively colder for two years (Fig.2).

To look into more closely the process of cold subsurface water extension, the case of 1974 is chosen as a good example. Figure 8 shows the vertical sections of temperature and salinity drawn along 07 stations (130° E longitude line) for the period February-June when the southward extension of cold subsurface water was remarkable. For reference, the BT section taken by MMO (1974) is also shown. In mid-February, FRDA sections show a deep front near observation line 105 which separates the cold/fresh and warm/saline waters. In BT section, taken 15 days earlier than those by FRDA in February, this deep front appears at about 100 Km to the north (see Fig.1 for location). This fact seems to indicate that vertical homogenization by surface cooling have continued during the first half of February. As time progresses, this front changes its form in such a way that the cold/fresh water penetrates into deeper layers including the subsurface layer toward the south. The southward extension of the subsurface front observed in Figures 4 through 6 is thus by the upward movement of the cold/fresh water from below. This type of variation is compatible with the southward propagation of baroclinic disturbances created by a sudden change in density structure. The northward spreading of warm/saline water over the thin surface layer seems to be a combined results of surface advection (in this case, surface current is expected northward as shown later) and surface heating. By June, most of the subsurface layer is occupied by the cold/fresh water. One thing which is not easily understood is that ,at this time, the surface layer is filled with the fresh (but not cold) water, on the contrary that it was once dominated by the warm/saline water. The possibility of the role of coastal trapped baroclinic waves will be further confirmed by a numerical experiment in the next section.

NUMERICAL EXPERIMENT

To perform the numerical experiment, we first idealize the problem by
assuming two regions of different density: one, in the north and the other, in
the south. In each region, vertical gradient of density is retained. This is
the "dam-breaking" experiment (e.g., Hsieh and Gill, 1984) in which a dam
initially separating two different water bodies is suddenly removed. The model
domain is a limited area with a horizontal extension from 128°E to 134°E and
from 33°N to 43°N (c.f., Fig.9). The dam is initially located along about 39°N
latitude line. Most interest is paid on Korean coastal region to the south of
the dam.

Initial density values are determined by taking the temperature and salinity
observed in February, 1974 (Fig.8) into consideration. The model employed is
the Semtner version (1974) of Bryan-Cox's General Circulation Model. Horizontal
grid size is taken as a quater degree of latitude and longitude and time step,

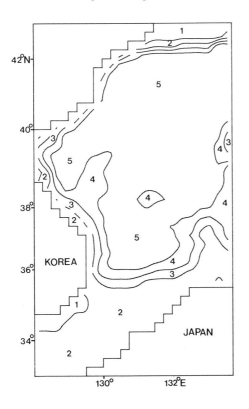

Fig. 9. Topography of the model basin. Bottom depth is expressed in number of
levels.

TABLE 1.

Thickness and initial density of each level.

level	thickness(m)	initial density(sigma-t) north/south
1	50	27.10/26.51
2	100	27.20/26.53
3	350	27.25/27.24
4	500	27.30/27.30
5	1000	27.30/27.30

as 30 minutes. Vertical resolution is such that the deepest part is divided into 5 levels. The thickness and initial density of each level are shown in Table 1 and the topography of model basin is shown in Figure 9. Horizontal eddy diffusivity is taken as 5×10^6 Cm^2/sec; horizontal eddy viscosity, as 2×10^7 Cm^2/sec; and both of vertical eddy viscosity and diffusivity, as 1 Cm^2/sec.

To minimize effects of artificial boundaries (to the north, south and east), a sponge layer of 10 grid intervals is placed adjacent to eastern boundary. Within the sponge layer, a linear friction is imposed such that its coefficient increases from zero at interior edge to 2×10^{-6}sec^{-1} at the boundary. Time integration is performed up to 40 days which is long enough to check the concept proposed earlier. Longer time integration is possible but effects of artificial boundaries may intervene. They trap disturbances, which otherwise freely leave the domain, and serve as wave guides.

Figure 10 shows the resulting density distribution at level 2 and the vertical excursion of water particles checked at level 2.5 (interface between level 2 and 3), after 20 days and 40 days have passed. As mentioned earlier, the initial density front lying along $39°$N propagates southward along the coast. These are coastal trapped baroclinic waves modified by bottom topography. As might be expected from the theory of baroclinic wave, the increase of subsurface density is seen due to the vertical advection of deeper waters. The subsurface density front represented by 26.87 in sigma-t unit, which is the average of initial densities in the north and south at level two (Table 1), moves at about 2 to 3 Km/day (Fig.10). This value is comparable to that observed during the period February-June (Fig.5 and 6). Further

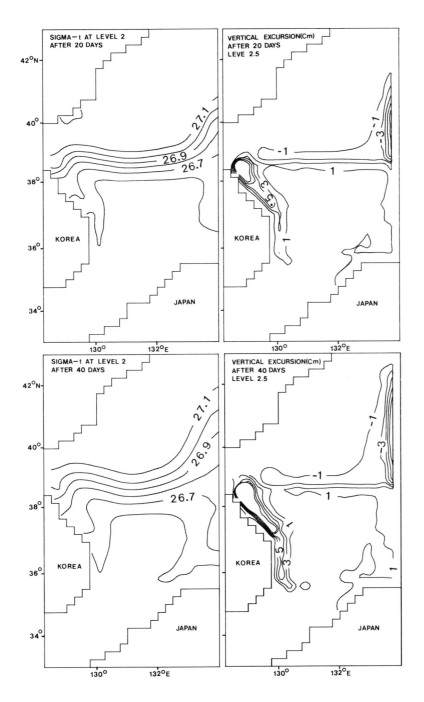

Fig. 10. Horizontal distributions of density at level 2 and vertical excursion of water particles measured at level 2.5 (interface between level 2 and 3) 20 and 40 days after the start of experiment.

discussions about the coastal trapped waves in this area, which are beyond the scope of present study, will be prepared in the future.

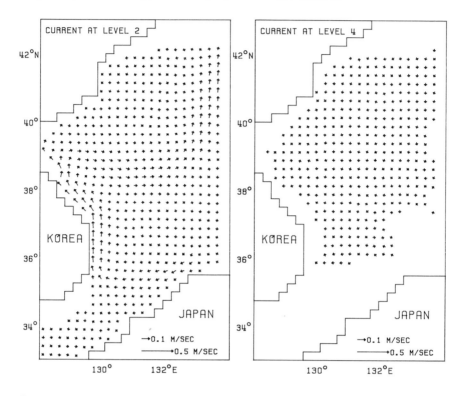

Fig. 11. Current fields at level 2 and 4 after 40 days.

The current field after 40 days (Fig.11) shows that the induced motions (barotropic + baroclinic) are noticible only in upper two levels. Current is trapped along the Korean coast and directs northward up to the point of initial density front where it then deflects eastward. Upper two levels are therefore continuously affected by northward advection while vertical advection affecting only level 2; no significant change in density is observed at the first level during the experiment (not shown here). This fact may explain the observation (Fig.8) that surface layer is progressively occupied by warm/saline water toward the north during the period February-April, though the appearence of fresh water in June is not properly explained.

CONCLUDING REMARKS

Long-term observations suggest that the winter cooling is closely related to the appearence of cold/fresh subsurface water off the Korean coast in the East Sea. The cold subsurface water usually occurs sometime after winter has passed indicating that the cold water is not formed by local cooling in this region. A simple numerical experiment confirms the idea that it is due to the vertical advection of deeper waters associated with the passage of coastal trapped waves. These waves are baroclinic disturbances generated at the initial front developed by winter cooling. The disturbances are so much strong as the winter is colder than the preceding one probably because this condition favors deeper, and therefore more intense, front.

We admit that the present study deals with the problem only in limited extent. Consideration of only subsurface layer may be incomplete. Observations have been performed only down to 500m depth. Even in observed layers below the subsurface, changes in temperature and salinity are too small to be accurately detected by traditional methods. The area covered by observations is not sufficient either. For example, we don't know yet how the surface/subsurface layers in other parts of the basin undergo the surface cooling. Also, the assumption of development of initial front by cooling should be further confirmed. Future studies with more accurate observations covering more extensive space, along with long-term current measurements, will certainly lead to much better results.

As for the numerical experiment, we are satisfied only with confirmation of the proposed idea. The most important defect is that we do not take the effect of the Tsushima Warm Current into consideration. The buoyancy advected by this current will certainly prevent the southward extension of cold subsurface water in such a way that the extension of cold subsurface water is determined by the competition between the two effects, namely, those of cooling and buoyancy advection by the warm current. The Tsushima Warm Current will also play an important role in the final stage of yearly cycle where the disturbed hydrography is rapidly restored to original state (c.f., FRDA, 1986). Inclusion of the inflowing Tsushima Warm Current into the model neccesitates the consideration of the whole basin and will be done in near future.

REFERENCES

Central Meteorological Office, 1961–1984. Monthly Weather Report,
    Jan., 1961 through Dec., 1984, Seoul, Korea.
Fisheries Research and Development Agency, 1964–1985. Annual Report of
    Oceanographic Observations, Vol. 10 through Vol. 33, Pusan, Korea.
Fisheries Research and Development Agency, 1986. Mean Oceanographic Charts of
    the Adjacent Seas of Korea. Dec., Pusan, Korea, 106pp.
Hsieh, W.W. and Gill, A.E., 1984. The Rossby adjustment problem in a rotating,
    stratified, with and without topography. J. Phys. Oceanogr., 14: 424–437.
Hong, C.H., Cho, K.D. and Yang, S.K., 1984. On the abnormal cooling phenomenon
    in the coastal area of East Sea of Korea in summer, 1981. J. Oceanol. Soc.
    Korea, 19: 11–17.
Japan Oceanographic Data Center, 1975. Marine Environment Atlas: Northwestern
    Pacific Ocean I (annual mean). Japan Hydrographic Association, Tokyo, 164pp.
Kim, K. and Legeckis, R., 1986. Branching of the Tsushima Current in 1981–83.
    Progress Oceanogr., 17: 265–276.
Maizuru Marine Observatory, Japan, 1974. Oceanographic Prompt Reports,
    Supplement for No. 266, March, Maizuru, Japan, 47pp.
Manabe S., 1957. On the modification of air mass over the Japan Sea when the
    outburst of cold air predominates. J. Meteor. Soc. Japan, Ser. II, 35:
    311–326.
Semtner A.J., 1974. An oceanic general circulation model with bottom
    topography. Numerical simulation of weather and climate Tech. Rep., No. 9,
    Univ. of California, Los Angeles, 99pp.
Seung Y.H., 1987. A buoyancy flux–driven cyclonic gyre in the Laborador Sea.
    J. Phys. Oceanogr., 17: 134–146.

AN OBSERVATION OF SECTIONAL VELOCITY STRUCTURES AND TRANSPORT OF THE TSUSHIMA CURRENT ACROSS THE KOREA STRAIT

A. KANEKO
Interdisciplinary Graduate School of Engineering Sciences, Kyushu University 39, Kasuga 816 (Japan)
S-k. BYUN
Korea Ocean Research and Development Institute, Ansan P. O. Box 29, Seoul 425-600 (Korea)
S-d. CHANG
Department of Ocean Engineering, National Fisheries University of Pusan, Daeyeon-Dong, Pusan 608 (Korea)
M. TAKAHASHI
Fukuoka Prefectural Fisheries Experimental station, 1141-1 Imazu, Nishi-ku, Fukuoka 819-01 (Japan)

ABSTRACT
Sectional velocity structures of the Tsushima Current were directly measured along a transect (W-line) in the western channel and two transects (E- and F-lines) in the eastern channel of the Korea Strait, by use of a fish-mounted Acoustic Doppler Current Profiler (ADCP). A comparison of the result of ADCP measurement with the geostrophic velocity fields shows that the geostrophic balance holds better in the western channel than in the eastern channel. The correlation coefficient between the ADCP velocity field, including tidal currents and the geostrophic velocity field with a level of no motion at the bottom is 0.908 for the W-line and 0.523 for the E-line. We get a strong evidence that the transport through the western channel is two to three times greater than that through the eastern channel.

1 INTRODUCTION

The Korea Strait, located between Japan and Korea, is known as a good fishing ground. Currents in the Korea Strait are characterized by the coexistence of the strong tidal currents and the Tsushima Current, a branching flow of the Kuroshio which transfers warm and saline water into the Sea of Japan. These situations stimulated Japanese and Korean oceanographers to study the current system in the Korea Strait (Yi, 1966; Lee, 1970; Moriyasu, 1972; Lee, 1974; Miita, 1976; Byun and Seung, 1984; Mizuno et al., 1986). The detailed structures of current across the strait, however, are still poorly understood because the strong fishing activities inhibit long-term current-meter moorings and the cooperation between Japan and Korea has been insufficient.

A fish-mounted Acoustic Doppler Current Profiler (ADCP) becomes a powerful tool to get the sectional structure of currents along a traverse line (Kaneko and Koterayama, 1988; Kaneko et al., 1990). For this study, a fish-mounted ADCP is operated under the close cooperation of Japan and Korea to measure the sectional structure of the Tsushima Current across the Korea Strait.

2 OBSERVATION SITE AND METHOD

A location map of the observation sites is presented in Fig.1. The Tsushima island divides the Korea Strait into two parts, western and eastern channels. Traverse surveys using a fish-mounted ADCP were carried out on 14 September 1987 along the W-line in the western channel, and on 15 September and 3 August 1987, respectively, along the E- and F-lines in the eastern channel. The ADCP fish was towed by the R/V Pusan 402 of the National Fisheries University of Pusan for the W- and E-line surveys and the R/V Genkai of the Fukuoka Prefectural Fisheries Experimental Station for the F-line survey.

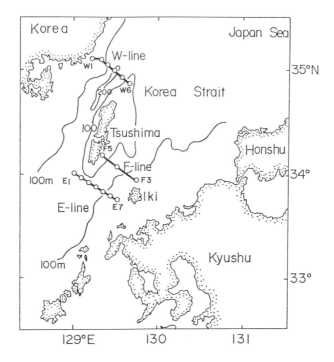

Fig. 1 Location map of the observation sites.

Fig. 2 Picture of the fish-mounted ADCP system used in this study.

Figure 2 shows the picture of the fish-mounted ADCP system. One may refer to Kaneko and Koterayama (1988) and Kaneko et al. (1990) for detailed description of the ADCP system. Velocity data were sampled every one minute at depth intervals of 8 m. Small-scale disturbances in the raw data were eliminated through a rectangular window of 10 minute x 16 m. Water depths along all the traverse lines were shallower than the ADCP bottom-tracking range of 480 m, so that the velocities obtained in the instrument coordinates could automatically be transformed to those in the earth coordinates by using bottom-tracking bins. The tidal components included in the ADCP data of the E- and F-lines is removed, using the values predicted by the the Hydrographic Department of the Japan Maritime Safety Agency (JMSA). The ADCP-measured bottom profiles and the positions where velocity data were collected are shown in Figs. 3a, b and c for the W-, E- and F-lines, respectively. At all the sections no data are available in a near-bottom layer. Data in this layer were produced through a linear interpolation with a level of no motion at the bottom. The bottom along the W-line slopes down toward a trough at the central part where the maximum depth is 225 m. The bottom is relatively flat along the E- and F-lines. The mean depth is about 120 m and 100 m for the E- and F- lines, respectively. The fish heading and motions were measured by a flux-gate compass and dual tilt sensors installed inside the ADCP vessel. The total number of velocity data and the mean values of fish speed and heading for each line are listed in Table 1.

The CTD casts were made at stations W1 to W6 on the W-ine and at stations E1 to E7 on the E-line. Due to bad recording on a magnetic tape, there were no data at station E1. The conductivity sensor was not stable electronically over the whole observation period. Therefore salinity data were reproduced using a temperature (T) - salinity (S)

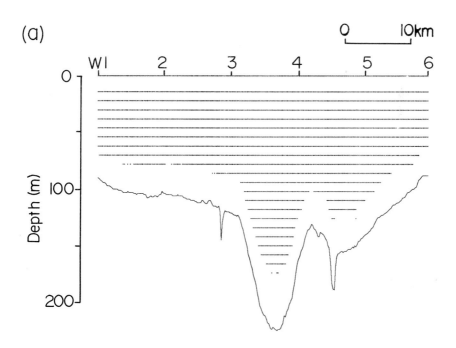

(b)

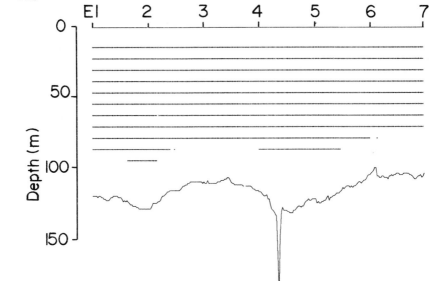

(c)

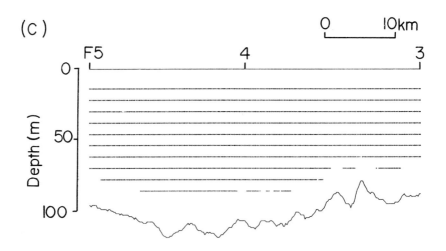

Fig. 3 Bottom profiles and positions of velocity measurements. (a) W-line (b) E-line (c)F-line

TABLE 1

Data number and the condition of ship operation for each traverse survey.

Traverse Line	Data Number	Ship Speed (cm / s)	Ship Heading (°)
W-line	8130	294.6	126.3
E-line	2553	301.1	317.8
F-line	2313	286.8	303.5

relationship determined from the data collected in the neighboring regions by the Fukuoka and Yamaguchi Prefectural Fisheries Experimental Stations during September and October of 1987. Data used in this procedure are plotted in Fig.4 with the least-squares fit

$$S = 33.4018 + 0.0840\ T + 0.00039\ T^2 - 0.0003\ T^3$$

where T and S are in °C and in %$_o$, respectively. We had the Nansen casts at stations F3 to F5 on the F-line.

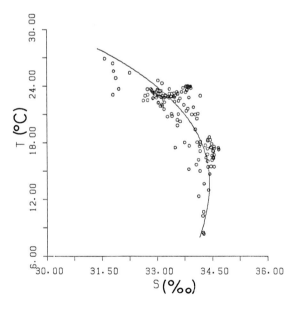

Fig. 4   T-S diagram from the data obtained in the neighborhood of the observation site by the Fukuoka and Yamaguchi Prefectural Fisheries Experimental Stations. A least-squares fit of the data is drawn with a solid line.

3 RESULTS

3.1 HYDROGRAPHIC DATA

Figure 5 shows the result of hydrographic observation for the W-line. A warm and less saline water which may be associated with the discharge of Chinese rivers (mainly, the Changjiang), occupies the upper 40 m of the section. A sharp thermocline and halocline exist just below the upper layer through stations W1 and W4. This stratification is weak between stations W4 and W6 downstream of the Tsushima island. Almost all isotherms between stations W1 and W4 sloped down toward the right (southeastward), implying the existence of the northeastward Tsushima Current. Cold water with the lowest temperature less than 4 °C is in the trough. There is a weak thermocline above the trough.

Figure 6 shows the result of hydrographic observation for the E-line. The isotherms of 18 to 24 °C, forming the sharp thermocline in Fig. 5a, are much more spaced in Fig. 6a. These suggest that the vertical shear of geostrophic currents is small in this region. The isohalines are also in the same situation as shown in Fig. 6b. The isotherms of 16 to 22 °C slopes downward to the right between stations E2 and E4 in contrast to the isotherms of 18 to 20 °C rising to the right between stations E5 and E6. This is an indication that the direction of geostrophic currents changes on the both sides of station E5. The near-surface warm and less saline water exists around station E3. The bottom cold water, as shown in Fig. 5a, is not found in Fig. 6a.

3.2 VELOCITY DATA

Figure 7a shows the vector plot of horizontal velocities obtained every one minute along the W-line. The NE

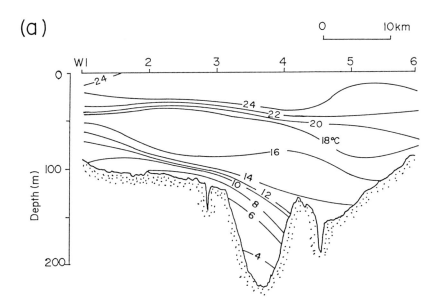

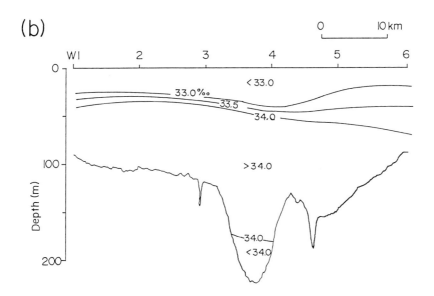

Fig. 5 Contour plot of the hydrographic data obtained along the W-line. (a) temperature (b) salinity

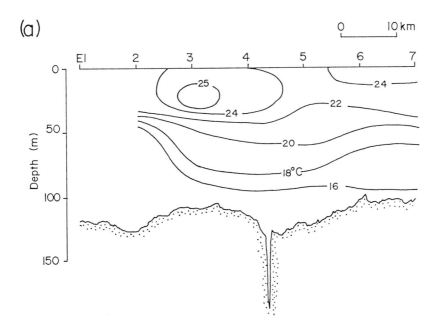

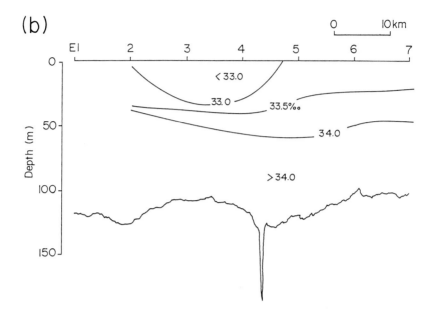

Fig. 6 Contour plot of the hydrographic data obtained along the E-line. (a) temperature (b) salinity

current is dominant over the whole section and its maximum speed is more than 100 cm/s between stations W3 and W4. The speed of the current decreases with horizontal distances from the peak, resulting in a jet-like profile. A weak SW current takes place at the surface layer around station W6. Water in the trough is almost at rest. Figure 7b shows the contour plot of the velocity component perpendicular to the W-line, smoothed through a 10 minute x 16 m rectangular window. The central part of the Tsushima Current with speeds greater than 50 cm/s is located between stations W2 and W5 of the western channel. Around the core of the Tsushima Current with a maximum velocity, the horizontal gradients of velocity are $3.9 \times 10^{-5} \mathrm{s}^{-1}$ on the left side of the core and $3.5 \times 10^{-5} \mathrm{s}^{-1}$ on its right side. The vertical gradient of velocity below the core is $5.0 \times 10^{-3} \mathrm{s}^{-1}$. The vertical shear of the current is greater by the order of 2 than the horizontal ones. The volume transport across the W-line is 2.2 SV ($1 \mathrm{SV} = 1 \times 10^{6} \mathrm{m}^{3} \mathrm{s}^{-1}$) estimated by integrating the velocities over the whole section. The tidal current predicted by JMSA for the same period is 10 cm/s to the east near station W4 and 7 cm/s to southeast near station W5. The JMSA has no sufficient data to predict the tidal currents between stations W1 and W4. Figure 8 shows the contour plot of the geostrophic current with a level of no motion at the bottom. Although the maximum speed is smaller by about 30 cm/s in Fig. 8 than in Fig. 7b, the overall flow pattern of the Tsushima Current is very similar in the both figures. The correlation coefficient between the ADCP and geostrophic flow fields is estimated as 0.908. The volume transport for the geostrophic current is 1.4 SV.

Figure 9a shows the vector plot of horizontal velocities obtained every one minute along the E-line. The NE current is confined between stations E1 and E3. The direction of the current is reversed in the rest of the section. Data acquired near the bottom of station E6 is contaminated by a large roll motion of the ADCP fish. Figure 9b shows the contour plot of the velocity component perpendicular to the E-line smoothed through a 10 minute x 16 m rectangular window. The core of the Tsushima Current with a maximum speed of 35 cm/s is seen at a depth of 30 m between stations E2 and E3. The core of the reverse flow forms in the surface layer around station E5. The velocity data

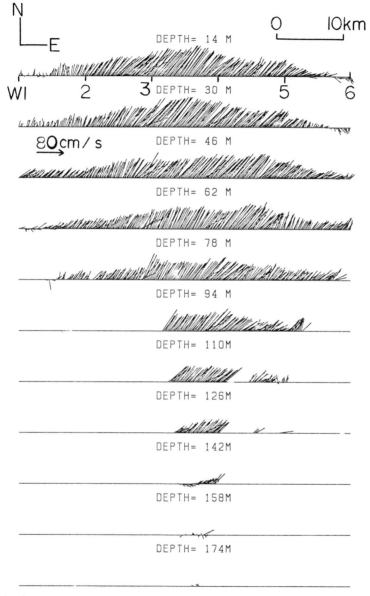

(a)

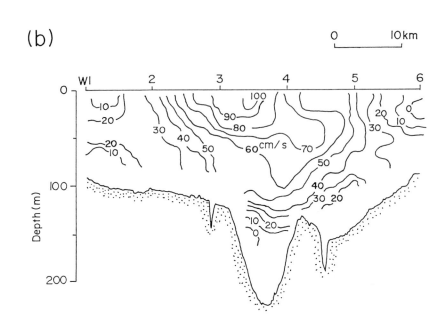

Fig. 7 ADCP data obtained along the W-line. (a) Vector plot of the horizontal velocities. The depth for each vector plot is indicated above the corresponding plot. (b) Contour plot of the velocity component perpendicular to the W-line.

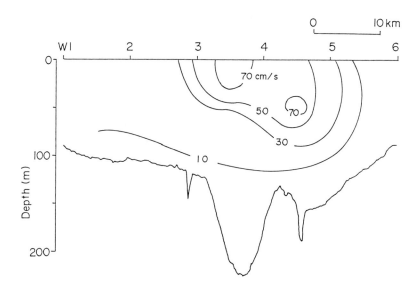

Fig. 8 Contour plot of the geostrophic current for the W-line. The level of no motion is taken at the bottom.

(a)

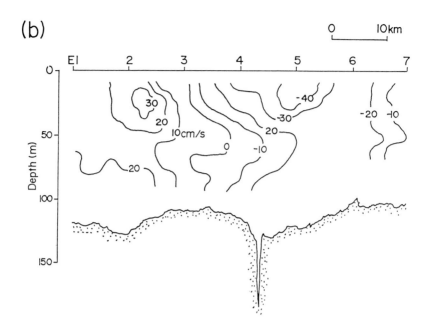

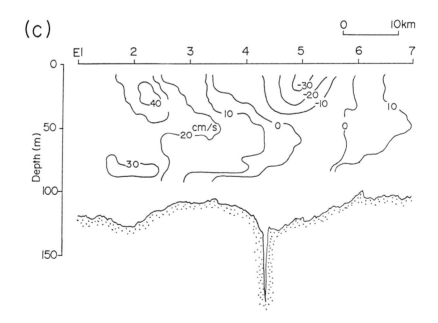

Fig. 9 ADCP data obtained along the E-line. (a) Vector plot of the horizontal velocities. The depth for each vector plot is indicated above the corresponding plot. (b) Contour plot of the velocity component perpendicular to the E-line with tidal currents. (c) Contour plot of the velocity component perpendicular to the E-line without tidal currents.

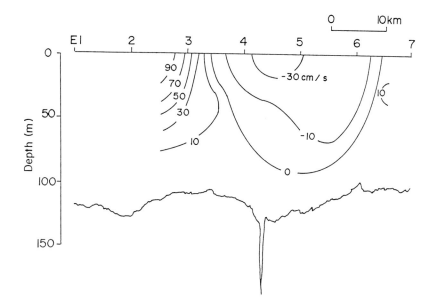

Fig. 10 Contour plot of the geostrophic current for the E-line. The level of no motion is taken at the bottom.

obtained by extracting the tidal components from the ADCP data are contoured in Fig. 9c. A comparison with Fig. 9b shows that the NE current is stronger by 10 cm/s while the reverse flow is weaker by 10 cm/s. In contrast to Fig. 9b there is a weak NE current on the right -hand corner of the section. As a result of the tidal correction, the area with the NE current is enlarged. The volume transports for Figs. 9b and c are estimated as -0.3 SV and 0.6 SV to the NE direction, respectively. Figure 10 shows the contour plot of the geostrophic current with a level of no motion at the bottom. The overall flow pattern is closer to Fig. 9c (without tidal current) than to Fig. 9b (with tidal current). The NE current between stations E1 and E3 is much stronger in Fig. 10 than in Figs.9b and 9c. The correlation coefficient is 0.523 between Figs. 9b and 10, and 0.758 between Figs. 9c and 10. The volume transport between stations E2 and E7 is 0.4 SV.

Figure 11a shows the vector plot of the horizontal velocities obtained every one minute along the F-line. The NE current between stations F5 and F4 diminshes significantly between stations F4 and F3. The NE current on left half of the section rotates the direction clockwise and decreases gradually the speed as it approaches the bottom. A weak reverse flow is seen in the mid and bottom layers around station F3. Figure 11b shows the contour plot of the velocity component perpendicular to the F-line smoothed through the rectangular window. The core of the Tsushima Current at a depth of 50 m between stations F5 and F4 has a maximum velocity of 36 cm/s. The boundary between the NE and the reverse flows slopes downward to the left from station F3. The contour plot of the velocity data obtained by extracting the tidal components from the ADCP data is shown in Fig. 11c. In contrast to Fig.11b, the speed of the NE current has no remarkable changes while the reverse flow disappears in the whole section. The volume transport through the whole section are 0.5 SV and 0.7 SV for Figs. 11b and c, respectively. The geostrophic

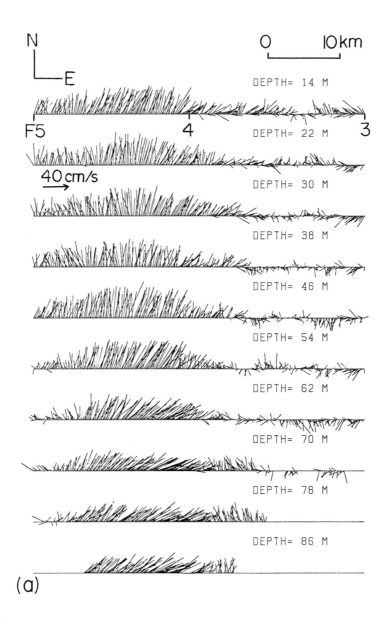

(a)

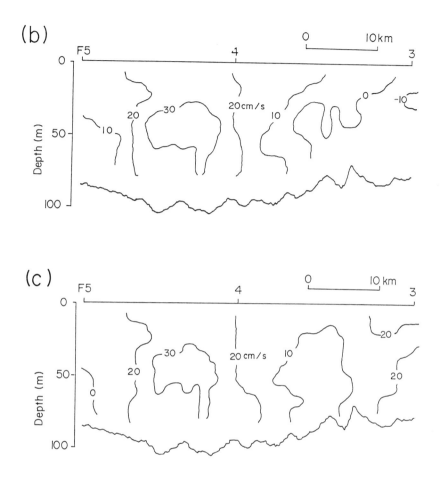

Fig. 11 ADCP data obtained along the F-line. (a) Vector plot of the horizontal velocities. The depth for each vector plot is indicated above the corresponding plot. (b) Contour plot of the velocity component perpendicular to the F-line with tidal currents. (c) Contour plot of the velocity component perpendicular to the F-line without tidal currents.

calculation is not attempted for F-line where the Nansen casts were done only at three stations.

Characteristic values of the flow fielde are listed in Table2 for all the cases of the W-, E- and F-lines.

4 DISCUSSION

The tidal current at the Korea Strait is stronger in the western channel than in the eastern channel of the strait although the former is deeper (Odamaki, 1989). The density stratification is much weaker in the eastern channel than in the western channel (Figs. 5 and 6). Some isotherms make a rise to the southeast on the Kyushu side of the eastern channel. Also the geostrophic balance becomes less satisfactory in the weakly stratified water of the eastern channel. Wind forces which generate ageostrophic currents and the mixing of water may be more effective in the

194

TABLE 2

Characteristic values of the velocity fields for each traverse line.

	W-line	E-line	F-line
Correlation coefficient between ADCP and geostrophic velocity fields			
with tidal current	0.908	0.523	-
without tidal current	-	0.758	-
Transport (SV)			
ADCP data			
with tidal current	2.2	-0.3	0.5
without tidal current	-	0.6	0.7
Geostrophic calculation			
referred to the bottom	1.4	0.4*	-

* This is a value for the section between E2 and E7.

eastern channel with shallower depths. In the eastern channel, the observed velocity field becomes better correlated with the geostrophic velocity field when the tidal component is removed from the data. The periods of tidal oscillation are likely to be shorter than the period of geostrophic adjustment. A further observation is required to clarify the reason why the geostrophic balance is poorly satisfied in the eastern channel.

From a conventional geostrophic calculation with a level of no motion at 125 m, Yi (1966) obtained the annual mean volume transport of 0.9 SV for the western channel and 0.3 SV for the eastern channel. The mean transport of 1.3 SV across the western channel was reported by Lee (1974) through a direct observation from an anchored ship. A difference of 0.4 SV between the estimates of the transport for the western channel may be explained in view of the existence of a barotropic transport. Miita (1976) analyzed the historical current-meter data collected by JMSA and other Japanese Agencies during 1924-1974. The volume transports estimated by him were 1.8 SV and 1.6 SV for the western and eastern channels, respectively. It should here be noticed that most of the data mentioned above were obtained with current meters suspended from the anchored ships for a day, and that only a small number of current meters were moored over longer periods. The transport through the eastern channel is much larger in the Miita's result than in the ADCP result of the E- and F-lines.

The present study suggests that the transport through the western channel is two to three times larger than that through the eastern channel. This result is consistent with that of Yi (1966). A cold water mass commonly exists along the northern coast of Kyushu (Fig. 6a). The geostrophic balance has a tendency to induce a SW current in the part of the eastern channel. This SW current effectively reduces the transport of the Tsushima Current through the eastern channel. The velocity structures and transport of the Tsushima Current may be determined more accurately by the repeated use of a fish-mounted ADCP along the same transect.

ACKNOWLEDGMENTS. We wish to thank Profs. H. Mitsuyasu, K. Kawatate, W. Koterayama and S. Mizuno of the Research Institute for Applied Mechanics, Kyushu University for their encouragements and stimulating suggestions during the course of this study. We greatly appreciate Mr. S. Sato of the Japan Maritime Safety Agency for the prediction of tidal currents in the Korea Strait. We also thank Dr. T. Hosoyamada and Mr. M.Ishibashi for their assistance in the Processing of ADCP data, and Mrs. Y. Arizumi for preparing the manuscript. The captains and the crews of R/V Pusan 402 (National Fisheries University of Pusan) and R/V Genkai (Fukuoka Prefectural Fisheries Experimental Station) are very much acknowledged for their active support in deployment and recovery of the ADCP fish. This study is supported by a grant from the Japan Ministry of Education, Culture and Science for International Scientific Research Programs and a grant from the Korea Ministry of Science and Technology for Korea-Japan Cooperative Research Programs which are greatly acknowledged. The data analysis was made by the FACOM M760/8 of RIAM.

5 REFERENCES

Byun, S-k. and Y-h. Seung, 1984. Description of current structure and coastal upwelling in the southwest Japan Sea- summer 1981 and spring 1982. In: T. Ichiye (Editor), Ocean Hydrodynamics of the Japan and East China Sea, Elsevier, Amsterdam: pp. 83-93.

Kaneko, A. and W. Koterayama, 1988. ADCP measurements from a towed fish. EOS Trans. AGU, 69: 643-644.

Kaneko, A., W. Koterayama, H. Honji, S. Mizuno, K. Kawatate and R. L. Gordon, 1990. A cross-stream survey of the upper 400-m of the Kuroshio by an ADCP on a towed fish. Deep-Sea Res., 37: 875-889.

Lee, C-k., 1970. On the currents in the western channel of the Korea Strait. Bull. Fish. Res. Dev. Agency of Korea, 6: 175-231 (in Korean).

Lee, C-k. 1974. A study on the currents in the western channel of the Korea Strait. Bull. Fish. Res. Dev. Agency of Korea, 12: 37-105 (in Korean).

Miita, T., 1976. Tsushima Current viewed in the short-term moored current-meter data. Proc. Jpn. Soc. Fish. Oceanogr., 28: 33-58 (in Japanese).

Mizuno, S., K. Kawatate and T. Miita, 1986. Current and temperature observations in the east Tsushima channel and the Sea of Genkai. Prog. Oceanogr., 17: 277-295.

Moriyasu, S., 1972. The Tsushima Current. In: H. Stommel and K. Yoshida (Editors), Kuroshio-Its Physical Aspects, Univ. Tokyo Press, Tokyo: pp. 353-369.

Odamaki, M., 1989. Tides and tidal currents in the Tsushima Strait. J. Oceanogr. Soc. Jpn, 45: 65-82.

Yi, S-u., 1966. Seasonal and secular variations of the water volume transport across the Korea Strait. J. Oceanol. Soc. Korea, 1 : 7-13.

MEASURING TRANSPORTS THROUGH STRAITS

D. PRANDLE

Proudman Oceanographic Laboratory, Bidston Observatory, Birkenhead, Merseyside L43 7RA, (England)

ABSTRACT

A range of techniques for measuring transports through Straits is reviewed, these include: conventional current meters, Acoustic Doppler Current Profilers (ADCP), H.F. Radar and sub-marine telephone cables. Indirect methods using tracer distributions are also described. While all such methods can be useful, no entirely accurate system is available. Combinations of methods are advocated with intensive (and costly) shorter period surveys supplementing long term (less costly and often shore-based) monitoring. While, ultimately, numerical models promise the best method, their performance must be examined by measurements of the kind reviewed here.

INTRODUCTION

When comparisons are made with meteorology, the lack of a systematic monitoring network in oceanography is often criticised. This difference can partly be explained by the high costs and technical difficulties of maintaining ocean monitoring stations. Shore-based measurements are much simpler and cheaper and a global network of tide gauge recordings has been established. Unfortunately, it is difficult to derive information on (sub-tidal frequency) circulation from coastal tide gauges. A compromise is often to try to determine transports through Straits since these can be critical indicators of large scale circulation and the technology can be partially shore-based. (Measurements of circulation in the Southern Oceans often focus on the Drake Passage for similar reasons.) This strategy has led to a study of such flows in Straits around the UK for over 40 years. This paper reviews these studies and describes future plans.

Section 2 reviews the many studies that made use of submarine telephone cables to measure the induced potential difference across Straits. Section 3 reviews the use of current meters and dispersed tracers while Section 4 describes new technologies.

2   TELEPHONE CABLE MEASUREMENTS

As early as 1832 Faraday suggested that tidal currents (in conductive sea water) flowing across the vertical component of the earth's magnetic field would induce a potential difference - the dynamo principle. Subsequent measurements by Wollaston (1881) across the river Thames in London confirmed this phenomenon. Practical interest was re-awakened in the 1940s when it was realised that the same principle could be used to detect submarines entering harbours. Quantitative scientific applications were stimulated by Longuet-Higgins (1949) who deduced theoretically the magnitude of the induced voltage across a channel.

Bowden (1956) used this theory to relate voltages recorded by the cross channel telephone cable across the Dover Strait to flow rates. The Strait is 30 km wide and with tidal currents up to $2ms^{-1}$, this potential difference varies by up to ± 1 volt between flood and ebb tide. His analysis of 15 months recordings accurately determined the tidal constituents and indicated close correlation of the non-tidal residual signal with wind forcing. Cartwright and Crease (1963) used similar recordings to estbalish the long term mean sea level gradient between the UK and France.

More detailed analysis of the long term residual flows deduced from cable measurements in the Irish Sea (Hughes 1969) showed that erratic 'contact potentials' limited the use of such measurements for long term transport calculations. Robinson (1978) showed theoretically how the measured potential difference represents a complicated weighted function of flow over a wide area surrounding the measurement locations. This theory was used by Prandle (1979) to explain the anomalous tidal signals measured across the Pentland Firth. Prandle (1978 & 1984) compared monthly mean values of flow through the Dover Strait from 9 years of cable records (Alcock & Cartwright 1978) with model predictions (figure 1). Correlations were also made with sea-level gradients from coastal tide gauge data. Monthly mean values extending for a period of 22 years were predicted with reasonable confidence.

Thus despite the problems of: magnetic field variability (Axe 1968) (figure 2), atmospheric tides (Malin 1973), internal waves (Sandford 1971), seasonally varying conductivity of both sea water and bed sediments (Hughes 1969); interesting long term measurement of integrated flows (tidal and non-tidal) have been made at minimal cost. Possibilities of using coastal discharge pipelines, offshore oil and gas pipes or even of laying specific cables have scarcely been explored. Moreover, with increased interest in long term changes in climate, intercomparison of cable recordings from strategic channels would be of interest to monitor oceanic impact. The complications of baroclinic flows and varying conductivity of sea water might be compensated using data from moorings. The capability of measuring 'catastrophic' storm events is particularly useful. Likewise for large scale phenomena such as Tsunamis, the timing of a recognisable signal is of interest regardless of the amplitude interpretation.

3   CURRENT METERS AND TRACERS

Van Veen (1938) carried out nearly 700 direct measurements of current profiles in the Dover Strait using a mechanical current meter. His work indicates how residual currents (and depth integrated transports) can vary significantly both temporally and spatially even within a confined channel. Further studies using moored and ship-based current meters by Cartwright and Crease (1963) and Prandle and Harrison (1975) highlight this problem. A recent analysis by Prandle et al. (1990) of current meter measurements in a straight kilometre-wide tidal channel emphasises that, even with recent developments in current meter technology, similar difficulties in achieving sufficient spatial

199

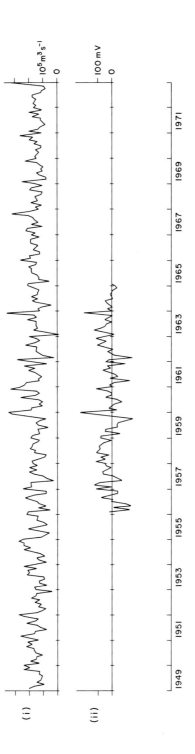

1. Transport through the Dover Strait, (top) model calculations, (bottom) telephone cable recordings.

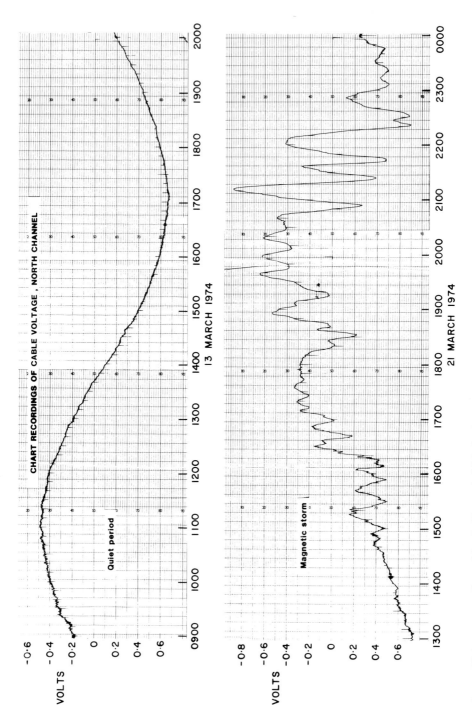

2.   Voltage recordings (North Channel, Irish Sea).

and temporal resolution persist. Instruments capable of spatial integration are necessary.

Indirect measurement from tracer distributions can be used to give first order estimates of net transports. Assuming the time and cross-sectionally averaged transport equation

$$\frac{u\partial c}{\partial x} + D\frac{\partial^2 c}{\partial x^2} = 0$$

both U and D (the longitudinal dispersion coefficient) can be determined if the temporal and spatial gradients of two (or more) tracer concentrations, c, can be measured (Wilson 1974). Alternatively, by separating the above equation into respective tidal constituents estimates of residual flows can be made (Prandle et al., 1990). Such calculations can determine the approximate magnitude of any large net drift but are unlikely to discern smaller drifts or the time variability in larger drifts.

By contrast the numerical simulation of the spread of $^{137}C_s$ discharged from Windscale (Prandle 1984) provided useful confirmation of the accuracy of the small residual currents involved (as little as 1 cm/sec in oscillating tidal flows of 1 m/sec i.e. beyond the accuracy of direct measurement). However the success of this simulation reflects the careful selection of the problem namely: (i) the tracer had been discharged at a known rate from (effectively) a single source over a time interval commensurate with the time over which the model was valid (i.e. much longer than the tidal periods - allowing tidal advection to be parameterised yet not so long that the open-boundary conditions significantly influence the interior region), (ii) excellent monitoring programmes at regular intervals with accurate concentration determinations, (iii) good forcing data were available (tidal and meteorological), (iv) extensive complementary data sets existed (long term tide gauge records), (v) the (large) spatial and (long) temporal scale of the problem and limited depths ensured bathymetric steering of residuals - to some extent reducing 3 dimensional dispersion to 1 dimension in which over-estimation of advection in one month can be compensated by underestimate the following (in the same way that a sequence of injections is maintained in a pipeline but smeared in an open sea).

4   NEW TECHNOLOGIES

The Proudman Oceanographic Laboratory is presently planning a new attempt to determine the net flow through the Dover Strait using two new instruments, bottom-mounted Acoustic Doppler Current Profilers for measuring vertical current profiles and shore-based H.F. Radar (Prandle 1989) for measuring surface currents. The year-long measurements will start in 1990, in addition a ship-mounted ADCP will be used for a two week intensive survey to provide comprehensive spatial interpolation (figure 3).

The use of HF radar for mapping near-shore surface currents has developed rapidly in the last 5 years. The UK's OSCR (Ocean Surface Current Radar) system has progressed from initial trials in '83, to comprehensive evaluation exercises in '84 including comparisons against current meters, moored and drifting buoys and detailed numerical models (Prandle & Ryder 1989).

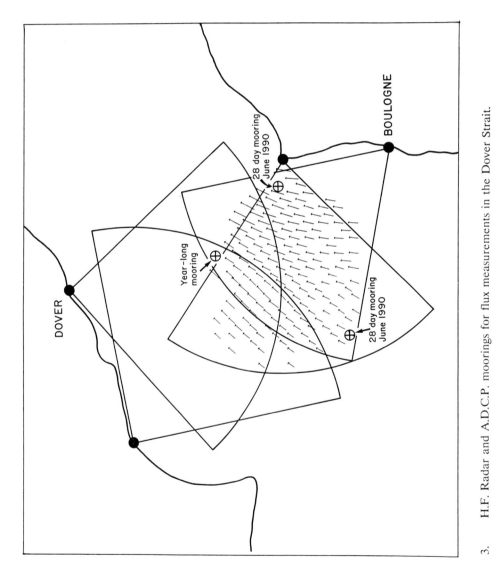

3. H.F. Radar and A.D.C.P. moorings for flux measurements in the Dover Strait.

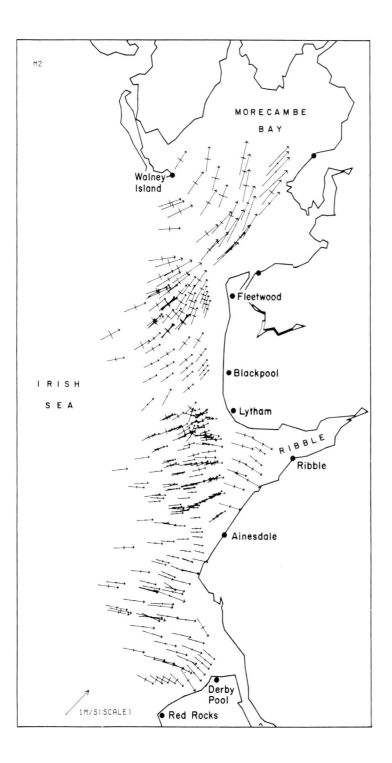

4. M$_2$ tidal current ellipses measured by H.F. Radar.

204

Figure 4 shows surface tidal current ellipses constructed from measurements made by HF radar from 10 stations between Liverpool Bay and Morecambe Bay. These measurements have been analysed to yield various tidal, net-residual and wind-driven components. For the predominant M_2 tidal constituent, the overall agreement between model and radar results for magnitude, phase and direction is excellent. Detailed inspection indicates the surface OSCR values are generally about 10% larger than the depth-averaged model values - this increase being consistent with both observations and theory of the vertical structure of tidal currents. The radar has also been used to reveal the nature of wind driven surface currents, indicating magnitude and veering increases in deeper water as predicted from Ekman's theory. The radar data also show that almost all coastal areas around the UK exhibit large time-averaged surface residual currents driven by (small) horizontal density gradients.

The development of 'purposeful tracers' (Watson 1987) offers new possibilities of deducing transports from the dispersion equation. (Moreover by using multiple tracers atmospheric ventilation rates can be determined). Research at the University of Southampton has shown that auto-correlation techniques used on successive satellite/aircraft sea-surface images (AVHRR and CZCS) can be used both to deduce residual currents and dispersion rates (I. Robinson, personal communication).

Recent developments in both satellite and Decca (navigational) self-tracking buoys have been used to determine long-term current drifts (Roberts et al., 1990). The Proudman Oceanographic Laboratory is presently investigating the use of acoustic tomography for measurements of integrated flows between shore-based stations.

Oceanography is becoming increasingly inter-disciplinary, while this paper has concentrated on water transports through Straits, the ultimate concern is often with fluxes of material. Where such material is adsorbed onto fine sediment, the concurrent cross sectional concentration distribution must be determined. Prandle et al. (1990) indicate the relative sensitivity of net fluxes to lateral and vertical variations in both concentration and currents.

5 REFERENCES

Alcock, G.A. and Cartwright, D.E., 1978. An analysis of 10 years' voltage records from the Dover-Sangatte cable. Deep-Sea Research (Supplementary volume: George Deacon 70th Anniversary Volume).
Axe, G.A., 1968. The effects of the earth's magnetism on submarine cables. P.O. elect. Engrs' J. 61, 37.
Bowden, K.T., 1956. The flow of water through the Straits of Dover related to wind and differences in sea level. Philos. Trans. r. Soc. (A) 248, 517.
Cartwright, D.E. and Crease, J., 1973. A comparison of the geodetic reference levels of England and France by means of the sea surface. Proc. Roy. Soc.   (A), 173, 558.
Faraday, M., 1832. Phil. Trans. pt. 1, p 175.
Faraday, M., 1966. Experimental researches in electricity, 2. Dover (reprint of 1932 edition).
Hughes, P., 1969. Submarine cable measurements of tidal currents in the Irish Sea. Limnol. Oceanogr. 15, 269.
Longuet-Higgins, M.S., 1949. The electrical and magnetic effects of tidal streams. Month. Not. R. astron. Soc. geophys. Suppl. 5, 285.
Malin, S.R.C., 1973. Worldwide distribution of geomagnetic tides. Philosophical Transactions of the Royal Society of London, (A), 174, 551.

Prandle, D., 1978. Monthly-mean residual flows through the Dover Strait, 1979-1982. Journal of the Marine Biological Association of the UK, 58, 965-973.

Prandle, D., 1979. Anomalous results for tidal flow through the Pentland Firth. Nature, 278, 541-542.

Prandle, D. and Harrison, A.J., 1975. Relating the potential differenc measured on a submarine cable to the flow of water through the Strait of Dover. Deutschen Hydrographischen Zeitschrift, 28(5), 207-226.

Prandle, D., 1984. Monthly-mean residual flows through the Dover Straits 1949-1980. Journal of the Marine Biological Association of the UK, 64, 722-724.

Prandle, D., 1984. A modelling study of the mixing of ^{137}Cs in the seas of the European continental shelf. Philosophical Transactions of the Royal Society of London, A, 310, 407-436.

Prandle, D., 1989. A review of the use of HF radar (OSCR) for measuring near-shore surface currents, pp. 245-258 in, Advances in water modelling and measurement, (ed. M. H. Palmer), Cranfield, Bedford: BMRA (Information Services), 402 pp.

Prandle, D., Murray, A. and Johnson, R., 1990. Analyses o flux measurement in the River Mersey. To be published in: Physics of Estuaries and Bays, Ed. R. Chery, Springer-Verlag.

Prandle, D. and Ryder, D.K., 1989. Comparison of observed (HF Radar) and modelled nearshore velocities. Continental Shelf Research, 9(11), 941-963.

Roberts, E., Last, D., Roberts, G., 1990. Precision current measurements using drifting buoys equipped with Decca Navigator and Argos. To be presented: 4th IEEE Conference: Current Measurement Technology: present and future trends. Washington DC.

Robinson, I.S.R., 1978. A theoretical model for predicting the response of the Dover-Sangatte cable to typical tidal flows. Deep-Sea Research (Supplementary volume: George Deacon 70th Anniversary Volume).

Sanford, T.B., 1971. Motionally induced electric and magnetic fields in the sea. J. geophys. Res. 76, 3476.

Veen, J. van, 1938. Water movements in the Straits of Dover. J. Cons. perm. int. Explor. Mer, 13, 7.

Watson, A.J., Liddicoat, M.I. and Ledwell, J.R., 1987. Perfluorodecalin and Sulphur hexafluoride as purposeful marine tracers: some deployment and analysis techniques. Deep Sea Research 34, 19-31.

Wilson, T.R.S., 1974. Caesium-137 as a water movement tracer in the St George's Channel. Nature, London, 248, 125-126.

Wollaston, C., 1881. J. Soc. Tel. Eng. 10, 51.

A CROSS-SPECTRAL ANALYSIS OF SMALL VOLTAGE VARIATION IN A SUBMARINE CABLE BETWEEN HAMADA AND PUSAN WITH SPEED VARIATION OF THE TSUSHIMA WARM CURRENT

Kazuo KAWATATE[*1], Akimasa TASHIRO[*1], Michiyoshi ISHIBASHI[*1], Takashige SHINOZAKI[*1], Tomoki NAGAHAMA[*1], Arata KANEKO[*1], Shinjiro MIZUNO[*1], Jyun-ichi KOJIMA[*2], Toshimi AOKI[*3], Tatsuji ISHIMOTO[*3], Byung Ho CHOI[*4], Kuh KIM[*5], Tsunehiro MIITA[*6], and Yasunori OUCHI[*7]

[*1] Research Institute for Applied Mechanics, Kyushu University, Japan
[*2] Meguro Research and Development Laboratories, Kokusai Denshin Denwa Company Limited, Japan
[*3] Hamada Cable Landing and Over Horizon Radio Relay Station, Kokusai Denshin Denwa Company Limited, Japan
[*4] Department of Civil Engineering, Sung Kyung Kwan University, Korea
[*5] Department of Oceanography, Seoul National University, Korea
[*6] Fukuoka Prefectural Buzen Fisheries Experimental Station, Japan
[*7] Fukuoka Prefectural Fukuoka Fisheries Experimental Station, Japan

ABSTRACT

We obtained a time series of electric voltage records by a submarine cable buried between Hamada and Pusan together with a set of speed records of the Tsushima warm current measured southeast of the Tsushima island. A cross-spectral analysis was made between small voltage variation and speed variation. A strong correlation exists between the two variations.

1 INTRODUCTION

Faraday's law tells us that the change of a magnetic flux yields an electric field. When a conductor moves in a magnetic field, a potential difference arises between the both ends of the conductor. Water is conductive. When it flows in the geomagnetic field, a potential difference is produced across the flow. In order to measure the potential difference Faraday conducted experiments: at the pond of the Kensington palace on 10 January 1832, and at the Waterloo Bridge in the Thames River on 12 January 1832 (Nakayama, 1984). He could not get expected results; however, his intention was succeeded. Wollaston showed in 1851 correctness of Faraday's idea by recording the potential difference induced by tides on a submarine cable set in the Strait of Dover (Alcock and Cartwright, 1977).

Several attempts have already been conducted to estimate a transport of flow through straits by measuring a small variation of the potential difference on the submarine cable (Prandle and Harrison, 1975; Larsen and Sanford, 1985; Baines and Bell, 1987).

Measurements of the Tsushima warm current have been made in summer since 1983 by use of moored meters through cooperation of Fukuoka Prefectural Fisheries Experimental

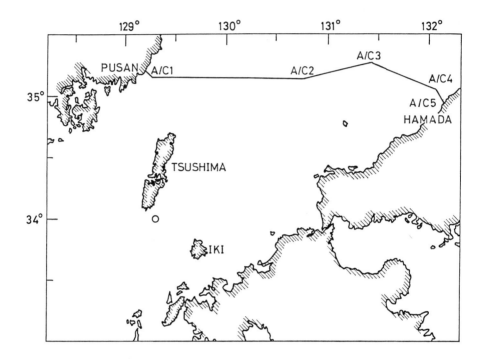

Fig.1a. Arrangement of the submarine cable and site of mooring line (open circle) A/C1, A/C2, A/C3, A/C4, A/C5 ··· turning points.

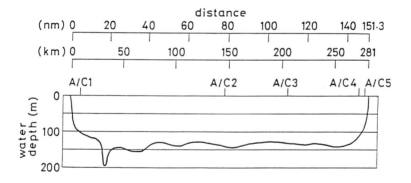

Fig.1b. Depth along the submarine cable.

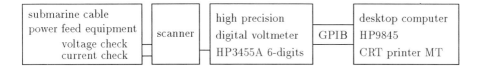

Fig.2. A scheme of voltage and current measurement.

Station and Research Institute for Applied Mechanics. In 1986 one of the authors Choi urged another one Kawatate to measure the electric potential difference for estimation of the flow, by introducing Faraday's idea and explaining the world tendency on the subject. Taking the opportunity the Kokusai Denshin Denwa Company has joined us. Thus a cooperative study has begun.

Japanese Ministry of Education has granted a fund to the study from 1988 to 1990 in the framework of Japan-Korea science and technology cooperation agreement.

2 OUTLINE OF MEASUREMENT AND ANALYSIS

A submarine cable is buried between Hamada and Pusan. The cable arrangement and the depth variation are shown in Figs.1a, b. A mooring line for measuring the ocean current was set 34°00'20"N 129°19'40"E at a depth of 110m, as shown by an open circle in Fig.1a.

Supply voltage and current on submarine cable power feed equipment were recorded every 15 minutes by use of a desktop computer HP9845, a digital voltmeter HP3455A of six digits, and a cassette type magnetic tape, as shown in Fig.2.

Three current meters were attached to the line. Its configuration is shown in Fig.3. R/V Genkai of 138 gross tonnage of Fukuoka Prefectural Fisheries Experimental Station deployed and retrieved the line. The ocean current speed, direction, water temperature, and depth were recorded on a cassette type magnetic tape every 10 minutes.

Among several methods for cross-spectral analysis, we adopted the following one. Let us assume that the data are obtained with time interval ΔT and the number of them is N, that is, the data are discrete and finite. We extend data cyclically, if necessary (Brigham, 1974). Denoting the speed variation along the mean flow axis by x and the voltage variation by y, we assume that

$$x(i\Delta T) = x((i \pm N)\Delta T),$$
$$y(i\Delta T) = y((i \pm N)\Delta T).$$

The cross-correlation function is defined by

$$R_{xy}(l\Delta T) = \frac{1}{N} \sum_{s=0}^{N-1} x((l+s)\Delta T)y(s\Delta T), \quad l = 0, 1, \ldots, N-1.$$

If y is substituted by x, the auto-correlation function of x is obtained. We also define the Fourier transform of x by

$$X(nf_0) = \frac{1}{N} \sum_{s=0}^{N-1} x(k\Delta T) \exp\left(-j'\frac{2\pi nk}{N}\right), \quad n = 0, 1, \ldots, N-1.$$

where $j' = \sqrt{-1}$ the imaginary unit, $f_0 = 1/(N\Delta T)$ the fundamental frequency. We also have a similar expression for y. The spectral density function, which is the Fourier transform of the correlation function, is written

$$S_{xy}(nf_0) = \frac{1}{f_0} X(nf_0)Y^*(nf_0), \quad n = 0, 1, \ldots, N-1.$$

where * signifies the complex conjugate. We use the physically realizable one-sided power spectral density functions (Bendat and Piersol, 1968), when we illustrate the results,

$$G_{xy}(nf_0) = \begin{cases} 2S_{xy}(nf_0), & n = 1, 2, \cdots, N/2 - 1, \\ \\ S_{xy}(nf_0), & n = 0, N/2, \end{cases}$$
$$= C(nf_0) - j'Q(nf_0),$$

where C and Q are, respectively, the co-spectrum and the quad-spectrum.

Let us further assume that the voltage variation y is an output and the speed variation x is an input; their relation is described by an impulse response function $h(t)$,

$$y(k\Delta T) = \Delta T \sum_{i=0}^{N-1} h(i\Delta T)x((k-i)\Delta T), \quad k = 0, 1, \ldots, N-1.$$

A frequency response function is defined by

$$H(nf_0) = \Delta T \sum_{l=0}^{N-1} h(l\Delta T) \exp\left(-j'\frac{2\pi nl}{N}\right), \quad n = 0, 1, \ldots, N-1$$
$$= |H(nf_0)| \exp(-j'\theta_H(nf_0)),$$

which is connected with the spectral density functions through a relation,

$$G_{xy}(nf_0) = H^*(nf_0)G_{xx}(nf_0).$$

The coherence function is given by

$$\gamma^2(nf_0) = \frac{|G_{xy}(nf_0)|^2}{G_{xx}(nf_0)G_{yy}(nf_0)}.$$

Computations are done by fast Fourier transform method (FFT) and auto-regression model method (AR). By FFT method, γ^2 is always 1 unless we filter results. In the present paper we shall show the results by AR method.

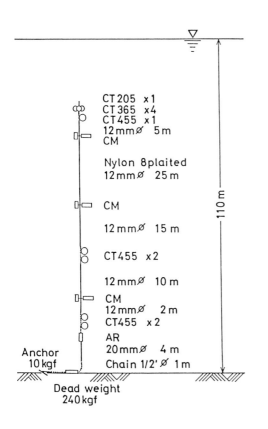

Fig.3. The mooring line.
CT205, CT365, CT455, floats with diameter 0.20, 0.36, 0.45m, respectively, and working depth
500m; CM, currentmeter; AR, acoustic release; SI conversion, 1kgf=9.8N, 1N=0.102kgf.

3 RESULTS

The voltage and current were obtained from 9 July 1987 to 16 February 1988. The
speed and direction were measured from 26 June 1987 to 21 September 1987. We shall show
the voltage and speed for 45 days (1080 hours) from 7 August 1987 to 20 September 1987:
simultaneous, continuous data sets of the voltage and speed by rejecting wrong data. The
voltage data were prepared every 1 hour by applying the Godin filter of $4 \times 4 \times 5$; and the
speed by $6 \times 6 \times 7$.

The voltage variation is shown in Fig.4. Data from the uppermost current meter showed
that the mean flow axis was at an angle of 70 degrees to the north, as shown in Fig.5a. Figures
5a, b show the speeds along, and perpendicular to, the mean flow axis, respectively.

A cross-spectral analysis was made for the variation of speed along mean flow axis and the
variation of voltage. For comparison we used the fast Fourier transform (FFT) method and

auto-regression model (AR) method. The FFT method was applied to data for 1024 hours from 09:00 on 7 August 1987 to 24:00 on 18 September 1987. The sampling interval ΔT was 1 hour (h), the number of data N 1024, the frequency resolution (fundamental frequency) $f_0 = 1/(N\Delta T) = 0.977 \times 10^{-3}$ cycles per hour (h^{-1}), and the Nyquist frequency $f_{N/2} = 1/(2\Delta T) = 0.5h^{-1}$. We sought trends $at + b$ for both data by use of least square method, subtracted them from the original data, and obtained series of variation. For the speed data we had $a = -0.9069 \times 10^{-4} ms^{-1}h^{-1}$ and $b = 0.3141 ms^{-1}$; for the voltage $a = -0.3281 \times 10^{-3} Vh^{-1}$ and $b = 551.1V$. On applying AR method to the both variations, Akaike's multiple final prediction error (Akaike and Nakagawa, 1975; Kawatate, 1978) took the minimum value provided that a datum is substituted by a linear combination of the past 25 hours data or the future 25 hours data. The above prediction error is given by $MFPE(M) = \{(N+2M)/(N-2M)\}^2 \times \det P(M)$, where $P(M)$ is a 2 by 2 matrix composed from auto- or cross-correlations of each difference between the data and auto-regressed values of the speed and voltage. In the present case $M = 25$ and $MFPE(25) = 1.031 \times 10^{-5} V^2 m^2 s^{-2}$.

The power spectrum of the speed along the mean flow axis G_{xx} is shown in Fig.6a by use of FFT method without filtering and in Fig.6b by use of AR method. Arrows below the abscissa designate the frequency of $0.0417h^{-1}$ and $0.0833h^{-1}$. Figures 6c, d show the power spectrum of the voltage variation G_{yy}, respectively, obtained by FFT method without filtering and AR method. Noise components are apparent in the results obtained by FFT method, because we did not filter the results along the frequency axis. Both FFT and AR methods give similar results on the whole. The power spectrum becomes almost zero as the frequency becomes high.

Now we shall show the results by AR method in the frequency range from zero to $0.14h^{-1}$. In other words, we put $f = nf_0$ and take n from 0 to 143.

The power spectrum of speed has the local maximum $23.7 m^2 s^{-2}h$ at a frequency of $0.081h^{-1}$ (period: 12h20min) and $19.5 m^2 s^{-2}h$ at $0.041h^{-1}$(24h23min), as given in Fig.7, which show semi-diurnal and diurnal motions. We took the length of data $N\Delta T = 1024h$ and $M = 25$ for expressing a datum by a linear combination of the past M and the future M data. Assuming the equivalent width of data $M\Delta T = 25h$, we derive the equivalent number of degree of freedom $k = 2N/M = 81.92 \doteq 80$. By using chi-square distribution we have the 90% confidence interval $18.6–31.3 m^2 s^{-2}h$ and the 80% confidence interval $19.6–29.5 m^2 s^{-2}h$ at the frequency $0.081h^{-1}$(12h20min). The other local maximum appears at $0.977 \times 10^{-3}h^{-1}$ (period: 1024h), which coincides with the length of data. It has no physical significance, being certainly brought either through the process of calculation in which we extended data cyclically when necessary, or by an effect of data truncation other than a multiple of the period.

The power spectrum of voltage has also the local maximum $6.95 V^2h$ at $0.081h^{-1}$(12h20min) and $1.16 V^2h$ at $0.040h^{-1}$(24h59min), as shown in Fig.8. The spectrum takes a value $1.1 V^2h$ at $0.041h^{-1}$(24h23min). It has another local maximum $0.434 V^2h$ at $0.125h^{-1}$(8h00min), which may be an indication of either one-third diurnal tide, or a complex tide caused by a non-linear interaction between waves. Their frequency is given by a sum (or a difference) of frequencies

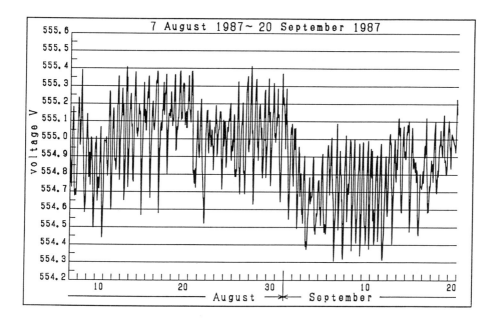

Fig.4. Time series of electric voltage.

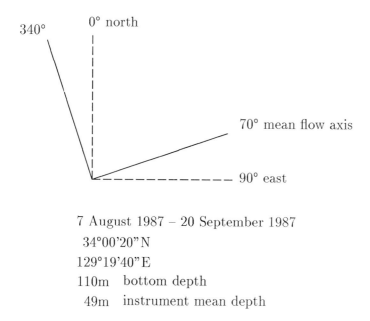

7 August 1987 – 20 September 1987
34°00'20"N
129°19'40"E
110m bottom depth
49m instrument mean depth

Fig.5a. Mean flow axis.

214

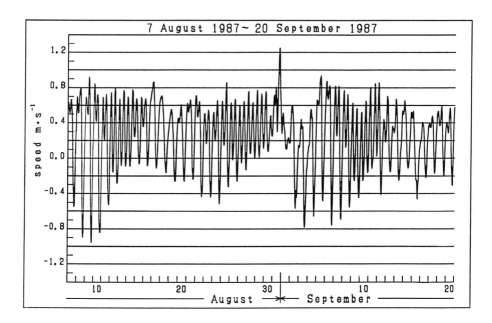

Fig.5b. Time series of speed along the mean flow axis.

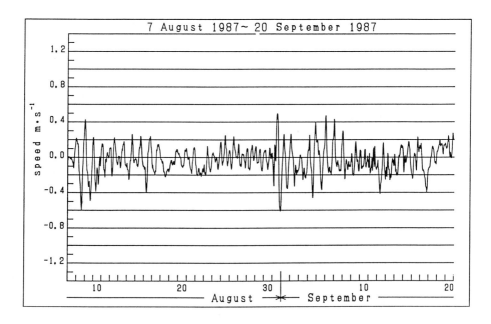

Fig.5c. Time series of speed perpendicular to the mean flow axis.

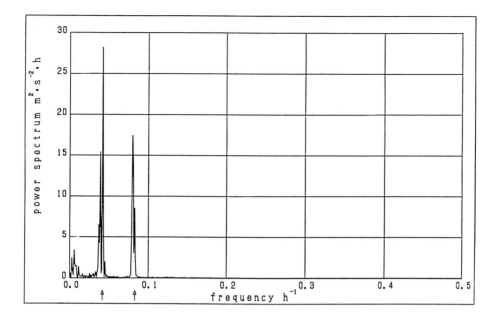

Fig.6a. Power spectrum of the speed along the mean flow axis by FFT without filtering.

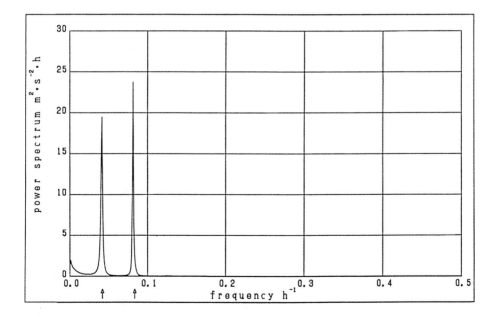

Fig.6b. Power spectrum of the speed along the mean flow axis by AR.

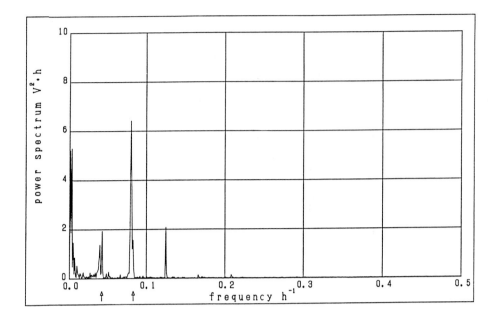

Fig.6c. Power spectrum of the voltage by FFT without filtering.

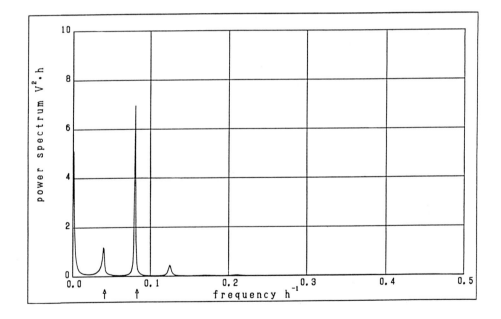

Fig.6d. Power spectrum of the voltage by AR.

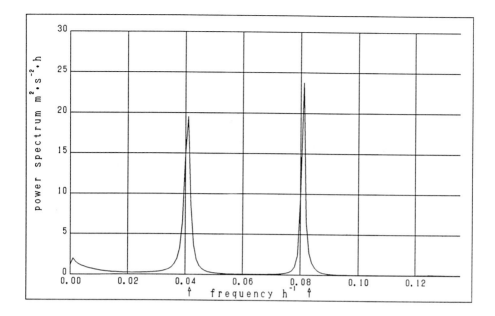

Fig.7. Power spectrum of the speed along the mean flow axis.

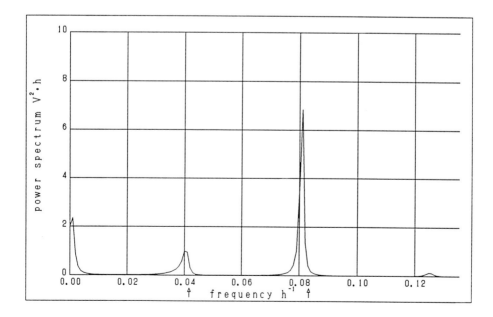

Fig.8. Power spectrum of the voltage.

218

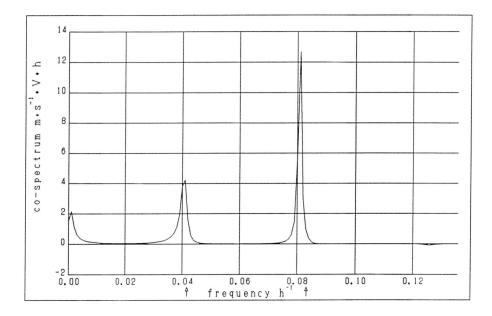

Fig.9a. Co-spectrum.

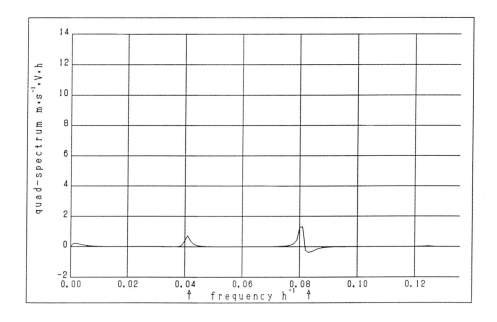

Fig.9b. Quad-spectrum.

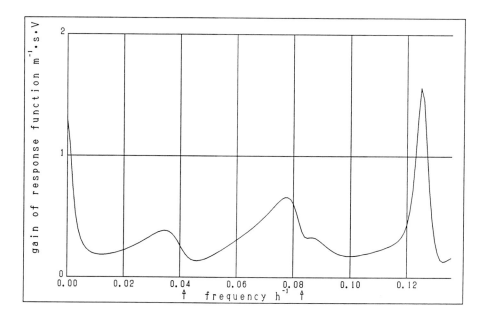

Fig.10a. Gain of response function.

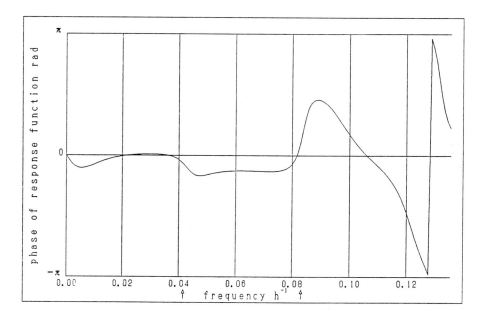

Fig.10b. Phase of response function.

220

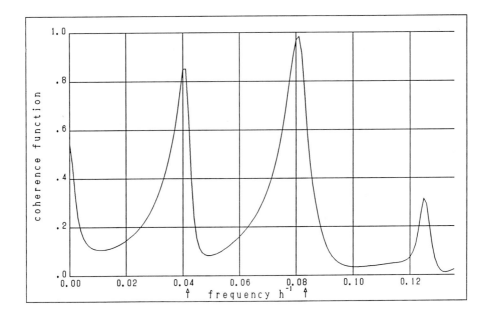

Fig.11. Coherence function.

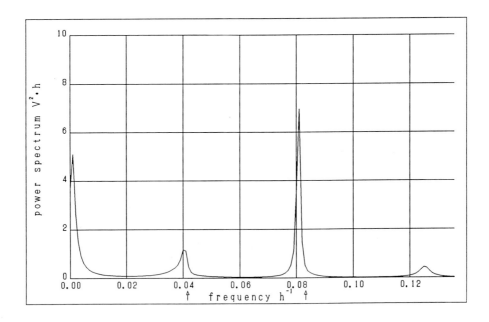

Fig.12. Power spectrum of the voltage.

of two waves: $1/12 + 1/24 = 1/8h^{-1}$ (period: 8h). However, no distinct indication of this component is in the current speed.

Figure 9 gives the co-spectrum and the quad-spectrum. The co-spectrum and the quad-spectrum were divided by the power spectrum of speed to get the response function, whose gain and phase are shown in Figs.10a, b. At frequencies where the value of the power spectrum of speed is very small the value of the response function loses confidence. The gain is high at long periods and at one-third diurnal motion, which however, has no confidence. The gain has a local maximum of $0.3840V/ms^{-1}$ at $0.0342h^{-1}$(29h15min), and another one of $0.6595V/ms^{-1}$ at $0.0703h^{-1}$(14h13min). These values are of the same order as the tidal conversion factors for various constituents derived by Prandle and Harrison (1975) in the Strait of Dover. At frequencies where the quad-spectrum is negative, the phase of response function is positive. The phase of speed (input) leads that of voltage (output). There are regions of negative phase in Fig.10b: the phase of voltage leads that of speed. It can not be true in the conventional linear model. To interpret this, more points for measuring speed are needed.

The coherence function is shown in Fig.11. It takes values 0.9826 at $0.081h^{-1}$(2h20min); 0.9698 at $0.080h^{-1}$(12h29min); 0.8519 at $0.041h^{-1}$(24h23min); 0.8512 at $0.040h^{-1}$(24h59min); and 0.3133 at $0.125h^{-1}$(8h00min). The coherence is high at the semi-diurnal and the diurnal motions. We calculated $\gamma^2 G_{yy} = |H|^2 G_{xx}$ and show it in Fig.12. After noise is eliminated, the power spectrum of voltage takes values $6.832V^2h$ at $0.081h^{-1}$(12h20min), $0.948V^2h$ at $0.040h^{-1}$(24h23min), $0.988V^2h$ at $0.040h^{-1}$(24h59min), and $0.136V^2h$ at $0.125h^{-1}$(8h00min). The value at the semi-diurnal motion is the maximum, and that of the diurnal motion the second maximum.

The semi-diurnal component at a frequency of 0.081cph (period: 12h20min) dominates over the spectrums of the speed and the voltage, which are highly correlated with each other. The voltage variation should be induced by the speed variation.

4   CONCLUDING REMARKS

A cross-spectral analysis of the speed and the voltage suggested that the voltage variation is induced by the speed variation. The result would be more conclusive if data of the geomagnetic flux variation were analyzed.

It is already reported that the wind and the water temperature influence the voltage variation. In their paper analyzing ten-year voltage records in the Strait of Dover, Alcock and Cartwright (1977) examined dependence of the fluctuating signal in the cable on the local weather and showed two situations: the southern wind produced the positive anomalies in the cable electromotive force, and the northern wind produced the negative ones. Murakami and Motomatsu (1979) reported two-year maintenance test results of the cable system between Reihoku, Japan and Shanghai, China. As the bottom temperature got high, the cable voltage loss became high. By installing a temperature automatic gain control device in the bottom relay they kept the variation of receiving voltage level small.

Various factors affect the voltage change. At least it would be necessary to investigate correlation of three kinds of data: voltage, speed, and magnetic flux.

Since only one mooring line was used for a term less than three months, the data used here are poor in space and time.

The present study is a first step toward the final goal which is to estimate the mean flow and the variability.

5 REFERENCES

Akaike, K., and Nakagawa, T., 1975. Statistical Analysis and Control of Dynamic System (in Japanese), Science Co., Tokyo, pp. 55-58.

Alcock, G.A., and Cartwright,D.E., 1977. An analysis of 10 year's voltage record from the Dover-Sangatte cable, A voyage of Discovery, George Deacon 70th Anniversary Volume, Supplement to Deep-Sea Research, M. Angel, Editor, pp. 341-365.

Baines, P.G., and Bell, R.C., 1987. The relationship between ocean current transports and electric potential differences across the Tasman Sea, measured using an ocean cable, Deep-Sea Research, Vol. 34, No. 4, pp. 531- 546.

Bendat, J.S., and Piersol, A.G., 1968. Measurement and Analysis of Random Data, John Wiley and Sons, Inc., Fourth Printing, pp. 82-83.

Brigham, E.O., 1977. The Fast Fourier Transform, Prentice Hall, Tokyo, pp. 99.

Kawatate, K., 1978. A comment on the auto-regression cross spectral analysis (Note in Japanese), Bulletin of Research Institute for Applied Mechanics, Kyushu University, No. 48, pp. 77-82.

Larsen, J.C., and Sanford, T.B., 1985. Florida Current Volume Transports from Voltage Measurements, Science, Vol. 227, pp. 302-304.

Murakami, Y., and Motomatsu, K., 1979. Results of two years activity of submarine cable system between Japan and China (in Japanese), Study of International Communication, pp. 395-405.

Nakayama, M., 1984. Electromagnetic Induction (in Japanese), Kyoritsu Publishing Co., Tokyo, p. 24.

Prandle, D., and Harrison, A.J., 1975. Relating the Potential Difference Measured on a Submarine Cable to the Flow of Water through the Strait of Dover, Sonderdruck aus der Deutschen Hydrographischen Zeitschrift, Band 28, Heft 5, pp. 208-226.

ACKNOWLEDGEMENTS

We express our sincere gratitude to Mr. Hideo Ishihara of the Kokusai Denshin Denwa Company for his kind considerations on performing the present investigation. We also owe a very large debt of gratitude to the management of the KDD Company, which made the present study possible. Our thanks go to Captain Mataichi Isobe and his crews of the Research Vessel Genkai for their efforts and skills in deploying and retrieving the mooring line. We take this opportunity to thank all friends who have helped us in various ways during the course of the present study.

The grant from the Japanese Ministry of Education is greatly acknowledged.

OUTFLOWS FROM STRAITS

Takashi Ichiye
Department of Oceanography, Texas A&M University
College Station, Texas 77843-3146, U.S.A.

ABSTRACT
Outflows from three straits, Tsushima, Soya and Tsugaru are treated from the law of potential vorticity conservation of a two-layer sea with the motionless lower layer. Water temperature profiles of these straits indicate a narrow geostrophic flow at a distance from the coast. Its width may be predicted from the transport through strait based on the law mentioned above.

1   INTRODUCTION

Among the four straits along the boundary of the Japan Sea, outflows from Tsushima, Tsugaru and Soya Straits are hydrographically well observed for many years by Japanese agencies. This paper deals with some common features of these outflows and proposes a simple model to explain them. Some algebraic derivations are shortened as much as possible to highlight a main point in dynamics. More elaborate models will be presented elsewhere.

2   EXAMPLES OF OUTFLOWS, IN TERMS OF TEMPERATURE DISTRIBUTIONS

Horizontal patterns of the flow around Japan can be represented by isotherms at 100m. The isotherms charts are routinely available from Japanese sources as "Ten Day Marine Report" of Japan Meteorological Agency, "Rapid Marine Report" of Hydrographic Department and "Fortnight Fishing Ground Report" of Fishery Agency.

Fig. 1 shows August 1984 isotherms at 100 m mainly from "Ten Day Marine Report." It shows the three outflows in general form a current parallel to the coast as seen from strong gradients, though its distance from the coast varies and it becomes unstable and departs from the coast after leaving the strait.

Three typical temperature cross sections A, B and C are selected for Tsushima, Soya and Tsugaru Strait as indicated in Fig. 1 and are shown in

224

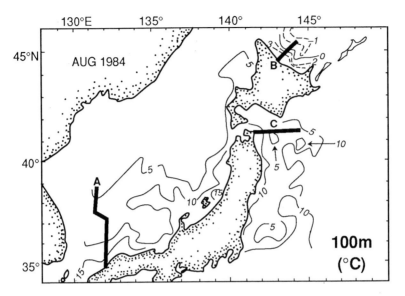

Fig. 1 Isotherms (°C) at 100 m depth in August, 1984. (Based on Ten Day Marine Report, Japan Meteorological Agency).

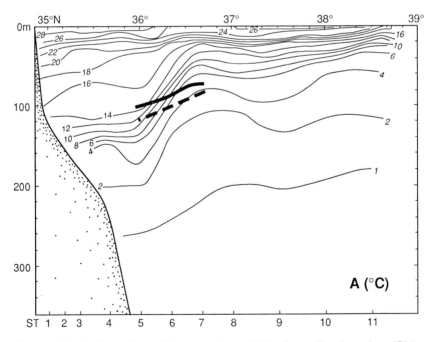

Fig. 2 Vertical section of temperature (°C) along Section A. (Shimane Pref. Fishery Station). (Location is shown in Fig. 1. Thick full and broken lines are explained in the text.)

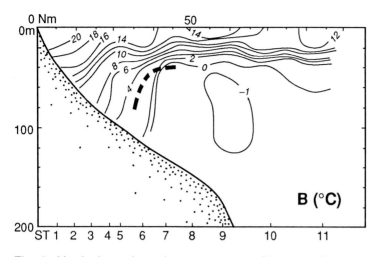

Fig. 3 Vertical section of temperature (°C) along Section B in August, 1984 (Wakkanai Fishery Station). (Location is shown in Fig. 1. The thick broken line is explained in the text.)

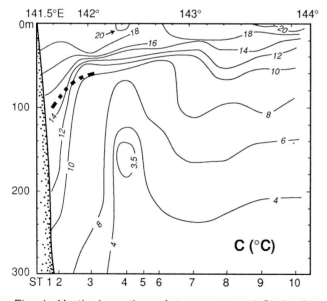

Fig. 4 Vertical section of temperature (°C) in August 1984 (Aomori Pref. Fishery Station). (Location is shown in Fig. 1. The thick broken line is explained in the text.)

Fig. 2, 3 and 4, respectively based on hydrographic data.(Fisheries Agency, 1988).

Section A indicates a strong eastward geostrophic flow between Stations 5 and 7 or between 36°N and 36.7°N, which is the first branch of Tsushima Current, but it also shows a weak westward countercurrent near the coast (Ichiye, 1984). Section B shows a south-westward flow called Soya Current (Oshima et al, 1990) between Stations 5 and 7. Section C indicates a southward flow termed Tsugaru Current between Stations 1 and 3, rather close to the coast. All the profiles indicate conspicuous thermocline that suggests applicability of a two-layer sea system. Temperature structures are almost uniform below 200 m and 100 m, respectively for A and B but for C horizontal gradient of temperature exists below 200 m mainly because of an isolated cold water separated from Oyashio Current (Kawai, 1972).

3 POTENTIAL VORTICITY CONSERVATION

Common features of the three sections are (1) presence of a strong thermocline and (2) a narrow geostrophic flow parallel to, but at some distance from, the coast after the outflow emerges from the strait. For such a flow, frictional effects of the coast may be negligible. Also the bottom friction is not significant because of the thermocline that shields its effects on the upper layer motion. Thus the potential vorticity may be conserved along the streamlines. Small dissipation guarantees the geostrophic relation across the flow as recognized upward slope of isotherms towards offshore in Sections A, B and C. A simple formulation of potential vorticity conservation combined with geostrophy is given by

$$(f + \Upsilon \, d^2 D \, / \, d \, x^2) \times D^{-1} = \text{constant along streamlines} \qquad (1)$$

(Gill, 1982, 232-233), where D and f are upper layer thickness and Coriolis parameter, respectively and

$$\Upsilon = g' / f \qquad (2)$$

with g' the reduced gravity. The x- axis is taken left to the coast with the sea being represented by negative x and with the nearshore boundary of the flow at x = o.

It is assumed that near the strait the relative vorticity is small compared to the local Coriolis parameter f_0 and depth is constant and equals D_0. Then equation (1) becomes

$$d^2 D / dx^2 = n^2 D - f^2 / g' \tag{3}$$

where

$$n^2 = f f_0 / g' D_0. \tag{4}$$

Equation (3) has a solution

$$D = A e^{nx} + B e^{-nx} - (f / f_0) D_0 \tag{5}$$

where A and B are constants to be determined by boundary and other conditions. We can specify the value of D at x = 0 as D_1 but the offshore boundary x = -a cannot be specified. Therefore we specify two conditions instead; (1) the transport M is specified, (2) at x = -a the x- derivative of D vanishes because of vanishment of the flow. These conditions lead to

$$2 M = \Upsilon (D_1^2 - D_a^2) \tag{6}$$

and

$$B = A e^{-2na} \tag{7}$$

where D_a is an unknown value of D at x = -a (a, unknown). The conditions at x = 0 and x = -a lead to

$$A (1 + e^{-2na}) = D_1 - C \tag{8}$$

$$2 A e^{-na} = D_a - C \tag{9}$$

where

$$C = D_0 f / f_0 \tag{10}$$

Elimination of A from (8) and (9) leads to

$$a = -n^{-1} \ell_n [R - (R^2 - 1)^{\frac{1}{2}}] \tag{11}$$

where

$$R = (D_1 - C) (D_a - C)^{-1} \tag{12}$$

equation (11) gives a when D_1 and M are specified.

4 APPLICATION OF THE MODEL AND ITS MODIFICATION

The above model is applied to the three sections A, B and C. From these profiles geographical location of the shoreside boundary of the current and D_1 are estimated. Also the Coriolis parameter f_o is taken equal to f, since the latitudinal differences of these sections from the respective strait are small. The reduced gravity is taken as 3×10^{-2} m/sec^2, from the observed density difference of the area.

TABLE 1 PARAMETERS

Section	x = 0 at	D_1 (m)	D_o (m)	M (10^6m/s)	f (10^{-5}s^{-1})	D_a (m)	D_a' (m)	a (km)	a' (km)
A	St 5	120	60	1.4	8.7	83	60-80	72	70-90
B	St 5-6	80	50	0.7	10.3	42	30-40	33	40-50
C	St 2	100	50	1.0	9.5	60	40-60	44	30-40

D_a' and a' indicate estimated values from temperature profiles in Figs. 2 to 4. The initial point x = 0 is estimated for each section from the same figures.

The assumed values of D_1, D_o, M and f and computed and estimated (observed) values of D_a, and a are listed in Table 1. The estimated values are not exact because the definition of the upper layer thickness is vague. However, the calculated values of a and D_a seem to be in agreement with their estimated values. The calculated profiles from solution(5) with the prescribed values of Table 1 are shown with a broken thick line in Fig. 2, 3 and 4.

When D_o varies with x, the solution is not expressed by (5). However, gradient of D_o with x is small within the strait for x = 0 to -a as is estimated, then the exponential terms of equation (5) can be replaced with exp ($\int$ ndx) and exp (- $\int$ ndx) for the first approximation under the condition (WKB method, Carrier et al, 1968).

$$(n)^2 \gg |n'| \tag{13}$$

when the prime indicates differential with x. For comparison with constant D_0 case, a functional form of D_0 is assumed as

$$D_0 = D_S (1 - x / x_0)^{\frac{1}{2}} \tag{14}$$

To determine D_S and X_0, two conditions are applied: (i) geostrophic transport M with (14) is the same as the constant case given in Table 1 and by equation (5) and (ii) the average of the profile (14) over the range (-a, 0) is the same as given in Table 1.

This approximation is applied to SectionA only, since both M and D_0 of Table 1 are more certain in A than in B and C. The profile of computed D with this approximation is shown in the thick full line in Fig. 2. The above two assumptions lead to $D_S = 129.1$ m and $x_0 = 99.7$ km. Fig. 2 shows the variable D_0 does not yield conspicuous change in D from the constant D_0 case.

5 CONCLUSION

An outflow from the strait becomes a jet-like flow at some distance from the coast, though the flow is trapped by the coast. The flow may conserve the potential vorticity near the strait. A simple model for this flow is proposed based on the potential vorticity conservation law. The width of the current and the depth of the upper layer calculated from the model seem to be within estimated values from the temperature profiles.

Elaboration of the model should include the deviation of streamlines from the straight lines as the flow takes its course and changes the flow direction near the outlet (Ichiye, 1984), both involving two-dimensional flow patterns with relative vorticity due to curvature and shear. Effects of dissipation processes due to coastal and bottom boundaries become important near the shore. Potential vorticity equation (1) may be modified following Ichiye et al (1984).

However, these elaboration needs more facts that can be provided by field experiments. The latter is different from the routine hydrographic observations which cannot provide the necessary parameters with any certainty even for this simple model. The author hopes that Japanese oceanography community would carry out such experiments in the future, because it has a large fleet of research vessels and high technology for ocean research and is favored by proximity of the area concerned. Perhaps this could be implemented as a JECSS joint program.

6 REFERENCES

Carrier, G.F. and C.E. Pearson (1968). ordinary Differential Equations. Blaisdell, Waltham, MA 229 pp.

Fisheries Agency (1988). The Results of Fisheries Oceanographic Observations, Jan.-Dec. 1984, 1155 pp.

Gill, A.E. (1982). Atmosphere-Ocean Dynamics. Academic Press, N.Y. 662 pp.

Ichiye, T. (1984). Some problems of circulation and hydrography of the Japan Sea and the Tsushima Current in "Ocean Hydrodynamics of the Japan and East China Seas" (edited by T. Ichiye) 15-54, Elsevier, Amsterdam.

Ichiye, T. and L. Li (1984). A numerical study of circulation in a northeastern part of the East China Sea. ibid 187-208.

Kawai, H. (1972). Hydrography of the Kuroshio Extension in "The Kuroshio" edited by H. Stommel and K. Yoshida). 235-352, Tokyo University Press.

Oshima, K.I. and M. Wakatsuchi (1990). A numerical study of baroclinic instability associated with the Soya Warm Current in the Sea of Okhotsk. J. of Physical Oceanography. 570-584.

DISPERSAL PATTERNS OF RIVER-DERIVED FINE-GRAINED SEDIMENTS ON THE INNER SHELF
OF KOREA STRAIT

(1)   (2)
S.C. PARK and K.S. CHU
(1) Department of Oceanography, Chungnam National University, Taejon 305-764,
    Korea
(2) Department of Oceanography, Korea Hydrographic Office, Inchon 400-037,
    Korea

ABSTRACT

   The inner shelf off the southeastern coast of Korea (Korea Strait) receives
large amounts of fine-grained sediments derived from Nakdong River. An
analysis of topmost sediment shows a progressive seaward decrease in grain size.
Sandy muds and muds accumulate within about 10 km of the river mouth, whereas
clay-size fractions prevail further offshore. Satellite image together with
tidal current data suggests that suspended sediments are transported both west-
and east-ward in the coastal area, strongly influenced by tidal currents. The
westward transport converges to an area east of Geoje Do, whereas the eastward
transport extends further northeastward along the Korean coast. About 14% of
the river-derived fine-grained sediments, approximately 0.64 million tons per
year, accumulate on the inner shelf of Korea Strait; the remaining 86% escape
the shelf and are transported to other parts of the sea.

INTRODUCTION

   The continental shelf off the southeastern coast of Korea (Korea Strait) is
divided into inner shelf, mid-shelf and outer shelf based on bathymetry and
bottom features (Park and Yoo, 1988). The inner shelf, which is shallower than
70 m, is an area of thick accumulations of modern sediments. Detailed surveys
(Park, 1985; Park and Choi, 1986; Suk, 1986; Park and Yoo, 1988; Park et al.,
1990) reveal that sediments of the Korea Strait are dominated by muds on the
inner shelf and sands on the mid- and outer shelves. The mud deposits in the
inner shelf are more recent in origin, whereas the sands on the mid- and outer
shelves are relict, formed close to the shoreline during low stand of sea level
(Park, 1985; Park and Yoo, 1988).

   The Nakdong River in the southeastern province of the Korean Peninsula is
the main source of fine-grained sediments to the shelf (Park and Yoo, 1988; Lee
and Chough; 1989; Park et al., 1990). This river contributes annually 10
million tons of sediments with its discharge concentrated during the rainy
season from July to August (Korea Ministry of Construction, 1974). Most coarse-
grained sediments are deposited in the river mouth, forming tidal sand ridges
and sandy shoals in water depths less than about 20 m (Kim and Lee, 1980). The
remaining fine-grained sediments are transported further offshore in a

suspended mode.

This paper describes distribution and dispersal patterns of fine-grained sediments on the inner shelf off the southeastern coast of Korea. Distribution patterns of fine-grained sediments are interpreted from textural characters of surface sediments and their thickness. Dispersal patterns of suspended sediment are interpreted from the Landsat image together with tidal current data. The sediment sink and budget are also estimated from the data of long-term accumulation rates based on the thickness of Holocene sediments as well as Pb-210 accumulation rates of sediment cores (Jang, 1990).

OCEANOGRAPHIC SETTINGS

The Nakdong River is the second largest fluvial system in Korea; the drainage basin of which comprises an area of 23,656 km^2 (Kim and Park, 1980). This river discharges about 63 billion tons of fresh water, of which about 70% occurs during summer. During summer floods the discharged water is turbid and contains more than 150 mg/l of suspended particulate matters in surface waters (Kim et al., 1986). The river plume extends about 10 km offshore from the river mouth. Tidal range at the river mouth is 1.7 m during flood tide and 0.4 m during neap tide. The tidal currents in the coastal area flow west- to southwest-ward during flood and east- to northeast-ward during ebb (Figs. 2 and 3). The maximum current velocity is higher than 100 cm/sec (Korea Hydrographic Office, 1982). Kim et al. (1986) report that the coastal current flows northeastward along the southeastern coast of Korea. A band of cold coastal water extends about 20 km offshore as a turbidity front in water depths of 40-60 m. Further offshore, the warm Tsushima Current flows northeastward; the typical speed is 30-90 cm/sec (Korea Hydrographic Office, 1982).

SEDIMENT DISTRIBUTION AND DISPERSAL

Sediment samples were collected from the inner shelf during the cruise aboard R.V. Busan 403 in March and May, 1989 using a gravity corer (Fig. 1). The sediments on the inner shelf can be classified texturally into three types: sandy mud, mud and clay (Fig. 4). The sandy muds and muds predominate the near-coastal area within about 10 km from the river mouth, whereas clays are present offshore over a wide area of the inner shelf. The mean grain size ranges from 7 to 9.5 phi, showing a progressive seaward decrease (Fig. 5). There is also a decreasing trend in sand and silt content with increasing water depth and distance from the river mouth, whereas the clay content increases seaward. Figure 6 shows the isopach map of offshore clays above the mid-reflector which is recognized as the pre-Holocene surface (Park et al., 1990). A depocenter is present in the area east of Geoje Do where the sediment thickness reaches up to 18 m.

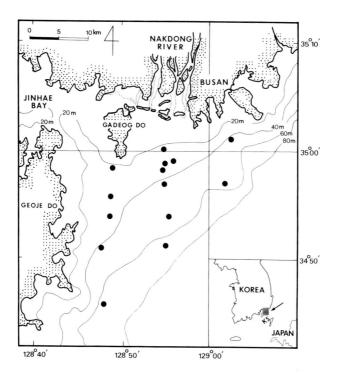

Fig. 1. Bathymetry of the study area and core locations (dots). KS= Korea Strait.

Wells (1988) reports that a residual coastal flow from the southeastern Yellow Sea, forced by monsoon wind from the north, could potentially transport large volumes of suspended sediments into the Korea Strait. However, Lee and Chough (1989) indicate that the muddy sediments on the inner shelf of the Korea Strait are largely derived from the Nakdong River, whereas sandy muds in the western South Sea result from transportation of suspended materials from the southeastern Yellow Sea. On the basis of high-resolution seismic profiles, Park et al., (1990) also suggest that the sediments accumulating on the inner shelf of the Korea Strait are derived from the Nakdong River. The sediments show a prograding deltaic wedge pattern, decreasing in thickness from the sediment source (Nakdong River) seaward. A seaward decrease in silt content accompanied by an increase in clay content in the present study also suggests that the sediments were mainly derived from the Nakdong River. The preponderence of silt and clay reflects the large quantity of fine-grained sediment supplied in suspension.

Figure 7 shows a Landsat image taken on July, 1979 after heavy rainfall in the southeastern Korean Peninsula. The image displays a seaward turbid water plume which extends about 10 km offshore from the river mouth. The image was

234

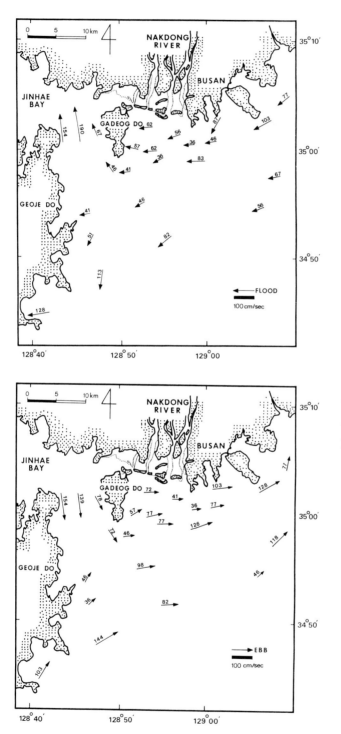

Fig. 2. Maximum velocity (cm/sec) and direction of flood tidal currents during spring tide. After Korea Hydrographic Office (1982) and Kim et al. (1986).

Fig. 3. Maximum velocity (cm/sec) and direction of ebb tidal currents during spring tide. The source of data is same as Fig. 2.

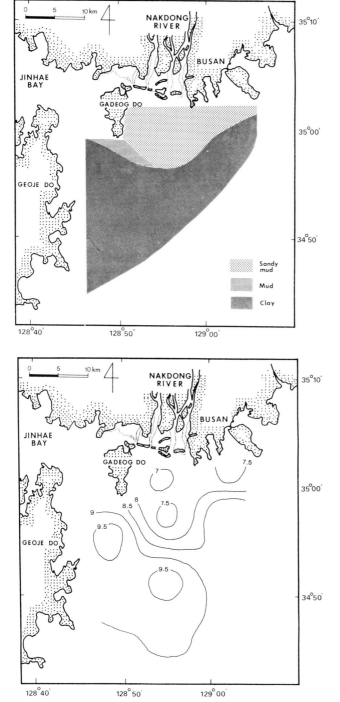

Fig. 4. Distribution
of bottom sediment
types on the inner
shelf of the Korea
Strait. The sediment
type is based on the
nomenclature
suggested by Folk
(1954).

Fig. 5. Grain size
distribution (in phi
scale) of sediment
samples.

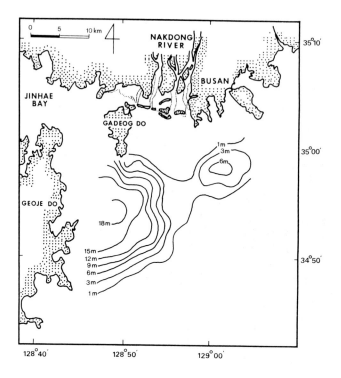

Fig. 6. Isopach map showing the thickness of clayey sediments. Note the maximum accumulation in the area east of Geoje Do where the sediment thickness reaches up to 18 m.

taken 30 minutes after high water during spring tide. Kim et al. (1986) indicate that the ebb current dominates at this time. It shows that one turbid plume extends westward toward Jinhae Bay, which appears to have been mainly affected by the flood current. However, the plume tends to turn to the south, which might be influenced by the ebb current. The ebb current is a return flow from Jinhae Bay, directed to the south (Fig. 3). Kim et al. (1986) indicate that this turbid plume converges in the area east of Geoje Do where thick accumulation of fine sediments is observed (Fig. 6). These fine-grained sediments reflect slow deposition of suspended sediments from the turbid plume and deposition of suspended sediments may be largely influenced by tidal currents. The other plume can be also observed on the Landsat image, which begins to flow northeastward(Fig. 7). Kim et al. (1986) suggest that this plume is transported further northeastward along the coast as a result of the combined influence of tidal and coastal currents. Park (1985) also indicates high concentrations of suspended matters in the coastal waters of the southern East Sea, which might be transported from the Nakdong River.

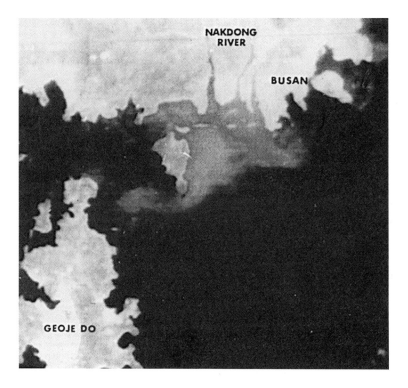

Fig. 7. Landsat image showing the high turbid river plumes that are discharged from the Nakdong River. The image was taken 30 minutes after high water on July 9, 1979 after heavy rainfall in the southeastern province of Korea. Note that one plume extends westward toward Jinhae Bay whereas the other tends to flow northeastward along the coast (for details see text).

SEDIMENT BUDGET

The Nakdong River discharges about 63 billion tons of fresh-water per year and about 70% of total discharge occurs during rainy summer season from July to September (Chu, 1978). The total amounts of annual sediment discharge by the Nakdong River is as much as 10 million tons. According to the Analyses by Kim et al. (1986), the concentration of total suspended matters near the river mouth is as much as 150 mg/l during summer flood, whereas it decreases to 30 mg/l during dry season. The concentration of suspended sediments is nearly uniform from surface to bottom and tends to decrease with distance. With these data, the total discharge of suspended sediments is estimated to be about 4.6 million tons per year which is about the half of the total sediment discharge by the Nakdong river (Table 1).

The mud deposit on the inner shelf off the Nakdong River covers approximately 400 km^2. On the basis of high-resolution (3.5 kHz) seismic profiles, the thickness of the mud above the distinct mid-reflector is in the range from 5 to

20 m. Park and Yoo (1988) suggest that sediment began to accumulate approximately 5000 years B.P., when sea level approached the present level. This implies an average long-term accumulation rate of 1-4 mm/yr during the last 5000 years. This rate agrees well with that of Pb-210 accumulation rates determined from the sediment cores (Jang, 1990). Jang (1990) also reports that the dry bulk density of the mud is 0.80 g/cm^3. These data indicate that the accumulation rate for the mud is about 0.64 million tons per year, which is about 14% of the annual discharge of suspended sediments by the Nakdong River. The remaining 86% of the river-derived fine sediments may escape the Nakdong shelf and is transported to other parts of the sea floor. The oceanic processes effectively disperse the fine-grained sediments on the shelf off the southeastern coast of Korea.

TABLE 1

Annual accumulation rate and budget of the river-derived fine sediments on the inner shelf of the Korea Strait (for details see text).

		References
Annual discharge of water	63×10^9 tons (70% are concentrated during summer flood times)	Kim and Park (1980)
Annual discharge of total sediments	10×10^6 tons	Ministry of Construction (1974)
Concentration of suspended sediments in waters	30-150 mg/l	Kim et al. (1986)
Annual discharge of suspended sediments	4.6×10^6 tons	Kim et al. (1986)
Average accumulation rate	2 mm/year	Isopach map (Park and Yoo, 1988) Pb-210 method (Jang, 1990)
Annual accumulation of suspended sediments on the inner shelf of Korea Strait	0.64×10^6 tons	Bulk density of sediments: 0.80 g/cm^3 (Jang, 1990; Kim and Suk, 1985) Total area studied: about 400 km^2

CONCLUSIONS

The river-derived fine sediments on the inner shelf of the Korea Strait are texturally characterized by sandy mud, mud and clay with a mean grain size between 7 and 9.5 phi. The sandy muds and muds accumulate within about 10km of the river mouth, whereas clays are dominant offshore over a wide area of the inner shelf. The Landsat image combined with the tidal current data suggest that the dispersal pattern of suspended sediments discharged from the Nakdong River is mainly west- and east-ward, which agrees well with the distribution pattern of bottom sediments. The westward plume, influenced by tidal currents, converges to the site of thick accumulation of fine sediments east of Geoje Do. However, the eastward plume appears to be transported further northeastward along the coast as a result of the combined influence of tidal and coastal currents.

About 14% of the annual discharge of suspended sediment by the Nakdong River, that is about 0.64 million tons per year, accumulates on the inner shelf off the southeastern coast of Korea (Korea Strait). The remaining 86% of the river-derived fine sediment escapes the Nakdong shelf and transported to other parts of the sea floor. The oceanic processes effectively disperse the fine-grained sediments on the shelf.

ACKNOWLEDGEMENT

This research was partly supported by Korea Science and Engineering Foundation (1989-1990). We thank K.M. Jang, S.D. Lee and S.K. Hong for their valuable assistance in sampling and data analysing. S.K. Chough and G.S. Chung are thanked for reviewing the manuscript.

REFERENCES

Chu, K.S., 1978. The correlation study between the sea condition and oceanic environment in the Nakdong River Bay area. Korea Hydrogr. Off. Techn. Rpt., 9-18.
Folk, R.L., 1954. The distinction between grain size and mineral composition in sedimentary rock nomenclature. Jour. Geol., 62: 334-359.
Jang, K.M., 1990. Seismic stratigraphy and accumulation of fine-grained sediments on the Korea Strait shelf. Unpubl. MS. thesis, Chungnam Nat. Univ., Taejon, Korea, 83 pp.
Kim,M.S., Chu, K.S. and Kim, O.S., 1986. Investigation of some influence of the Nakdong River water on marine environment in the estuarine area using Landsat imagery. Rpt. Korea Ministry Sci. Technology, 93-147.
Kim, S.R. and Suk, B.C., 1985. The sound velocity and attenuation coefficient of the marine surface sediments in the nearshore area, Korea. Jour. Oceanol. Soc. Korea, 20: 10-21.
Kim, W.H. and Lee, H.H., 1980. Sediment transport and deposition in the Nakdong Estuary. Jour. Geol. Soc. Korea, 16: 180-188.
Kim, W.H. and Park, Y.A., 1980. Microbiogenic sediments in the Nakdong Estuary, Korea. Jour. Oceanol. Soc. Korea, 15: 34-48.
Korea Hydrographic Office, 1982. Marine environmental atlas of Korean waters. Korea Hydrog. Off., Inchon, Korea, 38 pp.
Korea Ministry of Construction, 1974. Report on the Nakdong Estuary: Industrial site investigation. Unpubl. Rpt., 1-56.
Lee, H.J. and Chough, S.K., 1989. Sediment distribution, dispersal and budget in the Yellow Sea. Mar. Geol., 87: 195-205.

Park, S.C. and Yoo, D.G., 1988. Depositional history of Quaternary sediments on the continental shelf off the southeastern coast of Korea (Korea Strait). Mar. Geol., 79: 65-75.

Park, S.C., Jang, K.M. and Lee, S.D., 1990. High-resolution seismic study of modern fine-grained deposits: Inner shelf off the southeastern coast of Korea. Geo-Mar. Lettr., 10: 145-149.

Park, Y.A., 1985. Late Quaternary sedimentation on the continental shelf off the southeast coast of Korea, a further evidence of relict sediments. Jour. Oceanol. Soc. Korea, 20: 55-61.

Park, Y.A. and Choi, J.Y., 1986. Factor analysis of the continental shelf sediments off the southeast coast of Korea and its implication to the depositional environments. Jour. Oceanol. Soc. Korea, 21: 34-45.

Suk, B.C., 1986. Depositional environment of Late Quaternary sediments and suspended particulate matter on the southeastern continental shelf, Korea. Jour. Geol. Soc. Korea, 22: 10-20.

Wells, J.T., 1988. Distribution of suspended sediment in the Korea Strait and southeastern Yellow Sea: Onset of winter monsoons. Mar. Geol., 83: 273-284.

ESTIMATION OF ATMOSPHERIC VARIABLES AND FLUXES ON THE OCEAN SURFACE AROUND KOREAN PENINSULA

IN-SIK KANG and MAENG-KI KIM
Department of Atmospheric Sciences, Seoul National University, Seoul, 151 Korea

ABSTRACT

Daily-mean time series of the atmospheric variables observed by the buoys of Japan Meteorological Agency are compared with the corresponding time series of European Center for Medium Range Forecast (ECMWF) 1000 mb data for one winter of 1 Dec. 1983 - 29 Feb. 1984. On the basis of the comparison the formulas of converting the ECMWF data to that of the ocean surface are derived, and the distributions of winter-mean air temperature and specific humidity are obtained over the seas around Korean peninsula.

Using the atmospheric variables on the ocean surface thus obtained from the ECMWF 1000 mb data and National Meteorological Center (NMC) sea surface temperature, sensible and latent heat fluxes on the ocean surface are calculated using the bulk formula. Sea surface winds are also computed using the so-called Cardone model, and the distributions of surface wind stress and wind stress curl are obtained from the wind.

1 Introduction

Air-sea interaction typified by heat, moisture, and momentum exchanges between ocean and atmosphere has been agreed by ocean and atmospheric research communities to be an important subject to focus on for understanding the ocean and atmospheric circulations and certain phenomena such as ocean waves and air-mass modification (Fissel et al., 1976; Smith, 1980; Large and Pond, 1980; Park and Joung, 1984; Kang, 1986). However, investigations of air-sea interaction problems are hindered mainly because of the limitation of meteorological data on the open ocean surface. Although several studies (Matsumoto, 1967; Han, 1970; Bong, 1976; Kang, 1984) have dealt with heat exchange between air and ocean around Korean peninsula, their studies are confined to the coastal area.

In the present study, we examine the possibility of the use of European Center for Medium Range Forecast (ECMWF) analysis data at 1000 mb in the air-sea interaction problem over the seas around Korean peninsula. The regression formula of converting the ECMWF data to that of the ocean surface are derived using the Japan Meteorological Agency (JMA) buoy data and the ECMWF data nearest the buoy location. The formula thus obtained are applied to the ECMWF data over the seas around Korea. Using the estimated surface variables over the seas, heat fluxes, wind stress, and wind stress curl are calculated over the open ocean surface.

2 Data

The data sets used in this study are daily means of JMA buoy observations and ECMWF 1000 mb variables for the winter from 1 Dec. 1983 to 29 Feb. 1984 (hereafter this period is referred to as 1983-1984 winter), and monthly mean SST obtained from NMC. JMA buoy locations are marked by crosses in Fig. 4. The horizontal resolutions of ECMWF and National Meteorological Center (NMC) SST data, respectively, are 2.5° x 2.5° and 2° x 2° in longitude and latitude. To match the grid point, SST data is transferred to 2.5° x 2.5° grid data using an optimum interpolation method. Domain of analysis is the oceans inside the region of 20° N - 50° N and 110° E - 150° E. The variables used are air temperature, relative humidity, wind speed and direction, and sea surface temperature. The period of ECMWF data is chosen because buoy observations are more complete during the 1983-1984 winter than during other winters.

3 Estimation of air temperature and humidity at buoy locations

Air temperature and relative humidity observed by Buoy No. 3 and 6 are compared with corresponding variables of ECMWF 1000 mb data at the grid points nearest the buoys. The locations of Buoy No. 3 and No. 6 are 25° 40´ N, 135° 55´E and 37° 45´N, 134° 23´E, respectively, and the locations of corresponding grid points nearest are 25° N, 135° E and 37.5° N, 135° E, respectively. On the basis of comparison, the regression formula of converting ECMWF 1000 mb data to the surface variables are obtained at the buoy locations.

In Fig. 1a is shown scattogram of daily mean air temperature observed by Buoy No. 3 versus that of ECMWF 1000 mb. Since the period of data used is from 1 Dec. 1983 to 29 Feb. 1984, 91 data points are plotted. It is clear that buoy temperature is warmer than the 1000 mb temperature. The mean difference is 1.9 °C. During winter 1000 mb level is generally higher than the sea surface, therefore 1000 mb temperature should be cooler than that of the surface. Applying the lapse rate of standard atmosphere, 0.65 °C/100 m, the surface temperature is estimated by the use of 1000 mb temperature and geopotential height. As shown in Fig. 1b, the estimated temperature is still cooler than the buoy temperature. The mean difference is 1.0 °C, therefore 0.9 °C is corrected by the use of height information.

In addition to the height, the degree of mixing near the sea surface may be an important factor to determine the surface temperature. The mixing or surface air stability is often parameterized by the difference between the air and sea surface temperatures. Considering both effects, the surface air temperature is estimated using the following formula.

$$T_S = T_{E1000mb} + a \; GPH + b \; (SST - T_a)\tag{1}$$

where T_S is the estimated sea surface temperature, $T_{E1000mb}$ and GPH, respectively, ECMWF 1000 mb temperature and geopotential height, and T_a the surface air temperature. Since the SST data available is monthly mean, daily SST is obtained by interpolation of monthly means, and T_a is used with ECMWF 1000 mb temperature for simplicity. The coefficients a and b are determined by the use of regression method. For Buoy No. 3, a = 0.61 °C/100 m and b = 0.23.

Fig. 1c shows scattogram of Buoy No. 3 temperature versus the estimated surface temperature. The mean error is now zero and the root mean square (rms) of the error is 1.0 °C. Considering the uncertainty of ECMWF 1000 mb temperature and daily SST interpolated, the estimation mostly within the error of 1.0 °C may be acceptable.

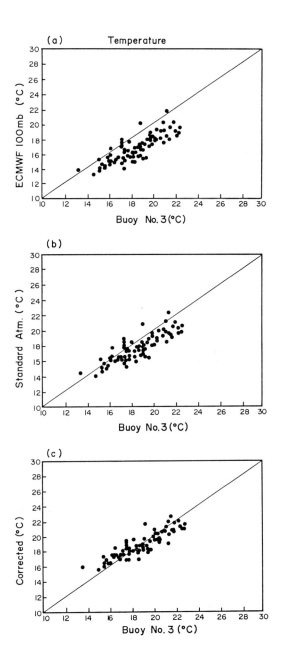

Fig. 1. (a) Scattogram of the temperature observed by Buoy No. 3 (x-axes) versus ECMWF 1000 mb temperature at the grid nearest the buoy (y-axes). (b) As in (a) except the surface temperature corrected from the ECMWF temperature with standard atmosphere lapse rate. (c) As in (a) except the surface temperature estimated using Eq. (1).

244

We now compare the surface humidity observed by buoys and that of ECMWF 1000 mb. Figs. 2a and 2b, respectively, are scattograms of relative humidity and virtual temperature. X- and Y-axes indicate Buoy No. 3 and ECMWF 1000 mb values, respectively. ECMWF and buoy datas provide relative humidity, and virtual temperature (T_V) is calculated using the following formula.

$$T_V = T_a (1 + 0.61 \, q_a) \tag{2}$$

where q_a is the specific humidity as calculated by the equation,

$$q_a = (0.622/P) \, (e_s \times RH/100) \tag{3}$$

P is the pressure, e_s the saturated water vapor pressure at the air temperature, RH the relative humidity.

In Fig. 2a, the 1000 mb RH is less variable than that of buoy and the relationship between the

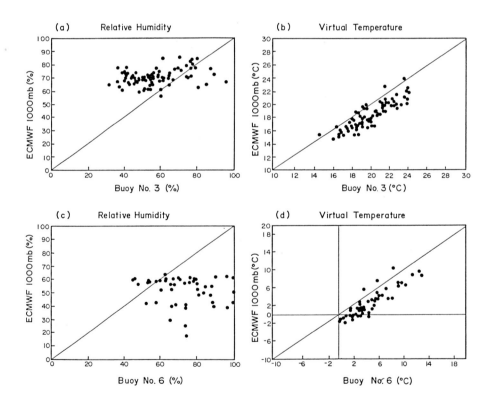

Fig. 2. (a) Scattogram of the relative humidity observed by Buoy No. 3 (x-axes) versus that of ECMWF 1000 mb at the grid nearest the buoy (y-axes). (b) As in (a) except the virtual temperature. (c) As in (a) except Buoy No. 6. (d) As in (b) except Buoy No. 6.

two variables is not clear. However, as shown in Fig 2b, buoy and ECMWF virtual temperatures have some relationship. The reason that virtual temperature has better relationship than relative humidity is that the height (pressure) information included in specific humidity and virtual temperature as shown in Eq. (3). The mean difference between the two temperatures is 1.6 °C. Figs. 2c and 2d, respectively, are as in Figs. 2a and 2b except for Buoy No. 6. At the buoy location RH at 1000 mb level is generally lower than that of the buoy, although revered situation is true at Buoy No. 3. However, both virtual temperatures at 1000 mb level is lower than those observed at Buoy No. 3 and No. 6, and the relationships between T_V's are clear and similar to that of temperature shown in Fig. 1a. Therefore the regression formula of converting ECMWF 1000 mb T_V to that of the surface is obtained with the same formula as expressed in Eq. (1). For Buoy No. 3, the coefficients a and b, respectively, are 0.56 °C/100 m and 0.18. Using the formula, T_V at the surface is calculated and the scattogram of the calculated T_V versus that of Buoy No. 3 is shown in Fig. 3. The mean error is zero and the rms is 1.0 °C.

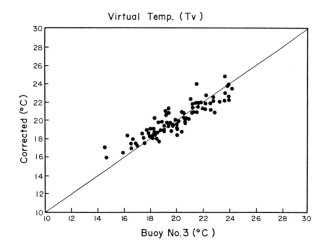

Fig. 3. Scattogram of the virtual temperature of Buoy No. 3 versus that of the ocean surface estimated based on the regression formula, Eq. (1).

4 Horizontal distribution of atmospheric variables on the ocean surface

The regression formulas obtained at Buoy No. 3 may be hardly applied to other regions in the domain, because their coefficients are not uniform over the oceans. For example, for the surface temperature the coefficients a and b at Buoy No. 6 are 0.68 °C/100 m and 0.24, respectively, which are different from those at Buoy No. 3. In this section, we derive the formulas, which can be applied to all grid points, using all buoy data available after 1980 in the domain and corresponding ECMWF data.

The coefficient a is associated with height correction, and is set to 0.65 °C/100 m, the value of standard atmosphere. In fact, the value is in between the coefficients at Buoy No. 3 and No. 6.

Then using all available data in the domain for winters of 1980-1984, the coefficients b of temperature and virtual temperature are calculated by minimizing rms of the error, the difference between the buoy and estimated surface temperatures. b is 0.30 and 0.35 for temperature and virtual temperature, respectively.

The regression formulas thus obtained are applied to the ECMWF grid data in the domain for one winter of 1 Dec. 1983 - 29 Feb. 1984. In Fig. 4a is shown the distribution of surface air temperature averaged for the winter. The range of winter mean temperature in the domain is from about 20 °C at the southern boundary to -5 °C at the northern tip. Zero line crosses northern part of the East Sea (Sea of Japan) and Yellow Sea. From the virtual temperature, specific humidity on the ocean surface is derived and its distribution is shown in Fig. 4b. The winter-mean specific humidity is between 2 - 4 g kg^{-1} near Korean peninsula, and more than 10 g kg^{-1} in the subtropics.

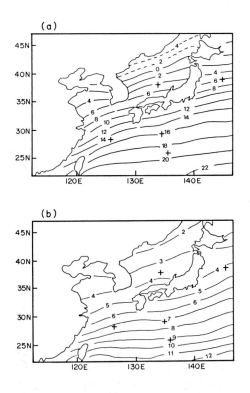

Fig. 4. (a) Distribution of the estimated temperature on the ocean surface averaged for the 1983-1984 winter. (b) As in (a) except the specific humidity. Contour interval is 2 °C.

In Fig. 5 is shown the sea surface temperature averaged over the 1983-1984 winter. SST varies from 26 °C in the southern boundary to 3 °C in the northern tip. In the South Sea of Korea the winter mean SST is about 12-14 °C, and over the East and Yellow Seas the SST decreases with latitude upto 4 °C in northern Korea.

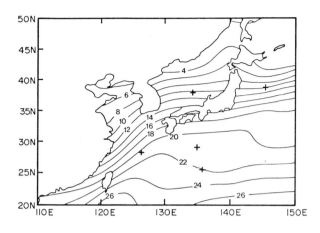

Fig. 5. Distribution of the sea surface temperature averaged for the 1983-1984 winter.

Sea surface wind is one of the most important variables for studying air-sea interaction problems. In this section, sea surface wind is derived using the model initially proposed by Cardone (1969, 1978), and developed by Kim (1989). In this model, atmosphere near the surface is devided by two layers; Ekman and surface layers. Ekman layer wind is determined by Ekman profile, and surface layer is characterized by turbulent mixing and a constant flux layer. The intensity of the mixing is determined by air stability near the surface, which is proportional to the difference between SST and air temperature. Thermal wind effect resulted from horizontal temperature distribution is also included in the model. For the top and bottom boundary conditions, the wind at the top of Ekman layer is assumed to be geostrophic and is determined by ECMWF 1000 mb geopotential height, and the wind is zero at the sea surface. At the layer between Ekman and surface layers, matching condition is applied to the Ekman and surface layer solutions. The sea surface wind is assumed to be the wind at the level of 10 m height; this height is usually within the surface layer. More details of the model and its associated dynamics can be inferred from Kim (1989) and Bong et al. (1989), respectively.

Using the model, horizontal distribution of sea surface wind averaged over the winter is obtained and displayed in Fig. 6. As shown in Fig. 6a, northerly wind is dominated around Korean peninsula and north-easterly wind over the south East China Sea. The magnitude of the winter-mean surface wind, shown in Fig. 6b, is not much varied in the domain; the range is between 5 m s^{-1} and 10 m s^{-1}.

5 Heat, moisture, and momentum exchanges on the ocean surface

Using the horizontal distributions of temperature, specific humidity, SST, and sea surface wind obtained in the previous section, we now estimate various fluxes on the ocean surface. The following bulk formula is used for the estimation of sensible and latent heat fluxes.

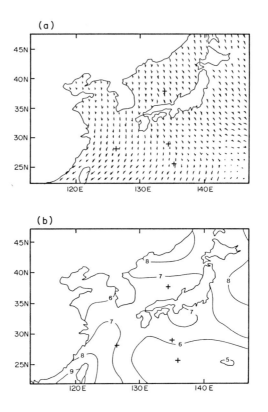

Fig. 6. (a) Distribution of the estimated sea surface wind averaged for the 1983-1984 winter. (b) As in (a) except the wind magnitude. The arrow of one grid length in (a) is 10 m s^{-1} and contour interval in (b) is 1 m s^{-1}.

$$F_S = \rho_a C_a C_H (T_S - T_a) |V| \qquad (4)$$

$$F_L = \rho_a L C_E (q_S - q_a) |V| \qquad (5)$$

where F_S and F_L are sensible and latent heat fluxes, respectively, ρ_a the surface air density 1.225 kg m^{-3}, q_S and q_a are specific humidity of ocean surface and of air above the ocean, respectively, and $|V|$ is the magnitude of sea surface wind. C_a the air specific heat 1005 J $^\circ$K^{-1}kg^{-1}, L the latent heat of evaporation 2.5 x 10^6 J kg^{-1}, C_H and C_E are sensible and latent heat exchange coefficients, respectively, and the value 0.0015 is used for both coefficients.

In Figs. 7a and 7b are shown distributions of sensible and latent heat fluxes, respectively, obtained by substituting the winter means into Eqs. (4) and (5). Sensible heat flux generally increases with latitude. Maximum value is about 100 W m^{-2} at the north-west coast of the East Sea.

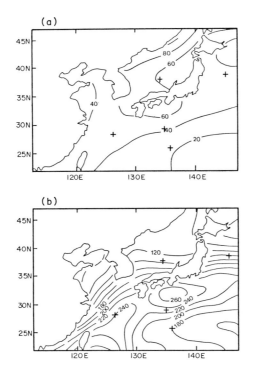

Fig. 7. (a) Distribution of the winter-mean sensible heat flux at the ocean surface. Contour interval is 20 W m^{-2}. (b) As in (a) except latent heat flux.

On the other hand, latent heat flux decreases with latitude near Korean peninsula, and the maximum appears off the south coast of Japan and the East China Sea. It is interesting that the region of maximum latent heat flux coincide with the region of Kuroshio current. Also noted is that over the northern part of the East Sea, the magnitude of sensible heat flux is comparable to that of latent heat flux. However, in most regions of the domain latent heat flux is much larger than sensible heat flux. The magnitudes of sensible and latent heat fluxes obtained in the present study are comparable to the counterparts of Esbensen and Kushnir (1981).

Wind stress and its curl are also computed using the sea surface wind model explained in the previous section. Wind stress (τ) is defined as $\rho_a u^{*2}$, where u* is the friction velocity determined by the surface layer dynamics of the model. Among them important are surface roughness and intensity of surface layer mixing. More details can be found in Kim (1989) and Kim (1990).

In Figs. 8a and 8b are shown distributions of wind stress and its curl. Relatively large wind stress appears over the northern part of the East Sea and off the east coast of China. The distribution of wind stress is similar to that of the surface wind shown in Fig. 6a. Wind stress curl represents a

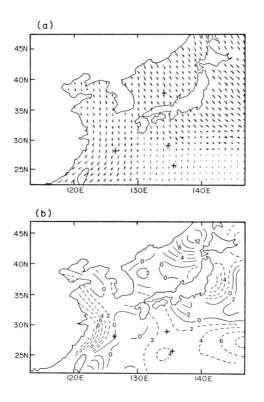

Fig. 8. (a) Distribution of the winter-mean surface wind stress. (b) As in (a) except the wind stress curl. Contour intervals for (a) and (b) are 0.02 N m^{-2} and 2 x 10^{-8} N m^{-3}. Negative values are denoted by dashed lines in (b).

rather complicated structure. In the East Sea, however, the sign of the curl is positive in most of the region. Since vertical motion is proportional to wind stress curl in a large-scale motion, upwelling can be induced by the wind over the East Sea. In Particular, significant upwelling can exist in the west of Hokkaido. On the other hand, negative wind stress curl is prominent off the east coast of China.

6 Summary and concluding remarks

In the present study, we investigate a possible way of obtaining atmospheric variables on the open ocean surface using ECMWF 1000 mb data. Daily-mean time series of temperature and virtual temperature of ECMWF 1000 mb are compared with the corresponding time series observed by JMA buoys for 1983-1984 winter. On the basis of the comparison, regression formulas of converting ECMWF 1000 mb data to the surface are derived. The formulas thus obtained are applied to the

ECMWF data over the seas around Korean peninsula. Sea surface wind is calculated using the so-called Cardone model. Using the estimated atmospheric variables on the ocean surface and NMC SST, distributions of sensible and latent heat fluxes, wind stress, and wind stress curl over the seas are obtained.

It should be mentioned that the data sets in deriving the regression formulas include several uncertainties. For example, daily SST data interpolated from monthly means and ECMWF 1000 mb data, which is actually the output of global atmospheric model, should be different from actual daily value. However, the fact that the error of the estimated surface temperature is about 1 °C is encouraging. Another uncertainty of this study is that in the formulas obtained in section 4, a is set to the lapse rate of standard atmosphere and b is determined using available buoy and ECMWF data. Atmospheric lapse rate near the surface can vary with time and location. Therefore, the model based on the surface layer dynamics should be developed to obtain objectively the atmospheric variables on the ocean surface. We now try to build up the model by adding an temperature profile equation in the equation set of the sea surface wind model. In the model, the wind, temperature, and heat flux on the ocean surface are coupled each other.

7 References

Bong, J.-H., 1976. Heat exchange at the sea surface in the Korean coastal seas. J. Oceanogr. Soc. Korea, 11, 43-50.

Bong, J.-H., Oh, I.-S., Kang, I.-S. and Choi, J.-B., 1989. The study of meteorological characteristics and marine forecasting over the seas around Korea (III). Research report, Meteor. Research Institute, Korea Meteor. Service, 355 pp.

Cardone, V. J., 1969. Specification of the wind distribution in the marine boundary layer for wave forecasting. Report GSL-TR69-1, College of Engineering and Science, New York Univ., 181 pp.

-------, 1978. Specification and prediction of the vector wind on the United States continental self for application to an oil slick trajectory forecast program. Contract T-36430, Institutes of Marine and Atmos. Sciences, New York Univ., 210 pp.

Esbensen, S. K. and Kushnir, Y., 1981. The heat budget of global ocean: An atlas based on estimates from surface marine observations. Report No. 29, Climate Research Institute, Oregon State Univ., 240 pp.

Fissel, D. B., Pond S. and Miyake, M., 1976. Computation of surface fluxes from climatological and synoptic data. Mon. Wea. Rev., 105, 26-36.

Han, Y.-H., 1970. On the estimation of evaporation and sensible heat transfer in the south-eastern part of the Yellow Sea in the month of January. J. Korean Meteor. Soc., 6, 83-87.

Kang, I.-S., 1986. Atmospheric linear responses to large-scale sea surface temperature anomalies. J. Korean Meteor. Soc., 22-2, 35-47.

Kang, Y.-Q., 1984. Atmospheric and oceanic factors affecting the air-sea thermal interactions in the East Sea (Japan Sea). J. Oceanogr. Soc. Korea, 19, 163-171.

Kim, K.-M., 1989. Sea surface wind model around the Korean peninsula. M. S. Thesis, Depart. Atmos. Sci., Seoul National University, 76 pp.

Kim, M.-K., 1990. A study on the estimation of meteorological variables on the ocean surface around the Korean peninsula. M. S. Thesis, Depart. Atmos. Sci., Seoul National University, 76 pp.

Large, W. G. and Pond, S., 1980. Open ocean momentum flux measurements in moderate to strong winds. J. Phys. Oceanogr., 11, 304-336.

Matsumoto, S., 1967. Budget analysis on the sea effect snow observed along the Japan sea coastal area. J. Meteor. Soc. Japan, 36, 123-134.

Park, S.-U., and Joung, C.-H., 1984. Air modification over the Yellow Sea during cold-air outbreaks in winter. J. Korean Meteor. Soc., 20, 35-50.

Smith, S. D., 1980. Wind stress and heat flux over the ocean in Gale force winds. J. Phys. Oceanogr., 10, 709-726.

IDENTIFICATION OF WATER MASSES IN THE YELLOW SEA AND THE EAST CHINA SEA BY CLUSTER ANALYSIS

K. KIM[*], K.-R. KIM[*], T.S. RHEE[*], H.K. RHO[+], R. LIMEBURNER[†] and R.C. BEARDSLEY[†]
* Department of Oceanography, Seoul National University, Seoul 151-742, KOREA
+ Department of Fishery, Cheju National University, Cheju 690-121, KOREA
† Woods Hole Oceanographic Institution, Woods Hole, MA 02543, U.S.A.

ABSTRACT

A statistical cluster analysis technique is applied to hydrographic data taken in the Yellow Sea and the East China Sea mostly during January and July, 1986. The technique clusters individual points on the temperature (T) – salinity (S) plane into groups which can then be examined for their T-S properties and spatial distributions. Horizontal maps of the 11 groups provide a clear synoptic picture of water mass distribution in the area. As a result of this analysis, we introduce East China Sea Water as a distinct water mass, which is characterized by $12.5°C < T < 16.5°C$ and $33.5‰ < S < 34.60‰$.

1 INTRODUCTION

The Yellow Sea and East China Sea are typical mid-latitude epicontinental seas. The surface temperature changes as much as 20°C seasonally and the salinity varies from 29‰ to 35‰ in these seas. The extremely large ranges of temperature and salinity variation make these seas very unique, and the characteristics of water masses and the circulation in the area have been subject to many investigations (Asaoka et al., 1966; Lim, 1971; Nakao, 1977; Mao et al., 1981; Nagata, 1981; Lie, 1984 and 1986; Park, 1985 and 1986).

It should be noted that only Yellow Sea (Bottom) Cold Water has been identified as a distinct water mass in the past(Nakao, 1977), as it was located clearly in the central part of the Yellow Sea. Coastal waters along the Chinese coast and the Korean coast were rather loosely defined(Gong, 1971; Kondo, 1985; Lie, 1984; Mao et al., 1981). This reflects that classification of water masses in the Yellow Sea and the East China Sea is a difficult task because of the extremely large range of property variation. One way to overcome this difficulty is to introduce an analysis technique which treats hydrographic data statistically.

Here we apply a cluster analysis to temperature and salinity data in an attempt to identify distinct water masses by the grouping of data points on the T-S diagram. Cluster analysis is a statistical technique which has been frequently used in biology to classify species into groups according to similarity in parameters among species (Clifford et al., 1973; Sneath et al., 1973). In the analysis done here, we chose the average linkage between groups method, often called UPGMA (unweighted pair-group method using arithmatic average), which combines groups so that the average distance between all cases in the resulting group is as small as possible (Norusis, 1986). The distance is defined as the squared Euclidean distance based on the normalized T and S difference between points. The program used for our analysis is from the statistical package, SPSS-X2.1.

The cluster analysis was applied for limited areas of the Yellow Sea (Li et al., 1984; Qiu et al., 1985). The purpose of our study is to examine water masses in the Yellow Sea and the East China Sea from synoptic data, which was not available in the past.

2 DATA

Most of the hydrographic data used in this study were obtained on a R/V Thomas G. Thompson (TT) cruise in January, 1986 and a R/V Thomas Washington (TW) cruise in July, 1986. Both cruises were parts of US-Korea-China and US-Korea cooperative research programs, respectively. The data from these two cruises are unique, as they are the most extensive and systematic data obtained in the Yellow Sea and the East China Sea in recent years.

Additional data came from cruises organized by Nagasaki Marine Observatory of Japan (NJ), Korean Fisheries Research and Development Agency (KF) and Cheju National University of Korea (CU) (Table 1). Although some data were taken in different months, observation periods are divided into January and July nominally. All the stations are shown in Fig. 1(a) and 1(b) for January and July, 1986, respectively.

During the TT and TW cruises temperature and salinity were measured with a Neil-Brown Instrument System's Mark III CTD and the hydrographic data from other sources were obtained from bottle casts with reversing thermometers. From the entire data set some representative data are selected for the cluster analysis. In January we chose CTD data at 4 m depth for the surface water value because this was the shallowest CTD data reported. Since waters were well mixed vertically in the water column shallower than 100 meters, any depth may represent the surface. However, very strong thermoclines with depths ranging from 5 m to 40 m occurred in July with mixed layers both above and below these thermoclines. We used CTD data at 2 m depth for the surface water values and CTD data at 50 m depth for deep waters in July. In both month, additional surface and 50 m data were obtained by bottle casts. Henceforth, the surface and deep waters refer to the waters at these selected depths.

TABLE 1

Hydrographic data used for the cluster analysis. Data are divided into
January and July, 1986 nominally.

Cruise	Research Vessel/Organization	Period	Cast	Stations
January 1986				
TT	R/V Thomas G. Thompson	10 JAN – 31 JAN	CTD	177
CU	Cheju National University	25 JAN – 14 FEB	BOTTLE	21
KF	Korean Fisheries Research and Development Agency[1]	29 JAN – 4 FEB	BOTTLE	31
NJ	Nagasaki Marine Observatory[2]	5 FEB – 11 FEB	BOTTLE	41
July 1986				
TW	R/V Thomas Washington	5 JUL – 19 JUL	CTD	144
KF	Korean Fisheries Research and Development Agency[1]	10 AUG – 16 AUG	BOTTLE	31
NJ	Nagasaki Marine Observatory[3]	16 JUL – 31 JUL	BOTTLE	39

1   Korean Fisheries Research and Development Agency(1987)

2   Nagasaki Marine Observatory(1986a)

3   Nagasaki Marine Observatory(1986b)

TABLE 2

Water groups by the cluster analysis.

Group	Number of Samples	Temperature(°C)	Salinity (‰)
1	9	1.5 – 5.6	29.98 – 30.84
2	116	3.2 – 9.7	31.03 – 32.32
3	148	6.4 – 13.2	32.12 – 33.60
4	222	11.3 – 17.0	33.53 – 34.80
5	14	7.2 – 10.1	33.78 – 34.34
6	102	16.8 – 23.7	34.00 – 34.88
7	44	24.3 – 29.4	33.69 – 34.75
8	29	22.7 – 29.3	31.92 – 33.41
9	117	18.1 – 26.5	30.09 – 32.41
10	27	25.8 – 29.7	30.49 – 31.71
11	19	21.6 – 27.0	29.13 – 29.94

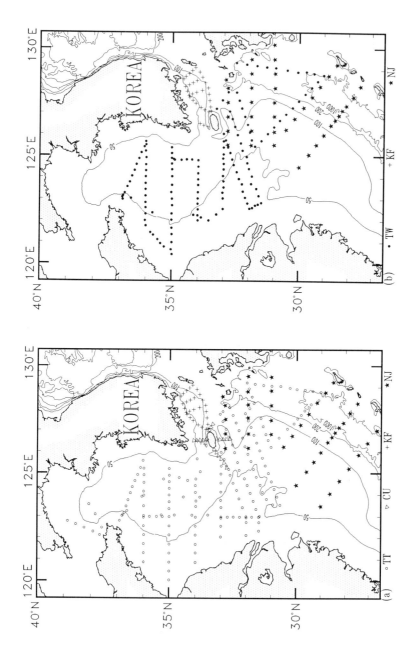

Fig. 1. Stations occupied in January, 1986 (a) and July, 1986 (b). Symbols in the bottom panel represent cruises and organizations listed in Table 1. Isobaths in meters.

3 ANALYSIS AND RESULTS

We used temperature and salinity as the two independent parameters for the cluster analysis. In order to see the relationships between water masses which may occur in winter and summer, we combined all data from January and July together for clustering.

Since temperature ranges from 1.5°C to 29.7°C and salinity varies from 29.130‰ to 34.876‰, all T and S data were normalized as follows before grouping:

$$XN = \frac{X - Xmin}{Xmax - Xmin}$$

where XN is a normalized parameter, X is a parameter value, and Xmin, Xmax are the minimum and the maximum values of each parameter, respectively. After normalization the T, S data were analyzed using the cluster technique. In a cluster analysis, it is rather arbitrary or subjective to determine how many groups are required to best define data; after experimentation we chose eleven groups in our analysis.

Results of our analysis are presented in terms of temperature and salinity rather than the normalized parameters for sake of convenience. The hydrographic characteristics of the 11 groups are summarized in Table 2. The groups are numbered, beginning with the coldest and relatively low salinity water (group 1) and finishing with the water which is characterized by the lowest salinity and relatively high temperature (group 11). The change of characteristics by groups can be seen in Fig. 2, where the group number increases counterclockwise except for group 5 and 9.

It should be pointed out that the T-S diagram in Fig. 2 is different from usual T-S curves. In Fig. 2 the vertical variation of temperature and salinity on the T-S plane is represented by two data points at the surface and 50 m. Otherwise the entire T-S plane is filled (Nakao, 1977, Fig. 5b). This is an important step to simplify the water mass analysis. It can be justified by the fact that water columns are vertically homogeneous because of strong vertical mixing in January and are made of two layers separated by a seasonal thermocline at mid-depth in July. The purpose of this procedure is to reduce the vertical structure as simple as possible, leaving the remaining variation of temperature and salinity subject to horizontal processes.

In general temperature increases linearly with salinity from group 1 to group 4 (Fig. 2). Then temperature increases at a much larger rate from group 4 through group 6 and group 7, while salinity changes little around 34.4‰. From group 7 to group 11 the relationship between temperature and salinity is again approximately linear, but it is not as tight as

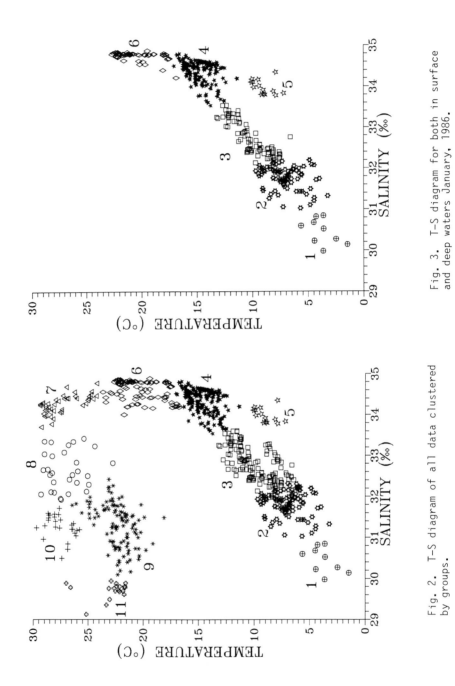

Fig. 2. T-S diagram of all data clustered by groups.

Fig. 3. T-S diagram for both in surface and deep waters January, 1986.

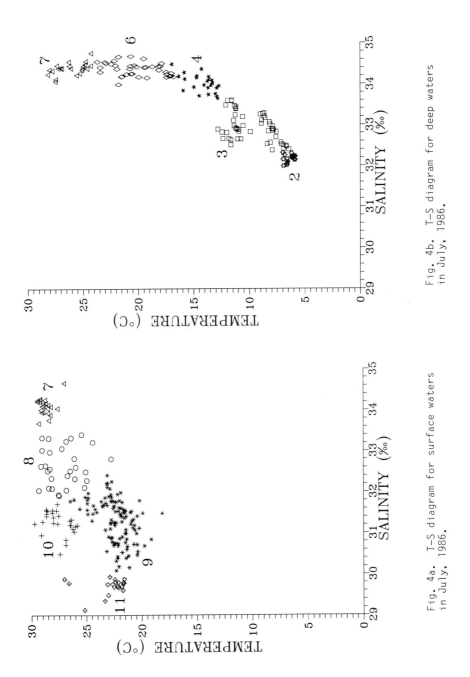

Fig. 4a. T-S diagram for surface waters
in July, 1986.

Fig. 4b. T-S diagram for deep waters
in July, 1986.

that from group 1 to group 4. It should be noted that the T-S relationship for the highest salinity of group 4 is close to the characteristics of the Kuroshio(Nitani, 1972).

As waters are well mixed vertically in January, the surface and deep waters are represented by the same groups in Fig. 3. It is rather striking that the T-S characteristics for waters in January are very similar to those of deep waters in July shown in Fig. 4b, where groups 3, 4 and 6 are common. The differences are that waters of group 1, 5 and most of group 2 appear only in January and group 7 appears only in July. From this comparison it seems that group 3, 4 and 6 are present in deep waters continuously year-round, while group 1, 2, 5 and 7 are replaced in season.

Surface waters in July (Fig. 4a) are made of group 7 through group 11, while the deep waters during the same period are composed of group 2 through group 7 (Fig. 4b). Furthermore, the surface waters in July (Fig. 4a) show no resemblance at all with those in January (Fig. 3). Obviously, seasonal heating raises the surface temperature, but additional analysis is required to determine if the change in T at the surface is entirely due to local heating or to other processes.

4 DISTRIBUTION OF WATER GROUP

One benefit of T-S cluster analysis is that it allows mapping the geographic distribution of different water property groups. Here we describe the spatial distributions of the water groups for January and July, 1986.

4.1 January

The distribution at 4 m (Fig. 5) also represents that at 50 m in January, since waters are well mixed vertically over the top 50 m in winter. Waters of group 1 and 5, which appears only in January in the T-S diagram, are observed only along the east coast of China and the south coast of Korea respectively. It should noted that stations belonging to group 1 are located separately either around the tip of Shandong Peninsula or off Jiangsu Province. The very low temperature and salinity of group 1 in the T-S diagram (Fig. 3) are probably due to the heat loss to the atmosphere and the discharge of fresh water into the very shallow coastal band around the western Yellow Sea. Group 2 prevails in waters shallower than 50 m on the Chinese side and the northern Yellow Sea. Group 3 is an continuation of group 2 in the Yellow Sea and part of the East China Sea west of 126°E and north of 30°N.

In January the surface water of the Kuroshio is represented by group 6 (Fig. 5). On the continental shelf of the East China Sea group 4 is abundant. It is interesting that group 4 also appears around Cheju Island and in the western channel of the Korea Strait. Group 5 is observed only in a limited area off the south coast of Korea, which is surrounded by waters of

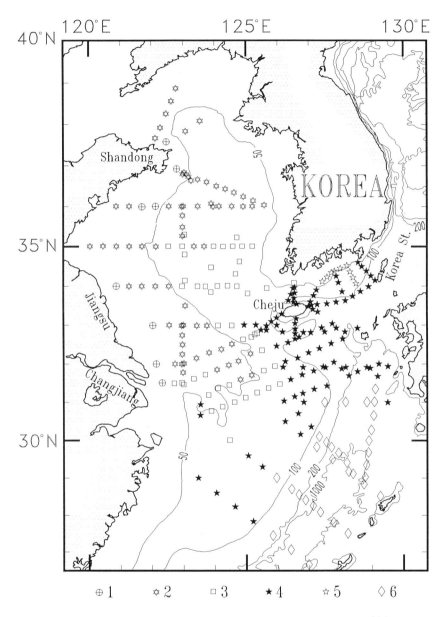

Fig. 5. Distribution of water groups for January, 1986.
Groups are indiated by symbols in the bottom panel.

group 4. In Fig. 3, temperature of group 5 is less than that of group 4 by about 5°C and salinity of the former is slightly less than that of the latter. This suggests that group 5 water is modified from group 4 water in winter by cooling. The discharge from small rivers of Korea may lower the salinity of group 5. From the T-S diagram and geographical distribution, we find that in January group 2 and group 3 make up the most of the shelf water, and group 4 water is a mixture between shelf water and the Kuroshio water, which is represented by group 6. Group 1 and 5 are modified shelf waters, which might be called coastal waters.

4.2 July

In July group 9 and group 10 dominate the surface water in the Yellow Sea and most of the East China Sea respectively (Fig. 6). Group 10 is warmer than group 9 by 5°C approximately(Fig. 4a) and appears mainly between Cheju Island and Tsushima Island, reflecting the effect of the Kuroshio. The Kuroshio region on the continental slope is represented by group 7, which replaced group 6 in January. From the Kuroshio towards the Korea Strait group 7, group 8 and group 10 appear in the order of decreasing salinity. It is notable that some stations occupied between the mouth of the Changjiang River and Cheju Island are combined as group 11, which is characterized by the lowest salinity of all data (S < 30‰). There is little doubt that the extremely low salinity observed there is a result of mixing with the discharge of fresh water from the Changjiang River (Beardseley et al., 1985).

The horizontal distribution of the groups at 50 m in July (Fig. 7) show clearly that they are the same groups as observed in January (Fig. 5). The same composition of groups at 50 m indicates persistence of deep water masses in the Yellow Sea and the East China Sea. It is, furthermore, worthwhile to examine the detailed distribution of each group beginning from the Kuroshio region. In July group 7 represents the upper water of the Kuroshio. In comparison with January, group 6 appears more extensively on the East China Sea shelf west of Kyushu at 50 m in July, while the area of group 4 has shrunk noticably. There could be two different explanations for this change. First, we can consider the Tsushima Current which carries warm and saline water northward towards the Korea Strait (Uda, 1934). It is possible that most waters of group 4 in January flow towards the Korea Strait as waters of group 6 gradually follow and fill in. Secondly, it is also possible that because of heating, characteristics of group 4 evolve in time to those of group 6 (Fig. 4b). Although the Tsushima Current seems more likely responsible for the changes, we need direct measurements of currents to verify this.

In the Yellow Sea, the area of group 3 at 50 m in July is larger than

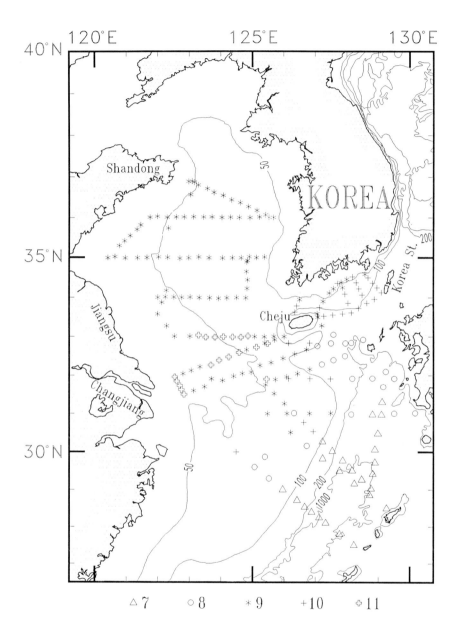

Fig. 6. As in Fig. 5 except for surface waters in July, 1986.

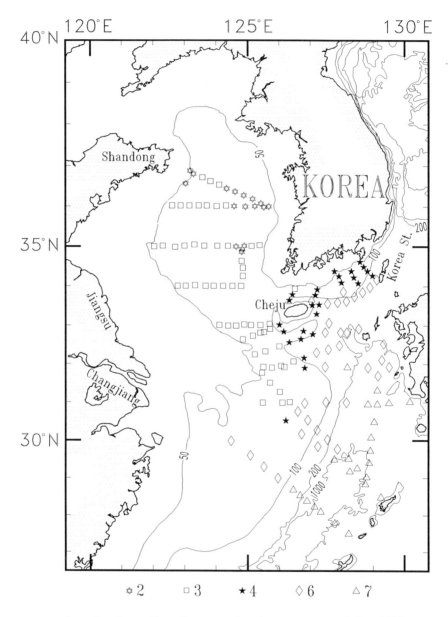

Fig. 7. As in Fig. 5 except for deep waters in July, 1986.

that in January, particularly southwest of Cheju Island. This may be an indication of the cold current flowing southward in spring through summer.

5 DISCUSSION

The T-S diagram in Fig. 2 shows that some groups may deserve names, although naming does not necessarily mean we understand the origin or dynamics of these water masses in the area. In the past the cold water found abundantly in the Yellow Sea was named the Yellow Sea (Bottom) Cold Water (Nakao, 1977). Individual investigator disagree slightly on the property limits of this water(Lie, 1984, 1986; Park, 1985, 1986), but it is usually characterized by 32.0‰< S <32.5‰ and T <10°C. This water corresponds to most of group 2 and the neighboring part of group 3 in Fig. 2. Kondo (1985) called this the Yellow Sea Central Cold Water. It is surprising that there is no other water mass names cited frequently except for this water mass.

Water masses usually refer to persistent T-S characteristics. Accordingly, group 4 which appears both in January and July on the East China Sea shelf may deserve a name. We suggest that group 4 water be called the East China Sea Water. Most of this water can be characterized by 12°C< T <16.5°C and 33.5‰< S <34.6‰ as shown in Fig. 2. In conventional water mass analysis, we might consider group 2, 3 and 4 as a single water mass in Fig. 2, since their T-S characteristics are linear approximately. However, only group 4 water is found in the East China Sea and the new name would be useful for identification at least. It is also important to note that this water may characterize the property of water entering the East (Japan) Sea though the Korea Strait(Lim, 1971; Nagata, 1981).

Water groups which were observed either in January (group 1 in Fig. 3) or in July (groups 7, 8, 9, 10 and 11 in Fig. 4a) only should be considered separately. The change of groups occur only at surface associated with the extreme change of temperature over 15°C. It is difficult to explain the change in water properties throughout the year in terms of currents or circulation. A study of conservative properties other than temperature and salinity, such as oxygen and hydrogen isotopic composition of waters, may resolve this difficulty. As physico-chemical tracers are included as new parameters, cluster analysis such as that described here may become a truly powerful method to analyze water masses and determine the spatial distribution of water properties.

Conventional water mass analysis utilizes T-S curves with a series of figures showing horizontal and vertical distributions of temperature and salinity. Although the T-S diagram and contoured figures are useful in their own sense, integration of these figures into a single picture associated with the T-S diagram is not an easy task when the T-S plot is

266

complex. Cluster analysis presented here can be viewed as a statistical way of extracting a single combined parameter from any number of parameters to be studied, simplifying the results as demonstrated in Fig. 5, 6 and 7.

Acknowledgment

This work has been supported in part by the Korean Science and Engineering Foundation (KK, KRK, TSR, HKR; 1988-1990). The supports for R. Limeburner and R.C. Beardsley came from the U.S. National Science Foundation under grants OCE85-01366 and OCE86-10937. We are thankful to J.S. Cho for preparation of this manuscript.

6 REFERENCES

Asaoka, O. and Moriyasu, S., 1966. On the circulation in the East China Sea and the Yellow Sea in winter(Preliminary report). Oceanogr. Mag., 18: 73-81.
Beardsley, R. C., Limeburner, H. Y. and Cannon, G. A., 1985. Discharge of the Changjiang(Yangtze River) into the East China Sea. Continental Shelf Res.,4: 57-76.
Choi, B. H., 1980. A tidal model of the Yellow Sea and the East China Sea. KORIDI Report 80-02, 72pp.
Clifford, H.T. and Stephenson, W., 1973. An introduction to numerical classification. Academic Press, Inc., New York, 229 p.
Gong, Y., 1971. A study on the South Korea coastal front. J. Oceanol. Soc. Korea, 6: 25-36.
Kondo, M., 1985. Oceanographic investigation of fishing grounds in the east china sea and the Yellow Sea - I (in Japanese). Bull. Seikai Reg. Fish. Res. Lab. No. 62. pp. 19-62.
Korean Fisheries Research and Development Agency, 1987. Annual Report of Oceanographic Observations. Vol. 35, 536 p.
Li, F., Su, Y. and Yu, Z., 1984. Application of cluster analysis method to modified water-masses in the shallow sea. Acta Oceano. Sinica, 3: 451-461.
Lie, H. J., 1984. A note on water masses and general circulation of the Yellow Sea(Hwanghae). J. Oceanol. Soc. Korea, 19: 229-242.
Lie, H. J., 1986. Summertime Hwanghae features in the southeastern Hwanghae. Prog. Oceanog., 17: 229-242.
Lim, D. B., 1971. On the origin of the Tsushima Current Water. J. Oceanol. Soc. Korea, 6:85-91.
Mao, H. and Guan, B., 1981. A note on circulation of the East China Sea. Proceedings of the Japan-China Ocean Study Symposium. pp. 1-24.
Nagasaki Marine Observatory, 1986a. Oceanographic prompt report of the Nagasaki Marine Observatory. No. 120, 24 p.
Nagasaki Marine Observatory, 1986b. Oceanographic prompt report of the Nagasaki Marine Observatory. No. 122, 35 p.
Nagata, Y., 1981. Oceanic conditions in the East China Sea. Proceedings of the Japan China Ocean Study Symposium. pp. 25-41.
Nakao, T., 1977. Oceanic variablity in relation to fisheries in the East China Sea and the Yellow Sea. J. Fac. Mar. Sci. Technol. Tokai University, Spec No., pp 199-367.
Nitani, H., 1972. Beginning of the Kuroshio. In: H. Stommel and K. Yoshida (Editor), Kuroshio-its physical aspects. University of Tokyo Press, pp. 129-164.
Norusis, M. J., 1986. Advanced statistics SPSS/PC+, SPSS Inc., pp. B-71 - B-89.

Park, Y. H., 1985. Some important summer oceanographic phenomena in the East China Sea. J. Oceanol. Soc. Korea, 20: 12-21.

Park, Y. H., 1986. Water characteristics and movements of the Yellow Sea Warm Current in summer. Pro. Oceanog., 17: 243-254.

Qiu, D., Zhou, S. and Li, C., 1985. Application of cluster analysis method in determing water-masses of the Huanghai Sea. Acta Oceano. Sinica, 4: 337-348

Sneath, P.H.A. and Sokal, R.R., 1973. Numerical taxonomy. W.H. Freeman and Co., San Francisco, 573 p.

Uda, M., 1934. The results of simultaneous oceanographical investigations in the Japan Sea and its adjacent waters in May and June, 1932(in Japanese). J. Imp. Fisher. Exp. St., 5: 57-190.

SYNOPTIC BAND WINTERTIME HEAT EXCHANGES IN THE YELLOW SEA

Y. HSUEH and JAMES H. TINSMAN, III*
Department of Oceanography, Florida State University, Tallahassee, FL 32306

ABSTRACT
Advection of heat into the Yellow Sea along the Yellow Sea trough is inferred from the imbalance at three current meter mooring locations between the independently calculated heat loss to the atmosphere and the amount of heat released from the drop in ocean temperature. The moorings were employed during January - April 1986 along the Yellow Sea trough in an experiment to study the dynamics of the return flow to the wind-driven currents during the cold front passage. Significant correlation in the synoptic band found betweent the northward trough flow and heat budget imbalance suggests that the return flow also advects heat into the Yellow Sea. The source of the heat is probably the Kuroshio water west of Kyushu.

1. INTRODUCTION

The Yellow Sea is a shallow embayment enclosed on three sides by land and open to the East China Sea to the south. It is bordered by China to the west and north and by the Korean Peninsula to the east (see Figure 1). The topography of the Yellow Sea is marked by a north-south oriented trough that is nearer to Korea than to China. The trough is the landward extension of a canyon formation that begins in deep waters west of Kyushu. It is well-known that the Kuroshio reaches these waters before it exits the East China Sea through the Tokara Strait (Nakao, 1977). The canyon thus provides the Yellow Sea with an access to the heat and salt carried by the Kuroshio. In this paper an effort is described that attempts to establish and document the transport of Kuroshio water, and thus Kuroshio heat and salt, along this canyon feature by examing the flow of heat in the Yellow Sea in winter.

The wintertime Yellow Sea is frequented by strong north wind pulses of one to several day duration. As a consequence of the forcing by these surface wind pulses, synoptic scale set-ups in coastal sea level are excited (Hsueh and Romea, 1983). The coastal sea-level set-up translates into a south-to-north pressure gradient force which has been shown to drive an upwind return flow in the trough (Hsueh et al., 1986). It is this return flow that is potentially capable of advecting Kuroshio heat from the south in winter. An opportunity to assess the significance of this heat advection presents itself when a current meter mooring program was implemented to document the existence of the return flow and its dynamics. Measurements of current and temperature at mid-depth and bottom were made at four locations along the trough, as well as at the bottom at two locations on the Korean shelf (see Fig. 1). In addition, bottom pressures were

* Now at the State of Florida Department of Health and Rehabilitative Services, Tallahassee, Florida

270

recorded at the two northernmost trough moorings. These measurements confirm the existence of a pressure gradient forced upwind flow in the trough in the synoptic frequency band and suggest a moderation in the rate of cooling in the trough region (Hsueh, 1988). The moderation in cooling necessarily imples a convergence in heat flow from the south. The purpose of the present paper is to establish the connection between the implied convergence of heat and the observed northward trough flow. In particular, for the argument to work, a statistically significant correlation must exist between (a) the northward flow bursts in the trough and (b) the peaks

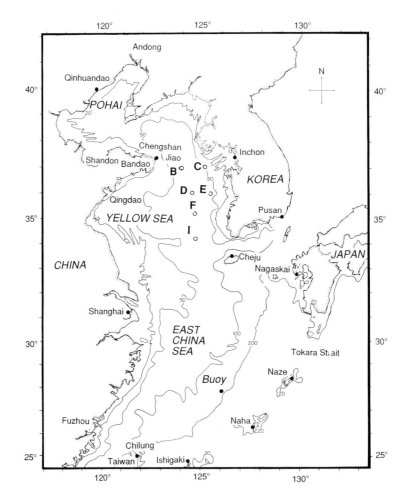

Figure 1. Map of the Yellow Sea and East China Sea region. Depths are in meters and those beyond 200 meters are not shown. Current meter moorings for the January - April 1986 experiment are marked by open circles with single letters. Six-hourly surface air temperature, dew point, and fractional cloud cover are obtained from weather maps for the fifteen named coastal locations.

in the amount by which the heat release from the drop in local water temperature falls short of the heat lost to the atmosphere.

2. DATA

In order to make estimates of this discrpancy in the heat budget, it is obviously necessary to construct the field of heat fluxes across the sea surface. Bulk aerodynamic formulae are the principal tools for calculating sensible and latent heat fluxes. Geostrophic winds for the period of the current meter mooring experiment (January - April, 1986) are calculated from six-hourly surface pressure charts for a 39x54 grid that covers the entire East China Sea and Yellow Sea region (see Fig. 2). These winds are a reasonable approximation to winds at the surface (Hsueh and Tinsman, 1987) and are used in the sensible and latent heat fluxes calculation. In addition, six-hourly values of surface air temperature, dew point, and cloud cover at a buoy and fourteen coastal stations (named in Figure 1) are taken from weather maps and interpolated over the same

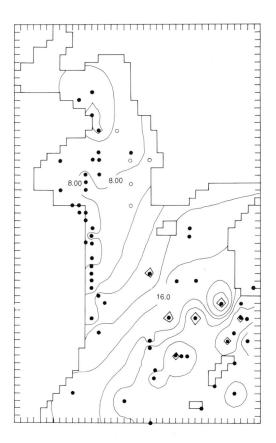

Figure 2. Distribution of Sea Surface temperature Reports for January 15, 1986. Contours show the temperature field obtained from interpolation. Numbers are in °C and contour interval is 2°C.

grid. The interpolation scheme weighs the influence of each observation station by the inverse of the square of distance to a given grid point. Mathematically, this scheme is given by

$$X_{i,j} = \left(\sum_{k=1}^{15} X_k / r_{i,j,k}^2 \right) \Big/ \left(\sum_{k=1}^{15} r_{i,j,k}^{-2} \right) , \tag{1}$$

where $X_{i,j}$ is the atmospheric variable to be interpolated at the grid point (i,j) and

$$r_{i,j,k} = \left([x(k) - i\Delta s]^2 + [y(k) - j\Delta s]^2 \right)^{\frac{1}{2}} .$$

Here, $r_{i,j,k}$ is the distance from the grid point (i,j) to the kth station and Δs is the size of the grid.

In a simlar manner, daily-compiled sea surface temperatures (SST's) from ship reports are also interpolated. A typical distribution of ship reported SST's over the 39x54 grid is shown in Figure 2. The grid-point values are then interpolated linearly in time to generate data at six-hourly intervals.

The six-hourly atmospheric variables and SST's are incorporated in the following formulae to calculate the sensible and latent heat fluxes (Q_s and Q_e, respectively) in units of Cal cm^{-2} day^{-1} (see Henderschott and Rizzoli, 1976):

$$Q_s = 5.27 \, (SST - T_a) w_s ,$$

$$Q_e = 1.44 \times 10^{-2} L \, (e_w - e_a) w_s ,$$

where T_a represents the surface air temperature in °C, w_s the wind speed in m sec^{-1}, L the latent heat of evaporation, e_w the saturation vapor pressure at the sea surface, and e_a the vapor pressure of the air calcualted from the dew point.

To complete the surface heat flux field, radiative components must be added. These are calculated according to the following formulae (see Talley, 1984):

$$Q_I = 0.9 Q_0 (1 - 0.62 n_c + 0.0019 A_n) ,$$

$$Q_b = \varepsilon \sigma T_s^4 (0.39 - 0.05 \sqrt{e_a})(1 - 0.8 n_c) ,$$

where A_n is the noon solar altitude, Q_0 incident radiation at the top of the atmosphere, ε emissivity of water, σ the Stefan-Boltzmann constant, T_s the SST in °K, and n_c the fractional cloud cover.

The incoming radiative flux Q_I is calculated only at noon and reduced linearly to zero at 0600 and 1800 hours. The back radiation Q_b is calculated at six-hourly intervals. The total flux of heat from the atmosphere to the ocean is thus given by

$$Q_T = Q_I - Q_b - Q_s - Q_e .$$

Figure 3 shows a typical distribution of the calculated Q_T.

3. BUDGET

A similar data base for the change in heat content of the Yellow Sea, however, is difficult to establish as the six-hourly change in heat content is too sensitive to temperature measurements to be estimated on the basis of ship-reported SST's. The only ocean temperature measurements that are of high enough quality for the heat content calculation are basically limited to those from the moored current meters at B, D, F, and I (with a resolution of 0.0002 °C and an absolute accuracy of 0.1 °C) (see Fig. 1). Since the Yellow Sea is very well mixed in winter, mid-depth temperatures from the moorings are used and the ocean heat content is thus calculated only for these mooring locations.

At a given point in the Yellow Sea in winter, the heat balance is described by the

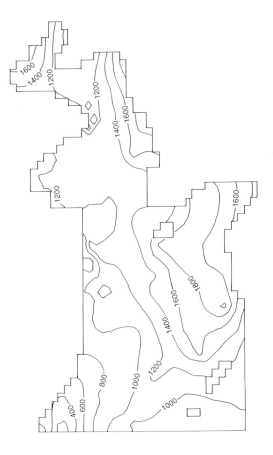

Figure 3. Distribution of Calculated Surface Heat Flux for 00 hours January 22, 1986. Numbers in Cal cm^{-2} day^{-1} and contour interval is 200 Cal cm^{-2} day^{-1}.

vertically integrated heat equation. Finite-differenced in time, this equation yields

$$\rho c_p H \Delta T = Q_T \Delta t - \rho c_p H \Delta t \nabla \cdot (\vec{v}T) \; , \tag{2}$$

where H is the local water depth, ΔT represents the change in water temperature in the time interval Δt, equal to 6 hours in the present analysis, and ∇ is the horizontal gradient operator.

With the data available, the convergence of heat at B, D, F and I, the last term on the righ-hand of (2), can be inferred from the imbalance between the heat content change (the left-hand side term) and the surface heat loss (the first term on the right). It thus makes sense to define a six-hourly discrepancy (in Cal cm^{-2}) such that

$$\text{Discrepancy} = \rho c_p H \Delta T - Q_T \Delta t \; . \tag{3}$$

The significance of (3) is that during an episode of strong northward return flow in the trough following a cold front passage, large positive discrepancy should appear as the advection of heat from the south makes up for the shortfall in heat supply to the atmosphere from the drop in water temperature ($\Delta T < 0$, $Q_T < 0$).

Time series analyses are carried out among the discrepancy defined in (3), the observed mid-depth northward velocities, and the observed mid-depth temperature at B, D, F, and I. Figure 4(a) shows the relevant time series at B. Except for the one event at the beginning, peaks

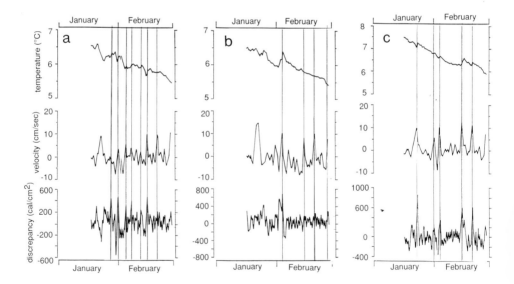

Figure 4. Plots of the temperature, northward component of the velocity and the discrepancy for mooring locations B, D, and F [(a), (b), and (c) respectively].

of discrepancy appear generally in phase with those of the northward velocity. The magnitude of the discrepancy during these events of northward flow bursts appears to be significantly above the level of noise in the data as indicated by the end portion of the time series which falls into the early spring period with weak winds and thus, weak trough flow.

Figures 4(b) and 4(c) show, respectively, similar plots for D and F. There is the expected phase locking among the time series. (Data from I cover too short a period to be useful for time series analysis.)

Figure 5 shows the coherency squared and phase between the discrepancy and northward velocity at B, D, and F. Significant correlation is found in the synoptic band and the phase is close to zero, a fact consistent with the advection scenario.

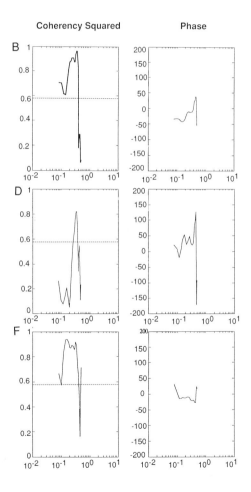

Figure 5. The coherency squared and phase between the northward component of the velocity and the discrepancy at mooring locations B, D, and F. The dashed line marks the 95% confidence level. A negative phase indicates that the northward component of the velocity leads the discrepancy.

4. CONCLUSIONS

According to data set from an experiment not originally designed to examine the heat transfer in the Yellow Sea in winter, it appears that significant amount of heat is advected into the Yellow Sea from the south by the episodic return flow along the Yellow Sea trough. This heat transfer process has a time scale of about 3 days and is probably confined in the trough in waters deeper than 50 m where the barotropic pressure gradient generated by the coastal sea-level set-up overcomes the surface wind stress and accelerates the flow to the north. The heat that is transferred to the north probably orignates in the Kuroshio water west of Kyushu.

5. ACKNOWLEDGEMENTS

Dr. A. Keneko kindly made available the ship reported SST data compiled by the Nagasaki Marine Observatory (NMO). Mr. David Legler supplied the authors with similar data from the Japan Meteorological Agency which complemented the NMO data. The research is supported by a grant from the Office of Naval Research, N00014-90-J-1820 and by the Institute for Naval Oceanography through UCAR contract N00014-86-CA001. The current meter mooring program from which temperature and velocity data were obtained is supported by an NSF grant, OCE-8500181.

6. REFERENCES

Hendershott, M.C., and P. Rizzoli, 1976: The winter ciruclation of the Adriatic Sea. Deep-Sea Res., 22, 353-370.
Hsueh, Y., and R.D. Romea, 1983: Wintertime winds and coastal sea-leve fluctuations in the northeast China Sea. Part I. Observations. J. Phys. Oceanogr., 13, 2091-2106.
Hsueh, Y., R.D. Romea, and P.W. deWitt, 1986: Wintertime winds and coastal sea-level fluctuations in the northeast China Sea. Part II. Numerical model. J. Phys. Oceanogr., 16, 241-261.
Hsueh, Y., and James H. Tinsman, III, 1987: A comparison between geostrophic and observed winds at a Japan Meteorological Agency buoy in the East Cina Sea. J. Oceanogr. Soc. Jpn., 43, 251-257.
Talley, L.D., 1984: Meridional heat transport in the Pacific Ocan. J. Phys. Oceanogr., 14, 231-241.

NUMERICAL PREDICTION OF THE VERTICAL THERMAL STRUCTURE IN THE BOHAI AND HUANGHAI SEAS--TWO-DIMENSIONAL NUMERICAL PREDICTION MODEL

Wang Zongshan, Xu Bochang, Zou Emei, Gong Bin, and Li Fanhua

First Institute of Oceanography, SOA, Qingdao, China

ABSTRACT

By using observed sea temperature data, we calculated the non−dimensional depth η and its rele vant non−dimensional temperature θ_T to construct a similarity function $\theta_T = f(\eta)$. Based on these, the thickness of the upper homogeneous layer $h = h(x,y,t)$, surface temperature $T_S = T_S(x,y,t)$, temperature in the thermocline $T_Z = T_Z(x,y,z,t)$, bottom temperature $T_H = T_H(x,y,t)$, the current velocity $u = u(x,y,t)$, $v = v(x,y,t)$ and the surface topography $\zeta = \zeta(x,y,t)$ were assumed to establish a numerical prediction model of the vertical thermal structure.

The solution of this two dimensional model was obtained by using "ADI" and "HN" methods. The results of a trial prediction are satisfactory.

1 INTRODUCTION

Until now, the studies of prediction on the vertical temperature structure have been focused on the characteristics of the upper ocean homogeneous layer (Kitaigorodskii, 1977; Nesterov, 1978; Resnanskii et al., 1980, 1983, 1986; Kraus et al., 1975, 1977). Few papers are concerned with the prediction of the vertical temperature structure. Early mathematical models are divided into three categories: (1). Models for solving the closure equations including the momentum equation, continuity equation, equation of state, thermal equation and salinity equation (Kitaigorodskii, 1977; Kraus, 1977; Kalazkii, 1978, 1980; Murakami et al., 1985; Omsteds et al. , 1983); (2). Forecasting models for describing the characteristics of the upper thermal structure according to the universal function derived by similarity theory (Wang et al., 1986); (3). Studies of the methods for calculating and simulating the vertical temperature structure in terms of the similarity function of the vertical temperature distribution (Xu et al., 1983, 1984) and thermal equation (Malkki et al., 1985). Category (1) is too difficult to put into effect because so many hydrographical and meterological factors have to be known for prediction. The coefficients in the equations are artificially taken as constants in most cases, which often produces errors in forecasting. As for category (2), though the model itself is simple, it describes only the mean characteristics of the vertical temperature structure. Category (3) need not some relevant coefficients in the equations, and the vertical temperature structure can be obtained somewhat objectively. Based on the category (3) and the available data of wind and air temperature field, a one−dimensional numerical prediction model for the vertical temperature structure in the Bohai and Huanghai Seas was developed (Wang et al., 1990). In the model, the effects of advection and lateral mixing are neglected, which holds good for the Bohai and Huanghai Seas. We could also verify it with the satisfactory trial prediction of the above model. However, with a strong cyclone passing, the simulated current velocities can reach 130 and 300 cm / s in some areas of the Bohai and Huanghai Seas, respectively (Zhang et al., 1983, 1988). Such huge wind−induced current will induce a large change in the three dimensional temperature structure. When a cyclone with wind speed 25−30m / s passes over the deep sea at mid−latitudes, the upper layer temperature in a large area decreases by more than 2℃ and the upper homogeneous layer thickness increases by more than 6 m. The depth influenced by wind could reach 100m (Japan Meteorol. Agency, 1985; Halpern, 1974; Nesterov,

1986). The upper homogeneous layer thickness in the Bohai and Huanghai Seas can increase by 5–8 m when heavy wind blows continuously over 24 hours (Ocean group of STC, PRC, 1964). The large change of the thermal structure makes effective operation of acoustic instrument and fishing activities difficult. Therefore, the effect of the wind–induced current on the thermal structure redistribution needs to be understood. In this context, we developed a numerical prediction model for the vertical temperature structure as associated with wind–induced current and lateral mixing.

2 PHYSICAL MODEL
2.1 SIMILARITY FUNCTION OF THE VERTICAL TEMPERATURE PROFILE

The vertical temperature distribution in the Bohai and Huanghai Seas in warm seasons is schematically shown in Fig. 1, which is divided into three layers: (1) the upper homogeneous layer, whose

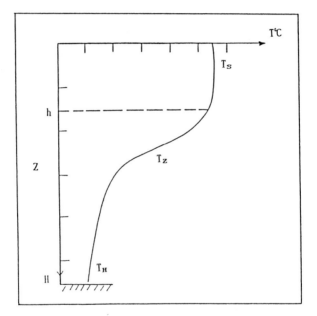

Fig. 1. Schematic represemtation of the vertical temperature profile.

thickness is h; (2) the thermocline; (3) the deep layer. Therefore, in the two–dimensional temperature structure, the non–dimensional depth η and its revelant non–dimensional temperature θ_T is written as:

$$\theta_T = \frac{T_s - T_z}{T_s - T_H}, \quad \eta = \frac{Z - h}{H - h} \tag{1}$$

where $T_s(x,y,t)$ denotes the surface temperature (or the temperature in the upper homogeneous layer); $T_z(x,y,z,t)$, the temperature which varies with depth z; $T_H(x,y,t)$, the temperature in the bottom layer; H, the water depth. Dimensionless variable θ_T is a function of η (Kitaigorodskii et al., 1970; Malkki et al., 1985; Xu et al., 1983, 1984). Based on the observed data of vertical temperature profile and meeting to boundary condictions:

$$\theta_T = \begin{cases} = 0 \\ = 1 \end{cases} \quad when \quad \eta \begin{cases} = 0 \\ = 1 \end{cases},$$

a similarity function has been constructed by using the least square method(solid line in Fig. 2):

$$\theta_T = a_1\eta + a_2\eta^2 + a_3\eta^3 + a_4\eta^4 \tag{2}$$

where a_1, a_2, a_3, a_4 are empirical constants determined by 5053 historical temperature profile data

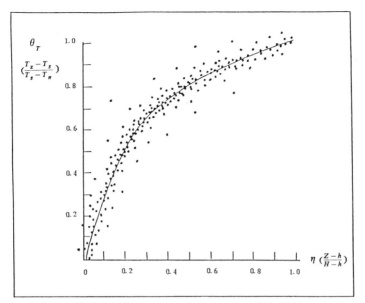

Fig. 2. The similarity profile of sea temperature in the Bohai and Huanghai Seas.

obtained by Nansen casts in the Bohai and Huanghai Seas during May–October from 1958 to 1982. We denote K_T and $\bar{K}_T$ by:

$$K_T = \int_0^1 \theta_T(\eta)d\eta, \quad \bar{K}_T = \int_0^{\prime}\int_0^{\eta}\theta_T(\eta)d\eta d\eta' \tag{3}$$

The average value of K_T and $\bar{K}_T$ for the Bohai and Huanghai Seas are 0.75 and 0.30, respectively.

According to the above formulae, the vertical temperature structure is expressed by following equations:

$$\begin{cases} T_s = T_s(x,y,t), & 0 \leqslant z \leqslant h \\ T_z = T_s(x,y,t) - \theta_T(\eta)[T_s(x,y,t) - T_H(x,y,t)], & h \leqslant z \leqslant H \end{cases} \tag{4}$$

2.2 GOVERNING EQUATIONS

To describe the two–dimensional current field, we use the following equations (Zhang et al., 1983, 1988):

$$\frac{\partial u}{\partial t} + u\frac{\partial u}{\partial x} + v\frac{\partial u}{\partial y} - fv + g\frac{\partial \zeta}{\partial x} - K_y\frac{\partial^2 u}{\partial x^2} = -\frac{1}{\rho_w}\frac{\partial P}{\partial x} + \frac{1}{\rho_w}\frac{\tau_{ax} - \tau_{bx}}{H} \tag{5}$$

$$\frac{\partial v}{\partial t} + u\frac{\partial v}{\partial x} + v\frac{\partial v}{\partial y} + fu + g\frac{\partial \zeta}{\partial y} - K_y\frac{\partial^2 v}{\partial y^2} = -\frac{1}{\rho_w}\frac{\partial P}{\partial y} + \frac{1}{\rho_w}\frac{\tau_{ay} - \tau_{by}}{H} \tag{6}$$

$$\frac{\partial \zeta}{\partial t} + \frac{\partial(HU)}{\partial x} + \frac{\partial(HV)}{\partial y} = 0 \tag{7}$$

in which $u = u(x,y,t)$, $v = v(x,y,t)$ are the velocity components, $\zeta = \zeta(x,y,t)$ the sea level, K_y the turbulent viscosity coefficient, f the Coriolis parameter, ρ_w the water density (taken as 1), H the water depth, P the pressure, g the gravitational acceleration, τ_a and τ_b the surface and bottom frictional stress given by:

$$\tau_{a_{x,y}} = \rho_a C_D |\overrightarrow{W}| W_{x,y} \tag{8}$$

and

$$\tau_{b_{x,y}} = \rho_w g K_b^{-2} |\overrightarrow{V}| V_{x,y} \tag{9}$$

with ρ_a (air density), C_D (the drag coefficient), $K_b = 1 / M \ \overline{H}^{1/6}$ (friction coefficient at the bottom), M (Manning coefficient), $\overline{H}$(mean depth in each mesh). In Eqs. (5) and (6), the first term of right side can be ignored in comparision with the second term.

Neglecting the relevant processes of the molecular diffusion, the heat conduction equation is:

$$\frac{\partial T}{\partial t} + u\frac{\partial T}{\partial x} + v\frac{\partial T}{\partial y} + w\frac{\partial T}{\partial z} - A_T \nabla^2 T = -\frac{\partial(\overline{W'T'} + R)}{\partial z} \tag{10}$$

where T is the temperature, A_T the horizontal turbulent heat conduction coefficient, $\overline{W'T'}$ the average vertical turbulent heat flux, R the penetrative component of the solar radiation divided by $C_p\rho_w$ (C_p is specific heat of the sea water).

Since u and v are independent of z, $w = 0$ at $z = 0$, and the solar energy is assumed to be entirely absorbed by the sea surface, integration of Eq. (10) over the upper homogeneous layer leads to:

$$\frac{\partial T_s}{\partial t} + u\frac{\partial T_s}{\partial x} + v\frac{\partial T_s}{\partial y} - A_T \nabla^2 T_s = \frac{Q_s - Q_H}{h} \tag{11}$$

where $Q_s = Q_L / (C_p\rho_w)$ is the surface heat flux, $Q_h = \overline{W'T'}|h$ the mean turbulent heat flux at the lower boundary of the upper homogeneous layer, and Q_L the heat budget at the sea surface.

In order to derive an equation for h and T_H, we substitute Eq. (4) into Eq. (10) and then integrate it with respect to z from h to H to get:

$$(1 - \theta_T)[\frac{\partial T_s}{\partial t} + u\frac{\partial T_s}{\partial x} + v\frac{\partial T_s}{\partial y} - A_T \nabla^2 T_s] + \theta_T[\frac{\partial T_H}{\partial t} + u\frac{\partial T_H}{\partial x} + v\frac{\partial T_H}{\partial y} - A_T \nabla^2 T_H] - (\eta$$

$$-1)\frac{\partial \theta_T}{\partial \eta}\{[\frac{T_s - T_H}{H - h}(\frac{\partial h}{\partial t} + u\frac{\partial h}{\partial x} + v\frac{\partial h}{\partial y} - A_T \nabla^2 h)] - \frac{2A_T}{H - h}[(\frac{\partial T_s}{\partial x} - \frac{\partial T_H}{\partial x})\frac{\partial h}{\partial x} + (\frac{\partial T_s}{\partial y}$$

$$- \frac{\partial T_H}{\partial y})\frac{\partial h}{\partial y}] - 2A_T \frac{T_s - T_H}{(H - h)^2}[(\frac{\partial h}{\partial x})^2 + (\frac{\partial h}{\partial y})^2]\} + A_T(\eta - 1)^2\frac{T_s - T_H}{(H - h)^2}\frac{\partial^2 \theta_T}{\partial \eta^2}[(\frac{\partial h}{\partial x})^2 + (\frac{\partial h}{\partial y})^2]$$

$$+ \frac{1}{H - h}\frac{\partial Q}{\partial \eta} = 0 \tag{12}$$

Integration of Eq. (12) with respect to ζ from 0 to 1 with the boundary condition $Q_H = 0$ gives:

$$(1 - K_T)[\frac{\partial T_s}{\partial t} + u\frac{\partial T_s}{\partial x} + v\frac{\partial T_s}{\partial y} - A_T \nabla^2 T_s] + K_T[\frac{\partial T_H}{\partial t} + u\frac{\partial T_H}{\partial x} + v\frac{\partial T_H}{\partial y} - A_T \nabla^2 T_H]$$

$$+ K_T\{\frac{T_s - T_H}{H - h}[\frac{\partial h}{\partial t} + u\frac{\partial h}{\partial x} + v\frac{\partial h}{\partial y} - A_T \nabla^2 h] - \frac{2A_T}{H - h}[(\frac{\partial T_s}{\partial x} - \frac{\partial T_H}{\partial x})\frac{\partial h}{\partial x} + (\frac{\partial T_s}{\partial y} - \frac{\partial T_H}{\partial y})\frac{\partial h}{\partial y}]$$

$$- 2A_T\frac{T_s - T_H}{(H - h)^2}[(\frac{\partial h}{\partial x})^2 + (\frac{\partial h}{\partial y})^2]\} + (2K_T - a_1)A_T\frac{T_s - T_H}{(H - h)^2}[(\frac{\partial h}{\partial x})^2 + (\frac{\partial h}{\partial y})^2] - \frac{Q_h}{H - h} = 0 \tag{13}$$

Double integration of Eq. (12) with respect to ζ, first from 0 to η, then from 0 to 1, yields:

$$(\frac{1}{2} - \overline{K_T})[\frac{\partial T_s}{\partial t} + u\frac{\partial T_s}{\partial x} + v\frac{\partial T_s}{\partial y} - A_T\nabla^2 T_s] + \overline{K_T}[\frac{\partial T_H}{\partial t} + u\frac{\partial T_H}{\partial x} + v\frac{\partial T_H}{\partial y} - A_T\nabla^2 T_H]$$

$$+ 2\overline{K_T}\{\frac{T_s - T_H}{H - h}(\frac{\partial h}{\partial t} + u\frac{\partial h}{\partial x} + v\frac{\partial h}{\partial y} - A_T\nabla^2 h)] - \frac{2A_T}{H - h}[(\frac{\partial T_s}{\partial x} - \frac{\partial T_H}{\partial x})\frac{\partial h}{\partial x}$$

$$+ (\frac{\partial T_s}{\partial y} - \frac{\partial T_H}{\partial y})\frac{\partial h}{\partial y}] - 2A_T\frac{T_s - T_H}{(H - h)^2}[(\frac{\partial h}{\partial x})^2 + (\frac{\partial h}{\partial y})^2]\} + [2\overline{K_T} - a_1$$

$$- 4\int_0^1(\eta - 1)\theta_T d\eta] \cdot A_T\frac{T_s - T_H}{(H - h)^2}[(\frac{\partial h}{\partial x})^2 + (\frac{\partial h}{\partial y})^2] - \frac{vQ_h}{H - h} = 0 \tag{14}$$

where $v = (Q_h - \int_0^1 Q d\eta)/Q_h$, $Q = (\overline{W'T'} + R)$. Simultanous equations (Eqs. (13) and (14)) together with Eq. (11) give:

$$\frac{\partial T_H}{\partial t} + u\frac{\partial T_H}{\partial x} + v\frac{\partial T_H}{\partial y} - A_T\nabla^2 T_H + C_3 A_T\frac{T_s - T_H}{(H - h)^2}[(\frac{\partial h}{\partial x})^2 + (\frac{\partial h}{\partial y})^2]$$

$$= C_1\frac{Q_h}{H - h} - C_2\frac{Q_s - Q_h}{h} \tag{15}$$

and

$$\frac{\partial h}{\partial t} + u\frac{\partial h}{\partial x} + v\frac{\partial h}{\partial y} - A_T\nabla^2 h - \frac{2A_T}{T_s - T_H}[(\frac{\partial T_s}{\partial x} - \frac{\partial T_H}{\partial x})\frac{\partial h}{\partial x} + (\frac{\partial T_s}{\partial y} - \frac{\partial T_H}{\partial y})\frac{\partial h}{\partial y}] - \frac{2A_T}{(H - h)}[(\frac{\partial h}{\partial x})^2$$

$$+ (\frac{\partial h}{\partial y})^2] - C_6 A_T\frac{1}{H - h}[(\frac{\partial h}{\partial x})^2 + (\frac{\partial h}{\partial y})^2] =$$

$$C_4\frac{Q_h}{T_s - T_H} - C_5\frac{(H - h)(Q_s - Q_h)}{h(T_s - T_H)} \tag{16}$$

in which

$$\begin{cases} C_1 = (2\overline{K_T} - K_T v)/(K_T\overline{K_T}) \\ C_2 = (2\overline{K_T} - K_T\overline{K_T} - \frac{1}{2}K_T)/(K_T\overline{K_T}) \\ C_3 = [2a_1\overline{K_T} + 2K_T\overline{K_T} - a_1 K_T + 4K_T\int_0^1(\eta - 1)\theta_T d\eta]/(K_T\overline{K_T}) \\ C_4 = (K_T v - \overline{K_T})/(K_T\overline{K_T}) \\ C_5 = (\frac{1}{2}K_T - \overline{K_T})/(K_T\overline{K_T}) \\ C_6 = [a_1(K_T - \overline{K_T}) - 4K_T\int_0^1(\eta - 1)\theta_T d\eta]/(K_T\overline{K_T}) \end{cases} \tag{17}$$

According to a processing method presented by Wang et al. (1990), Q_S, Q_h and γ in Eqs. (11), (15) and (16) are defined as follows. Q_S is written by a formula (Wang, 1983):

$$Q_S = [e_1 + e_2(T_S - T_a)]/(C_p\rho_w) \tag{18}$$

where T_a is the air temperature, and e_1, $e_2$empirical constants. It is difficult to get accurate value of the heat flux (Q_h) through the lower boundary of the upper homogeneous layer, because $Q_h$is related to the entrainment of the lower cold water to the upper homogeneous layer rather than the simple change in h. Based on turbulent kinetic energy equation, Resnanskii (1980, 1983, 1986) showed that the vertical salinity variation in the mid– latitude areas has almost no effect on the buoyance flux, and entirely depends on the vertical turbulent heat flux $\overline{W'T'}$. He also neglected the horizontal variations in the mean velocity, turbulent kinetic energy and their flux. Integration of the turbulent kinetic energy equation from 0 to h and use of Eqs. (10) and (11) gives:

$$Q_h = -Q_s - \frac{2F}{\beta h} \tag{19}$$

Here $\beta = g\alpha_T$is the buoyance coefficient, $\alpha_T$the heat expansion coefficient of sea water, F the conversion rate of turbulent kinetic energy in the upper homogeneous layer to potential energy, which is subject to the following conditions: the predicted h is close to the measured h under the wind mixing, and$Q_h$and entrainment velocity decrease with increasing h. Parameter F is written as (Resnanskii, 1983):

$$F = \Lambda b_1 (u_* - b_2 h |f|)u_*^2 + \frac{1}{4}\beta h(Q_s - |Q_s|)[1 - (1 - \frac{h}{H})^m] \tag{20}$$

in which, b_1, b_2, m are empirical constants taken as 2, 5, and 2, respectively.$\Lambda = \Lambda(x)$ is the unit function of independent variable $x = (U_* - b_2 h|f|)$ with$\Lambda = 1$ when $x>0$ and $\Lambda = 0$ when $x < 0$. $U_* = (\tau_a / \rho_w)^{1/2}$the friction velocity of sea water. In order to determine v, Eqs. (13) and (14) are transformed into:

$$d[1 - K_T + K_T n + K_T B + (2K_T - a_1)\alpha] = 1 \tag{21}$$

$$d[\frac{1}{2} - \overline{K}_T + \overline{K}_T n + 2\overline{K}_T B + (2\overline{K}_T - a_1 - 4\int_0^1 (\eta - 1)\theta_T d\eta)\alpha] = v \tag{22}$$

in which

$$\begin{cases}
d = \frac{H - h}{Q_h} \cdot \frac{Q_s - Q_h}{h} \\[2mm]
n = [\frac{\partial T_H}{\partial t} + u\frac{\partial T_H}{\partial x} + v\frac{\partial T_H}{\partial y} - A_T \nabla^2 T_H]/\frac{dQ_h}{H - h} \\[2mm]
\alpha = \frac{A_T(T_s - T_H)}{(H - h)^2}[(\frac{\partial h}{\partial x})^2 + (\frac{\partial h}{\partial y})^2]/\frac{dQ_h}{H - h} \\[2mm]
B = \{\frac{T_s - T_H}{H - h}[\frac{\partial h}{\partial t} + u\frac{\partial h}{\partial x} + v\frac{\partial h}{\partial y} - A_T \nabla^2 h] - \frac{2A_T}{H - h}[(\frac{\partial T_s}{\partial x} - \frac{\partial T_H}{\partial x})\frac{\partial h}{\partial x} \\[2mm]
\quad + (\frac{\partial T_s}{\partial y} - \frac{\partial T_H}{\partial y})\frac{\partial h}{\partial y}] + \frac{A_T(T_s - T_H)}{(H - h)^2}[(\frac{\partial h}{\partial x})^2 + (\frac{\partial h}{\partial y})^2]\}/\frac{dQ_h}{H - h}
\end{cases} \tag{23}$$

Eqs. (21) and (22) give:

$$v = [\frac{1}{2} - \overline{K}_T(1 - n - 2B - 2\alpha) - (a_1 + 4\int_0^1 (\eta - 1)\theta_T d\eta)\alpha]/[1 - K_T(1 - n - B - 2\alpha) - a_1\alpha]. \tag{24}$$

By use of historical temperature profile data and meteorological data, and corresponding u, v,

Q_S and Q_h calculated from Eqs. (5)–(7), (18), and (19), γ can be determined with Eq. (24). The average value of γ is 0.06 in the warming period and 0.55 in the cooling period in the Bohai and Huanghai Seas.

Then the numerical prediction model is constructed by using Eqs. (5) to (7), (11), (15) and (16).

At the closed boundary, $V_n = 0$, $T_S = T_H$, $h = H$, $Q_n = Q_h = Q_H = 0$. At the open boundary from the Changjiang river mouth to Chejudo island, $\partial u / \partial X = 0$, $\partial V / \partial y = 0$. The sea level ζ at the closed boundary is specified by:

$$\zeta = \sum_{i=1}^{n} H(x,y)_i Cos[\sigma_i t - g(x,y)_i]$$ (25)

where $i = 1, 2, \ldots \ldots, n$ denotes the serial number of tidal constituent, H_i and g_i are the harmonic constants of ith tidal constituent, σ_i is angular velocity.

The initial conditions are given by:

$$u = v = 0, \qquad \zeta = 0, \qquad h = h_0, \qquad T_S = T_{S0}, \qquad T_H = T_{h0}, \qquad \text{when } t = t_0.$$

3 NUMERICAL MODEL

The parallel of 36 ° N is taken as x axis (positive eastward), and the meridian of 120 ° E as y axis (positive northward), the mean sea level as $z = 0$ (positive upward).

3.1 DIFFERENCE SCHEME

In order to get a finite difference analog of the prediction model in differential form, we chose Platzman's alternative grid, and use alternating direction implicit method (ADI) (Leendertse, 1967; Zhang et al., 1983, 1988). The features of this method are: (1) the calculation is simple because an implicit differential form is alternatively used in one direction when estimating the variables in x and y directions. (2) an implicit and an explicit differential form are alternatively used in x and y directions that make the calculation stable and quick convergent. In the meantime a hydrodynamic numerical method with temporal forward difference and special central difference (HN) is used to disctetize the Eqs. (11), (15) and (16) describing the vertical temperature structure.

We evaluate ζ, h, T_S, T_h at a grid point (i, j), u, w_x at a grid point (i+1 / 2, j), v, w_y at a grid point (i, j+1 / 2) and H at a grid point (i+1 / 2, j+1 / 2).

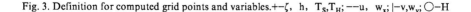

Fig. 3. Definition for computed grid points and variables.+–ζ, h, T_S,T_H; ––u, w_x; |–v,w_y; ○–H

3.2 DIFFERENCE FORM

Variables U^k, V^k, ζ^k are obtained from Eqs. (5) to (7) at $k\triangle t$ intervals. The appendix gives the implicit differential form for ζ and u, and explicit differential form for v from $k\triangle t$ to $(k+1/2)\triangle t$; the implicit differential form for ζ and v, and explicit differential form for u from $(k+1/2)\triangle t$ to $(k+1)\triangle t$; and the differential form of Eqs. (11), (15) and (16) from $k \triangle t$ to $(k+1)\triangle t$.

4 APPLICATION TO TIDAL PREDICTION

The wind speed w and air temperature T_a are known. The grid spacing is 20 km. The time step is 900 seconds. The basic parameters are specified as in Table 1.

TABLE 1

The relevant parameters in the model (cgs)

ρ_a	1.229×10^{-3}	C_D	$(0.8719+0.000704W_{10}^*) \times 10^{-3}$
ρ_w	1.023	A_T	10^6
C_p	0.938	K_γ	10^7
α	0.25×10^{-3}	K_b	$1/M\overline{H}^{1/6}$, $M = 0.016$—0.018
g	980		

* W_{10} means the wind speed at 10 m above the sea surface.

In the forecasting procedure, first of all, we calculate the current field from the given wind field, then put the current field and the given air-temperature field into the governing equations to predict h, T_S and T_H. Computation was made on IBM−4381 computer.

5 THE RESULTS AND DISCUSSION OF TRIAL PREDICTION

Because of the limited observations in this area, we only take the wind, air temperature and sea surface temperature data during FGGE as known variables to forecast the vertical temperature structure. The period of validity is about 4 days. The surface temperature (T_S), thickness of the upper homogeneous layer (h), bottom temperature (T_H), meridional distribution of temperature along 123.5 ° E and vertical temperature profiles on July 8, 1979 are presented in Figs. 4 to 8. These figures show that the sea surface temperature in the study area is 22~23℃ in most areas except that the SST in Haizhou Bay is higher than 25℃ and there is a cold eddy north of Cheng Shantou (Fig. 4). There are four areas where the thickness of the upper homogeneous layer is great (Fig. 5): the first one is located in the central Bohai Sea (greater than 5 m), the second one lies in the north of the northern Huanghai Sea Cold Water (greater than 15 m), the third one is at about 34.5 ° N, 123 ° E (greater than 10 m) and the last one in the Huanghai Trough area has the greatest thickness (20 m~25 m). The thickness of the upper homogeneous layer reaches 10 m off the coast of northern Jiangsu province due to strong mixing and is about 5 m in the other areas. The temperature profile from the trial prediction (Fig. 8a) clearly shows the upper homogeneous layer, thermocline and deep layer. In order to verify the reliability of the trial prediction, we roughly compare the calculated results with observed data obtained from a simultaneous hydrographic survey (Fig. 8b). Comparison shows that: (1) the temperature profile of trial prediction coincides with the observation north of 36 ° N, but there is a discrepancy between them in the deep layer south of 36 ° N, which might be due to a large meridional distance and asynchronous observational data between stations. (2) the thermocline obtained from the trial prediction is thicker than the observed one and the predicted temperature is vertically homogeneous in the thermocline, which may be due to the assumption that the current velocity is vertically homogeneous that induces the strong mixing between thermocline and its lower boundary. So three dimensional numerical prediction method is needed to put forward in the future.

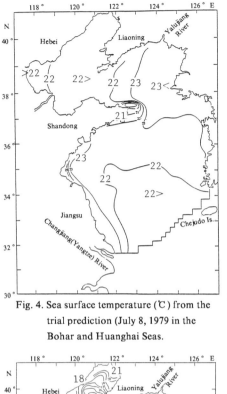

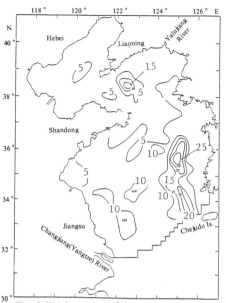

Fig. 4. Sea surface temperature (℃) from the
trial prediction (July 8, 1979 in the
Bohar and Huanghai Seas.

Fig. 5. Thickness (m) of the upper homogeneous
layer from the trial prediction (July 8, 1979)
in the Bohai and Huanghai Seas.

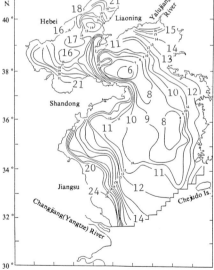

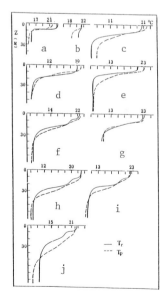

Fig. 6. Bottom temperature (℃, July 8, 1979) in
the Bohai and Huanghai Seas.

Fig. 7. The vertical temperature profile from
trial prediction(T_p) and observation (T_r)
at some grid points in the Bohai and
Huanghai Seas. a–j are the station
numbers shown as in Table 2.

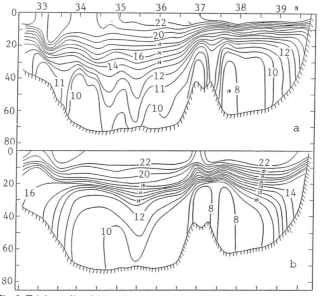

Fig. 8. Trial–predicted (a) and observed (b) meridional temperature (℃) at 123.5 ° E.

Table 2

Characteristics of predicted (T_p) and observed (T_r) at 10 grid points in the Bohai Hunghai Seas

station No	a		b		c		d		e		f		g		h		i		j	
Positions of grid points	40 ° 00′ N		39 ° 00′ N		38 ° 25′ N		38 ° 12′ N		37 ° 37′ N		36 ° 43′ N		36 ° 00′ N		36 ° 00′ N		35 ° 00′ N		35 ° 00′ N	
	121 ° 00′ E		120 ° 00′ E		122 ° 00′ E		123 ° 00′ E		123 ° 30′ E		123 ° 30′ E		122 ° 00′ E		124 ° 00′ E		123 ° 00′ E		124 ° 00′ E	
The data for predicted observed	T_p	T_r	T_p	T_r	T_p	T_r	T_p	T_r	T_p	T_r	T_p	T_r	T_p	T_r	T_p	T_r	T_p	T_r	T_p	T_r
	8 / 7	17 / 7	8 / 7	17 / 7	8 / 7	14 / 7	8 / 7	12 / 7	8 / 7	11 / 7	8 / 7	9 / 7	8 / 7	8 / 7	8 / 7	7 / 7	8 / 7	7 / 7	8 / 7	7 / 7
T_s	21.64	23.21	21.38	21.55	22.00	20.60	21.20	20.12	23.00	21.25	23.06	22.70	22.94	2252	22.88	23.01	22.29	22.59	22.03	22.55
T_H	15.00	15.51	18.36	20.80	7.00	5.59	6.50	6.45	8.00	7.82	8.50	8.98	11.00	10.37	8.10	8.60	9.42	9.92	9.40	11.45
h	4.4	5	5.0	5	4.3	5	5.0	7	4.5	5	4.0	6	7.0	5	10.1	8	6.4	6	10.0	6

ACKNOWLEDGEMENTS

The authors wish to thank Dr. Takano for his valuable suggestions and thanks go to Ms. Zhang Hongnuan for drawing the figures . This work was supported by the National Planing Committee, PRC.

Appendix:

The implicit differential from for ζ and u ,and explicit differential from for v from $k\triangle t$ to $(K+1 / 2)\triangle t$ for Eqs . (5)–(7) are as follows :

$$A_{i-\frac{1}{2},j}U_{i-\frac{1}{2},j}^{k+\frac{1}{2}} + B_{i,j}\zeta_{i,j}^{k+\frac{1}{2}} + C_{i+\frac{1}{2},j}U_{i+\frac{1}{2},j}^{k+\frac{1}{2}} = D_{i,j} \tag{26}$$

where

$$A_{i-\frac{1}{2},j} = -\frac{\triangle t}{4\triangle x}d_{i-\frac{1}{2},j}^{k}, \quad B_{i,j} = 1, \quad C_{i+\frac{1}{2},j} = \frac{\triangle t}{4\triangle x}d_{i+\frac{1}{2},j}^{k} \quad ,$$

$$D_{i,j} = \zeta_{i,j}^{k} - \frac{\triangle t}{4\triangle y}(d_{i,j+\frac{1}{2}}^{k}V_{i,j+\frac{1}{2}}^{k} - d_{i,j-\frac{1}{2}}^{k}V_{i,j-\frac{1}{2}}^{k}) \quad ,$$

$$A_{i,j}\zeta_{i,j}^{k+\frac{1}{2}} + B_{i+\frac{1}{2},j}U_{i+\frac{1}{2},j}^{k+\frac{1}{2}} + C_{i+1,j}\zeta_{i+1,j}^{k+\frac{1}{2}} = D_{i+\frac{1}{2},j} \tag{27}$$

where

$$A_{i,j} = -\frac{\triangle t}{2\triangle x}g, \quad B_{i+\frac{1}{2},j} = 1 + \frac{\triangle t}{4\triangle x}(U_{i+\frac{3}{2},j}^{k} - U_{i-\frac{1}{2},j}^{k}), \quad C_{i+1,j} = \frac{\triangle t}{2\triangle x}g \quad ,$$

$$D_{i+\frac{1}{2},j} = U_{i+\frac{1}{2},j}^{k} + \overline{V}_{i+\frac{1}{2},j}^{k}[\frac{f\triangle t}{2} - \frac{\triangle t}{4\triangle y}(U_{i+\frac{1}{2},j+1}^{k} - U_{i+\frac{1}{2},j-1}^{k})]$$

$$- K_{b}\triangle t\frac{U_{i+\frac{1}{2},j}^{k}[(U_{i+\frac{1}{2},j}^{k})^{2} + (\overline{V}_{i+\frac{1}{2},j}^{k})^{2}]^{\frac{1}{2}}}{d_{i+\frac{1}{2},j}^{k}} + \frac{K_{v}\triangle t}{2}[\frac{U_{i+\frac{3}{2},j}^{k} + U_{i-\frac{1}{2},j}^{k} - 2U_{i+\frac{1}{2},j}^{k}}{(\triangle x)^{2}}$$

$$+ \frac{U_{i+\frac{1}{2},j+1}^{k} + U_{i+\frac{1}{2},j-1}^{k} - 2U_{i+\frac{1}{2},j}^{k}}{(\triangle y)^{2}} \quad] + \rho_{a}C_{D}\triangle t\frac{(W_{x})_{i+\frac{1}{2},j}^{k+\frac{1}{2}}\{[(W_{x})_{i+\frac{1}{2},j}^{k+\frac{1}{2}}]^{2} + [(\overline{W}_{y})_{i+\frac{1}{2},j}^{k+\frac{1}{2}}]^{2}\}^{\frac{1}{2}}}{\rho_{w}d_{i+\frac{1}{2},j}^{k}}$$

$$V_{i,j+\frac{1}{2}}^{k+\frac{1}{2}} = \{V_{i,j+\frac{1}{2}}^{k} - \overline{U}_{i,j+\frac{1}{2}}^{k+\frac{1}{2}}[\frac{\triangle t f}{2} + \frac{\triangle t}{4\triangle x}(V_{i+1,j+\frac{1}{2}}^{k} - V_{i-1,j+\frac{1}{2}}^{k})] - \frac{\triangle t g}{2\triangle y}(\zeta_{i,j+1}^{k} - \zeta_{i,j}^{k})$$

$$+ \frac{K_{v}\triangle t}{2}[\frac{V_{i+1,j+\frac{1}{2}}^{k} + V_{i-1,j+\frac{1}{2}}^{k} - 2V_{i,j+\frac{1}{2}}^{k}}{(\triangle x)^{2}} + \frac{V_{i,j+\frac{3}{2}}^{k} + V_{i,j-\frac{1}{2}}^{k} - 2V_{i,j+\frac{1}{2}}^{k}}{(\triangle y)^{2}}]$$

$$+ \rho_{a}C_{D}\triangle t\frac{(W_{y})_{i,j+\frac{1}{2}}^{k+\frac{1}{2}}\{[(W_{y})_{i,j+\frac{1}{2}}^{k+\frac{1}{2}}]^{2} + [(\overline{W}_{x})_{i,j+\frac{1}{2}}^{k+\frac{1}{2}}]^{2}\}^{\frac{1}{2}}}{\rho_{w}d_{i,j+\frac{1}{2}}^{k}} \} / \{1 + \frac{\triangle t}{4\triangle y}(V_{i,j+\frac{3}{2}}^{k} - V_{i,j-\frac{1}{2}}^{k})$$

$$+ K_{b}\triangle t\frac{[(V_{i,j+\frac{1}{2}}^{k})^{2} + (\overline{U}_{i,j+\frac{1}{2}}^{k+\frac{1}{2}})^{2}]^{\frac{1}{2}}}{d_{i,j+\frac{1}{2}}^{k}} \} \tag{28}$$

the implicit differential form for ζ and v, and explicit differential form for u from $(k+1/2)\triangle t$ to $(k+1)\triangle t$ are as follows :

$$A_{i,j-\frac{1}{2}} V^{k+1}_{i,j-\frac{1}{2}} + B_{i,j} \zeta^{k+1}_{i,j} + C_{i,j+\frac{1}{2}} V^{k+1}_{i,j+\frac{1}{2}} = D_{i,j} \tag{29}$$

where

$$A_{i,j-\frac{1}{2}} = -\frac{\triangle t}{4\triangle y} d^{k+\frac{1}{2}}_{i,j-\frac{1}{2}}, \quad B_{i,j} = 1, \quad C_{i,j+\frac{1}{2}} = \frac{\triangle t}{4\triangle y} d^{k+\frac{1}{2}}_{i,j+\frac{1}{2}}, \quad D_{i,j} = \zeta^{k+\frac{1}{2}}_{i,j} - \frac{\triangle t}{4\triangle x}(d^{k+\frac{1}{2}}_{i+\frac{1}{2},j} U^{k+\frac{1}{2}}_{i+\frac{1}{2},j}$$
$$- d^{k+\frac{1}{2}}_{i-\frac{1}{2},j} U^{k+\frac{1}{2}}_{i-\frac{1}{2},j});$$

$$A_{i,j} \zeta^{k+1}_{i,j} + B_{i,j+\frac{1}{2}} V^{k+1}_{i,j+\frac{1}{2}} + C_{i,j+1} \zeta^{k+1}_{i,j+1} = D_{i,j+\frac{1}{2}} \tag{30}$$

where

$$A_{i,j} = -\frac{\triangle t}{2\triangle y} g, \quad B_{i,j+\frac{1}{2}} = 1 + \frac{\triangle t}{4\triangle y}(V^{k+\frac{1}{2}}_{i,j+\frac{3}{2}} - V^{k+\frac{1}{2}}_{i,j-\frac{1}{2}}),$$
$$C_{i,j+1} = \frac{\triangle t}{2\triangle y} g,$$

$$D_{i,j+\frac{1}{2}} = V^{k+\frac{1}{2}}_{i,j+\frac{1}{2}} - \overline{U}^{k+\frac{1}{2}}_{i,j+\frac{1}{2}}[\frac{f\triangle t}{2} + \frac{\triangle t}{4\triangle x}(V^{k+\frac{1}{2}}_{i+1,j+\frac{1}{2}} - V^{k+\frac{1}{2}}_{i-1,j+\frac{1}{2}})] - K_b \triangle t \frac{V^{k+\frac{1}{2}}_{i,j+\frac{1}{2}}[(V^{k+\frac{1}{2}}_{i,j+\frac{1}{2}})^2 + (\overline{U}^{k+\frac{1}{2}}_{i,j+\frac{1}{2}})^2]^{\frac{1}{2}}}{d^{k+\frac{1}{2}}_{i,j+\frac{1}{2}}}$$

$$+ \frac{K_v \triangle t}{2}[\frac{V^{k+\frac{1}{2}}_{i+1,j+\frac{1}{2}} + V^{k+\frac{1}{2}}_{i-1,j+\frac{1}{2}} - 2V^{k+\frac{1}{2}}_{i,j+\frac{1}{2}}}{(\triangle x)^2} + \frac{V^{k+\frac{1}{2}}_{i,j+\frac{3}{2}} + V^{k+\frac{1}{2}}_{i,j-\frac{1}{2}} - 2V^{k+\frac{1}{2}}_{i,j+\frac{1}{2}}}{(\triangle y)^2}]$$

$$+ \rho_a C_D \triangle t \frac{(W_y)^{k+\frac{1}{2}}_{i,j+\frac{1}{2}}\{[(W_y)^{k+1}_{i,j+\frac{1}{2}}]^2 + [(\overline{W}_x)^{k+1}_{i,j+\frac{1}{2}}]^2\}^{\frac{1}{2}}}{\rho_w d^{k+\frac{1}{2}}_{i,j+\frac{1}{2}}},$$

$$U^{k+1}_{i+\frac{1}{2},j} = \{U^{k+\frac{1}{2}}_{i+\frac{1}{2},j} + \overline{V}^{k+1}_{i+\frac{1}{2},j}[\frac{\triangle tf}{2} - \frac{\triangle t}{4\triangle y}(U^{k+\frac{1}{2}}_{i+\frac{1}{2},j+1} - U^{k+\frac{1}{2}}_{i+\frac{1}{2},j-1})]$$

$$- \frac{\triangle tg}{2\triangle x}(\zeta^{k+\frac{1}{2}}_{i+1,j} - \zeta^{k+\frac{1}{2}}_{i,j}) + \frac{K_v \triangle t}{2}[\frac{U^{k+\frac{1}{2}}_{i+\frac{3}{2},j} + U^{k+\frac{1}{2}}_{i-\frac{1}{2},j} - 2U^{k+\frac{1}{2}}_{i+\frac{1}{2},j}}{(\triangle x)^2}$$

$$+ \frac{U^{k+\frac{1}{2}}_{i+\frac{1}{2},j+1} + U^{k+\frac{1}{2}}_{i+\frac{1}{2},j-1} - 2U^{k+\frac{1}{2}}_{i+\frac{1}{2},j}}{(\triangle y)^2}] + \rho_a C_D \triangle t \frac{(W_x)^{k+1}_{i+\frac{1}{2},j}\{[(W_x)^{k+1}_{i+\frac{1}{2},j}]^2 + [(\overline{W}_y)^{k+1}_{i+\frac{1}{2},j}]^2\}^{\frac{1}{2}}}{\rho_w d^{k+1}_{i+\frac{1}{2},j}}\}$$

$$/ \{1 + \frac{\triangle t}{4\triangle x}(U^{k+\frac{1}{2}}_{i+\frac{3}{2},j} - U^{k+\frac{1}{2}}_{i-\frac{1}{2},j}) + K_b \triangle t \frac{[(U^{k+\frac{1}{2}}_{i+\frac{1}{2},j})^2 + (\overline{V}^{k+1}_{i+\frac{1}{2},j})^2]^{\frac{1}{2}}}{d^{k+1}_{i+\frac{1}{2},j}}\}. \tag{31}$$

Velocity comonents $\bar{u}, \bar{v}, \overline{w_x}$, and $\overline{w_y}$ in (26)–(31) denote the average values at the calculated points, and d_{ij} is two times average depth of the point (i, j).

The differential forms of Eqs. (11),(15) and (16) from $K\triangle t$ to $(k+1)\triangle t$ are as follows :

$$(T_s)_{i,j}^{k+1} = (T_s)_{i,j}^k - \frac{\triangle t}{4\triangle x}(U_{i-\frac{1}{2},j}^{k+1} + U_{i+\frac{1}{2},j}^{k+1}) \cdot [(T_s)_{i+1,j}^k - (T_s)_{i-1,j}^k]$$

$$- \frac{\triangle t}{4\triangle y}(V_{i,j-\frac{1}{2}}^{k+1} + V_{i,j+\frac{1}{2}}^{k+1}) \cdot [(T_s)_{i,j+1}^k - (T_s)_{i,j-1}^k] + A_T\triangle t[\frac{(T_s)_{i-1,j}^k + (T_s)_{i+1,j}^k - 2(T_s)_{i,j}^k}{(\triangle x)^2}$$

$$+ \frac{(T_s)_{i,j+1}^k + (T_s)_{i,j-1}^k - 2(T_s)_{i,j}^k}{(\triangle y)^2}] + \triangle t\frac{(Q_s)_{i,j}^{k+1} - (Q_h)_{i,j}^k}{h_{i,j}^k}, \tag{32}$$

$$(T_H)_{i,j}^{k+1} = (T_H)_{i,j}^k - \frac{\triangle t}{4\triangle x}(U_{i-\frac{1}{2},j}^{k+1} + U_{i+\frac{1}{2},j}^{k+1}) \cdot [(T_H)_{i+1,j}^k - (T_H)_{i-1,j}^k]$$

$$- \frac{\triangle t}{4\triangle y}(V_{i,j-\frac{1}{2}}^{k+1} + V_{i,j+\frac{1}{2}}^{k+1}) \cdot [(T_H)_{i,j+1}^k - (T_H)_{i,j-1}^k]$$

$$+ A_T\triangle t[\frac{(T_H)_{i-1,j}^k + (T_H)_{i+1,j}^k - 2(T_H)_{i,j}^k}{(\triangle x)^2}$$

$$+ \frac{(T_H)_{i,j+1}^k + (T_H)_{i,j-1}^k - 2(T_H)_{i,j}^k}{(\triangle y)^2}] - C_3 A_T\triangle t\frac{(T_s)_{i,j}^k - (T_H)_{i,j}^k}{(H_{i,j}^k - h_{i,j}^k)^2}$$

$$\cdot [(\frac{h_{i+1,j}^k - h_{i-1,j}^k}{2\triangle x})^2 + (\frac{h_{i,j+1}^k - h_{i,j-1}^k}{2\triangle y})^2] + C_1\triangle t\frac{(Q_h)_{i,j}^k}{H_{i,j}^k - h_{i,j}^k} - C_2\triangle t\frac{(Q_s)_{i,j}^{k+1} - (Q_h)_{i,j}^k}{h_{i,j}^k} \tag{33}$$

$$h_{i,j}^{k+1} = h_{i,j}^k - \frac{\triangle t}{4\triangle x}(U_{i-\frac{1}{2},j}^{k+1} + U_{i+\frac{1}{2},j}^{k+1}) \cdot (h_{i+1,j}^k + h_{i-1,j}^k) - \frac{\triangle t}{4\triangle y}(V_{i,j-\frac{1}{2}}^{k+1} + V_{i,j+\frac{1}{2}}^{k+1}) \cdot (h_{i,j+1}^k + h_{i,j-1}^k)$$

$$+ A_T\triangle t[\frac{h_{i-1,j}^k + h_{i+1,j}^k - 2h_{i,j}^k}{(\triangle x)^2} + \frac{h_{i,j-1}^k + h_{i,j+1}^k - 2h_{i,j}^k}{(\triangle y)^2}]$$

$$+ \frac{2A_T\triangle t}{(T_s)_{i,j}^k - (T_H)_{i,j}^k}\{[\frac{(T_s)_{i+1,j}^k - (T_s)_{i-1,j}^k}{2\triangle x}$$

$$- \frac{(T_H)_{i+1,j}^k - (T_H)_{i-1,j}^k}{2\triangle x}]\frac{h_{i+1,j}^k - h_{i-1,j}^k}{2\triangle x} + [\frac{(T_s)_{i,j+1}^k - (T_s)_{i,j-1}^k}{2\triangle y}$$

$$- \frac{(T_H)_{i,j+1}^k - (T_H)_{i,j-1}^k}{2\triangle y}]\frac{h_{i,j+1}^k - h_{i,j-1}^k}{2\triangle y}\}$$

$$+ (2 + C_6)\frac{A_T\triangle t}{H_{i,j}^k - h_{i,j}^k}[(\frac{h_{i+1,j}^k - h_{i-1,j}^k}{2\triangle x})^2 + (\frac{h_{i,j+1}^k - h_{i,j-1}^k}{2\triangle y})^2]$$

$$+ \frac{C_4\triangle t(Q_h)_{i,j}^k}{(T_s)_{i,j}^k - (T_H)_{i,j}^k} - C_5\triangle t\frac{[(Q_s)_{i,j}^{k+1} - (Q_h)_{i,j}^k](H_{i,j} - h_{i,j}^k)}{h_{i,j}^k[(T_s)_{i,j}^k - (T_H)_{i,j}^k]}, \tag{34}$$

where

$$(Q_s)_{i,j}^{k+1} = \{e_1 + e_2[(T_s)_{i,j}^{k+1} - (T_a)_{i,j}^{k+1}]\} / C_p\rho_w, \tag{35}$$

$$(Q_h)_{i,j}^{k} = -(Q_s)_{i,j}^{k+1} + \frac{2}{\beta h_{i,j}^{k}} \{ \wedge b_1 [\sqrt{\frac{\rho_a C_p}{\rho_w}} [((W_x)_{i,j}^{k})^2 + ((W_y)_{i,j}^{k})^2]$$

$$- b_2 |f| h_{i,j}^{k} \frac{\rho_a C_D}{\rho_w} [((W_x)_{i,j}^{k})^2 + ((W_y)_{i,j}^{k})^2] + \frac{1}{4} \beta h_{i,j}^{k} [(Q_s)_{i,j}^{k+1} - |(Q_s)_{i,j}^{k+1}|] \cdot$$

$$[1 - (1 - h_{i,j}^{k} / H_{i,j})^m] \} \tag{36}$$

REFERENCES

Halpern, D., 1974, Observation of the deepening of the wind—mixed layer in the northeast Pacific Ocean. J. of Phys. Oceanogr., 4, 454--466.

Japan Meteorological Agency, 1985, The results of marine meteorological and oceanographical observations. No. 78.

Kalazkii, V. E., 1978, Simulation of vertical thermal structure for the ocean active layer. Leningrad (in Russian), 1978.

Kalakzkii, V. E. and E. S. Nesterov, 1980, Numerical prediction of the ocean thermal structure with the influence of atmospheric process in weather scale. Tr. GMC SSSR (in Russian), No. 229; 37--44.

Kitaigorodskii, S. A. and V. Z. Miropolskii, 1970, On the theory of the ocean active layer. Izv. Akad. Nauk SSSR, Ser. FAO (in Russian), No. 6, 177--188.

Kitaigorodskii, S. A. 1977, Dynamics of seasonal thermocline. Okeanologiya (in Russian), No. 4, 6--34.

Kraus, E. R. and J. S. Turner, 1975, A one—dimensional model of the seasonal thermocline II : the general theory and its consequences . Tellus, No. 19,98--106

Kraus, E. R., 1977, Modelling and prediction of the upper layers of the ocean . Pergmon Press. 1977.

Leendertse, J. J., 1967, Aspects of a computional for long period water—wave propagation. Memorandum, RM--5394--PR, Rand Corporation, May, 1967.

Malkki, P. and R. Tamsalu, 1985, Physical features of the Baltic sea. Finnish Marine Res., No. 252, 50--67.

Murakami, M. et al., 1985, A numerical simulation of the distribution of water temperature and salinity in the Seto Inland Sea. J. of Oceanogr. Soc. of Japan, No.41, 213--224.

Nesterov, E. S. 1978, Numerical prediction for the thermal features of the upper layer in the north Atlantic Ocean. Tr. GMC SSSR (in Russian) , No. 200, 22--29.

Nesterov, E. S., 1986, Response of upper ocean to the temperate zone cyclone. Tr. GMC SSSR (in Russian), No. 281, 24--34.

Ocean Group of the Science and Technology Commission, PRC, 1964, Comprehensive marine investigation report . No. 2 and 5

Omsteds, A. et al., 1983, Measured and numerically—simulated autumn cooling in the Bay of Bothnia. Tellus, No. 35a, 231--240.

Resnanskii, U. D. and E. V. Trosnikov, 1980, Parametrization in the ocean active layer developed from the simulating zonal atmospheric circulation. Tr. GMC SSSR (in Russian), No. 229, 18--31.

Resnanskii, U. D., 1983, The influence of current on the evolution of features in the ocean active layer. Tr. GMC SSSR (in Russian), No.282, 23--33.

Resnanskii, U. D., 1986, Numerical experiment in the ocean active layer considering the term of space variation. Tr. GMC SSSR (in Russian) No. 291, 3--14.

Wang Zongshan, 1983, Asimplified method of computing heat budget at the sea surface. Mar. S. Bull. (in Chinese), No., 2, 22--25.

Wang Zongshan and Zou Emei, 1986, A parametric modle for thermal structure features of the ocean upper layer. Acta Oceanol. sinica, No. 5, 16—21.

Wang Zongshan, Xu Bochang, Zou Emei, Yang Keqi and Li Fanhua, 1990, A study on the numerical prediction model for the vertical thermal structure in the Bohai and huanghai seas (I)---One–dimensional numerical prediction model. Acta Oceanol. Sinica (in Chinese), 14(2), in Press.

Xu Bochang et al., 1983, Simulation of vertical distribution of sea–water density (σ_t) in the shallows. Mar. S. Bull. (in Chinese), No. 1, 9—12.

Xu Bochang et al., 1984, Simulation of vertical distribution of water temperature in the shallows. Mar. s. Bull. (in Chinese), No. 3, 1—6.

Zhang Yangting and Wang Yijiao, 1983, Simulation of wind field and numerical computation of storm surge in the Bohai sea. Acta Oceanol. Sinica (in Chinese), 5(3), 261—272.

Zhang Yangting and Wang Yijiao, 1988, The properties and the numerical modelling of the Typhoon surge of Huanghai sea. Preceedings of third science meeting of Chinese Oceanol. and Limnol. Soc. (in Chinese),57—63.

DEVELOPMENT OF TOWED VEHICLE SYSTEMS FOR ACOUSTIC DOPPLER CURRENT PROFILER

W. KOTERAYAMA, A. KANEKO, M. NAKAMURA and T. HORI

Research Institute for Applied Mechanics,Kyushu University, Kasuga 816 (Japan)

ABSTRACT

Two different types of towed vehicle, EIKO and DRAKE have been developed to house an ADCP (acoustic Doppler current profiler). EIKO is a simple and stable vehicle without any mechanism for the motion control. The depth and roll of DRAKE are controlled by the main wings and horizontal tail wings. The contours of EIKO and DRAKE are designed so as to increase the hydrodynamic damping force and to decrease the unstable roll moment acting on them. Structures of the two vehicles and results of on-site experiments for confirming their performances are described. The experiments showed that EIKO and DRAKE with ADCP are suitable for measurements of the ocean current.

1. INTRODUCTION

The acoustic Doppler current profiler (ADCP) measures the detailed vertical profile of ocean currents. Its accuracy is assured when mounted on the sea bottom (Lhermitte, 1982) or a stable platform.

Some oceanographers have attempted to use a shipboard ADCP (Joice et al., 1982) which enables to collect spatially high dense data for drawing the detailed map of the velocity distribution in an ocean current. They reported that motions of the ship due to surface waves, the cavitation noise generated by the ship's propeller and bubbles entrained under the hull had serious adverse effects on the ADCP data quality. These effects may be reduced by using the underwater vehicle. The underwater vehicle which could completely exclude these negative elements would be a non-tethered free-swimming one, but the speed and the radius of operation of the free-swimming vehicle are restricted owing to a difficulty in the power supply. Since we wanted to collect ocean data over wide areas of the sea we selected a towed vehicle system because of its high speed mobility.

Two towed vehicles, EIKO and DRAKE were developed. EIKO is simple and light and has no motion control system . Its maximum towing speed is 8 knots and its maximum working depth is 10m. EIKO fulfills its function when used behind a small ship. DRAKE is developed for high speed towing of over 10 knots and its maximum towing speed is basically unlimited; the set point of submerged depth can be varied from 0 to 300m. The main wings and tail wings are controlled automatically to keep a desired depth without roll. The towing point is determined on the basis of theoretical analyses so that the trim and pitch angles are minimized under any conditions of the submerged depth and towing speed.

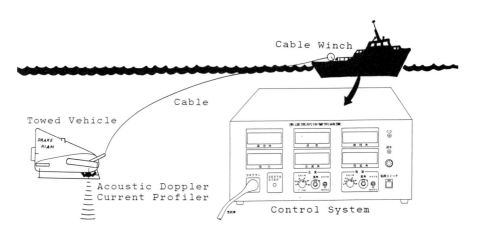

Fig.1. Schematic diagram of towed vehicle-ADCP system

In this report, the structures of EIKO and DRAKE and the results of on-site experiments for confirming their performances are described.

2. CONCEPT OF THE TOWED VEHICLE-ADCP SYSTEM

Figure 1 is a schematic diagram of the towed vehicle-ADCP system for ocean measurements. A CTD-sensor and an ADCP are housed in the vehicle.

We chose a product of RD Instruments (MODEL RD-DR0150) as the current profiler. The length is about 1.8m and the weight in the air is 80kg.

When mounted on the sea bottom or on a stable platform, the ADCP produces accurate ocean measurements (Lhermitte, 1982). When housed in a vehicle which oscillates by external forces, the correct measurement of ADCP's attitudes is essential because the backscattered acoustic signals are transformed into three dimensional velocity relative to the ADCP in the instrument on the basis of the measured data of attitudes and direction. Pendulum type tilt meters are used to measure the attitudes. By this type of tilt meter the distinction between the surge acceleration and the pitch or the sway acceleration and the roll is basically impossible. Such accelerations are induced by tension variations of the towing cable caused by the motions of the towing ship or shedding vortices from members of the towed vehicle. Therefore the accelerations of motions of the vehicle must be minimized.

We have developed towed vehicle systems for the purposes of ;
· measuring ocean currents accurately even under severe sea states,
· collecting data at a wider range of the depth than the ADCP normally covers,
· reducing the influence of cavitation noise generated by the ship's propeller,
· eliminating interference caused by bubbles entrained under the ship's hull and
· being able to use ADCP with any research vessel.

All purposes but the first can be achieved simply by utilizing a cable of proper length to tow the vehicle. To successfully achieve the first purpose, the vehicle's hull shape and control system were designed to minimize the oscillations of the towed vehicle in water. In addition, mother ship speeds should be measured accurately to calculate the velocity relative to the Earth by subtracting the vehicle's forward speed, of which the time average can be considered to be equal to that of mother ship speed, from the velocity relative to the towed vehicle. The ADCP can measure its own velocity relative to the Earth by bottom tracking in shallow water, but in deep water other navigational methods must be used. The Loran-C system was adopted in these experiments. We use currently the Global Positioning System, which is much more accurate than the Loran-C.

3. STRUCTURE OF EIKO

EIKO is pictured in Fig.2; the hull is made of fiber reinforced plastic (FRP). Its principal features are shown in TABLE 1.

TABLE 1
Features of EIKO

Operating depth	10m
Dimensions	L=2m,W=0.78m,H=0.55m
Weight in air	160kg (with ADCP)
Weight in water	0
Towing speed	0 ~ 8 knots
Instrumentation	ADCP, CTD sensor

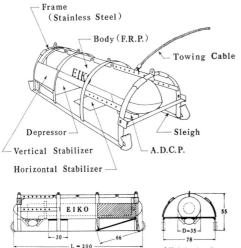

Fig.2 Figure of EIKO.

The submerged operating depth (10m) is chosen considering that the draft of the largest observation ship in Japan is not deeper than 7m. It is enough to avoid the effects of the mother ship on measurements. The lighter EIKO is, the easier it handles, but some bumping of EIKO against the mother ship is inevitable in deployment and retrieval. Therefore, EIKO should be of sufficient strength to withstand this without damage. Frames of EIKO are counted as longitudinal strength members and also expected to act as a shock absorber for the ADCP. While the deployment and retrieval are easy with excess buoyancy, the excess buoyancy would cause the coupling motion of the surge and heave, so that it is made neutral in water. In the body of EIKO, the buoyant material, polyurethane foam, was filled. This has now been replaced by high pressure syntactic foam because EIKO sometimes goes down deeper than the designed depth during deployment and retrieval. The lower part of the frontal area of EIKO is designed so that it does not obstruct acoustic transmission of ADCP. A depressor is set at its tail end for balancing the heavy head of the ADCP. It makes the trim of EIKO minimum by its downward lift force. The downward lift force can be controlled by changing the angle of the depressor. A horizontal stabilizer is set to increase hydrodynamic damping force for the pitching motion and minimize the trim angle. Vertical stabilizers are expected to increase the hydrodynamic damping forces for the roll and sway motion. The FRP body with the buoyant material protects the ADCP against the impulsive force induced by the bumping of EIKO against the ship. The body shape is streamlined and serves to diminish the drag force acting on the vehicle.

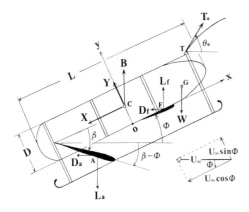

Fig.3 Forces acting on EIKO

The dimension and angle β of the depressor and other properties are determined considering the static balance of the forces acting on EIKO as shown in Fig.3, in which T_0 is the towing cable tension, L_f and D_f are the lift and drag force acting on the horizontal stabilizer, L_a and D_a are the lift and drag force acting on the depressor, W is the weight of the ADCP in water, B is the excess buoyancy of EIKO body, and X and Y are the x- and y-components of the drag forces acting on EIKO with the ADCP. At first, the static balance of the moment of L_f, D_f, L_a, D_a , W, B, X, and Y at the towing point T are considered. These values except W and B are functions of the towing speed U_∞ and the trim angle Φ. In addition to that, D_a and L_a depend also on the set angle of the depressor. When U_∞ and β are given, we obtain all these forces and the trim angle Φ through an iterative procedure. Next from the balance of the force at the towing point T, the value of the tension T_0 and the angle θ_0 of the towing cable at the towing point T are calculated. The profile of the towing cable and submerged depth of the vehicle are determined by substituting T_0 and θ_0 into the inverse catenary theory. Until the depth reaches to the desired one, the calculation is repeated by changing the value of β.

The detail of the calculation method is shown by Hori et al., (1988).

4. STRUCTURE OF DRAKE

Figure 4 and Table 2 show DRAKE and its principal feature. The hull is made of fiber reinforced plastic (FRP). The shape is streamline contour with its deep vertical contour designed to increase the damping- and added mass-force, thereby to reduce high frequency roll which is hard to control with the horizontal tail wings. The main wings have a symmetrical aerofoil profile. The lower part housing the ADCP and CTD sensor is designed so as to not obstruct the acoustic beam of the ADCP or the flow inlet for the CTD sensor.

Table 2

Principal features of DRAKE and the towing system

DRAKE	
Operating depth	$0 \sim 300$m
Dimensions	L=2m, W=2m,H=1.5m
Weight in air	360kg
Towing velocity	$5 \sim 12$knots
Depth control	By main wing
Roll control	By horizontal tail wing
Instrumentation	ADCP, CTD sensor
TOWING CABLE	
Length	800m
Diameter	12.9mm
Breaking tension	9 tons
Conductors	Power conductors 1 pair
	Signal conductors 10
CABLE WINCH	
Weight	2060kg
Dimensions	$1.8m \times 1.8m \times 2.0m$
Maximum reeling tension	1.5 tons
Maximum stopping tension	2.3 tons

The towing cable is a double-armored one with ten signal conductors and a pair of electric conductors for power supply to the sensors. The cable dynamics is very important

Fig.4 Picture of DRAKE.

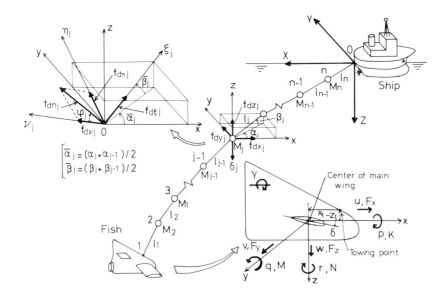

Fig.5 Concept of calculation method for DRAKE system.

for the DRAKE system because the towing cable is very long and heavy. We developed a three-dimensional lumped mass method for cable dynamics and six-degree-freedom equation for the motions of the DRAKE (Koterayama et al., 1988), of which the concept is shown in Fig.5. In the lumped mass method the cable is modeled as N-discrete masses interconected by springs. All forces such as the drag force, the added mass force, the inertia force, the buoyancy and weight acting on the cable are considered to be concentrated load on each mass, then the motions of the cable can be represented by simultaneous differential equations. The equations for the motions of DRAKE are similar to those for the airplane. The motions of the towing vessel, cable and towed vehicle are represented by a set of simultaneous equations. The hydrodynamic coefficients used in the calculation were obtained from model experiments and theoretical estimations. The body and control system were designed by numerical simulations with this calculation scheme. Koterayama et al.(1988) described details of the calculation method.

The perspective (Fig.6) shows that the impeller set at the tail end is coupled to a hydraulic pump. This provides power for the actuation of the main- and horizontal tail- wings. The idea of using a stream-driven impeller to generate hydraulic power was introduced in the development of the Batfish by the Bedford Institute of Oceanography (Dessureault, 1976).

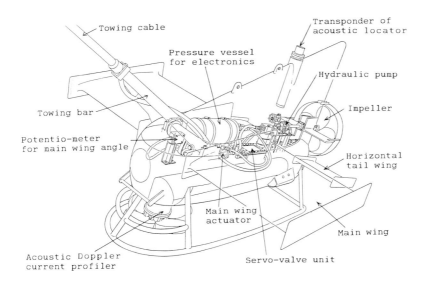

Fig.6 Perspective of DRAKE

5. ON-SITE EXPERIMENTS AND DISCUSSION

On-site experiments (Fig.7) of EIKO and DRAKE were separately carried out to confirm the performances of EIKO and DRAKE and to show the capability of the towed vehicle-ADCP system in ocean measurements.

The maximum towing speeds tested were 8 knots for EIKO and 12 knots for DRAKE; the latter speed was the maximum possible by the mother ship and the maximum towing speed of DRAKE is basically unlimited. When the towing speed of EIKO exceeded 8 knots, the angle of the heel became greater than 30 degree which is the maximum allowable tilt angle of the ADCP.

Figure 8 shows the static characteristics of DRAKE obtained from on-site experiments (Koterayama et al., 1990). Circles indicate the experimental results of main wing angle and the solid line is the calculated result. The squares and broken line are the results of the trim of DRAKE. The double circles and one-dash chain line are the attack angle of the main wings relative to the uniform flow. The triangles and two-dash chain line are the tension at the towing winch. This figure indicates the accuracy of theoretical estimations for the static ploblem.

The motions of EIKO and DRAKE were found to be much less than those of the mother ship. A comparison between the power spectra of motions of the mother ship and DRAKE (Fig.9) shows that the heave, roll and pitch of DRAKE are much less than those of the mother ship. We did not measure the surge, sway or yaw, but the theoretical analysis (Koterayama et al., 1988) suggests that the sway and yaw are much less than those of the mother ship; the surge of DRAKE is also less though greater than the sway or yaw. The theoretical analysis also suggests that the ratios of motions of DRAKE and the ship decrease with the increase of incident wave height.

Figure 10 compares data by the EIKO-ADCP system with data by a mooring system. Both are in fairly good agreement.

Figure 11 shows the velocity distribution of the Kuroshio taken by an ADCP mounted on DRAKE. It is presumed that the disturbances were caused by small islands. There was no loss of measured data, which is often caused by the noise generated by the ship's propeller and bubbles entrained under the ship's hull when the ADCP is set on a surface ship.

Velocity distribution in an ocean current is usually obtained from geostrophic calculation using the measured vertical profile of water temperature and salinity or direct measurement using the mooring system, but with these methods it is very difficult to obtain such detailed data as are shown in Fig. 11. The mooring system has an advantage over the ADCP carried by a vehicle or ship from the point of view of continuous measurement (Takematsu et al., 1986) while the ADCP boarded on the ship or vehicle is excellent in the point of spatial continuity. These two systems of measurement thus complement each other, and combined use of the two should become popular in ocean measurements. The system of ship-boarded ADCP has weak points on the data quality because the motions of the ship due to surface waves, the cabitation noise generated by ship's propeller and bubbles entrained under the

EIKO (a)   DRAKE (b)

Fig.7 Views of on-site experiments of EIKO (a) and DRAKE (b).

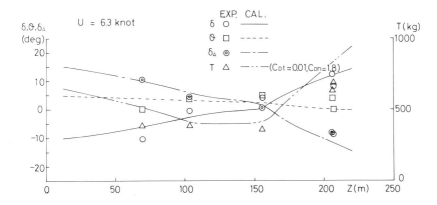

Fig.8 Static characteristics of DRAKE (towing speed is 6 knots).

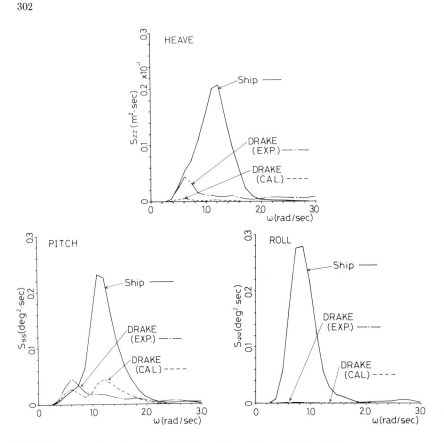

Fig.9 Power spectra of motions of the mother ship and DRAKE
(towing speed is 11 knots).

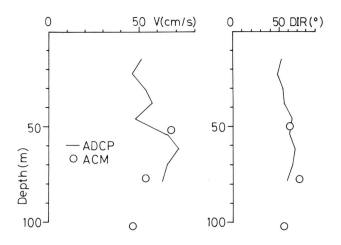

Fig.10 Comparison of data obtained by the EIKO-ADCP system and mooring system with
Aanderaa current meters on the current speed and direction at 34°04′N, 129°32′E.
(water depth is about 100m, towing speed is 6 knot).

hull have serious adverse effects. The results of on-site experiments shown in Figs.10 and 11 indicated that the well-designed towed vehicle can exclude these negative elements, but more deeper studies by means of numerical simulations and on-site experiments (Kaneko et al., 1990) are needed for the quantitative evaluation of the accuracy of measurements by an ADCP-towed vehicle system.

Acknowlegement

This work is a part of the Ocean Research Project of the Research Institute for Applied Mechanics financed by the Ministry of Education, Science and Culture, Japan.

The authors thank members of the Ocean Research Group of R.I.A.M.

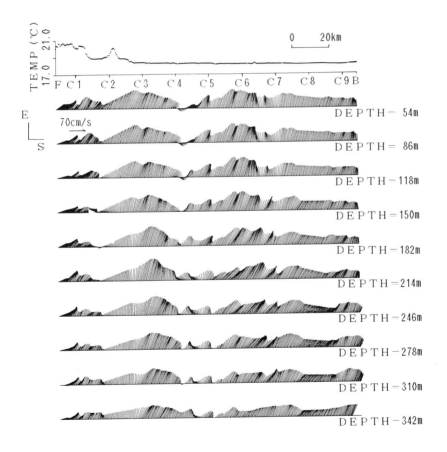

Fig.11 Velocity distribution of the Kuroshio obtained by the DRAKE-ADCP system at 30°5′N, 131°3′E (C1) ~ 28°57′N, 130°15′E (C9).
(towing speed is 9 knot, submerged depth of DRAKE is 50m, December, 1988).

304

6. REFERENCES

Dessureault, J.G., 1976. -Batfish- A depth controllable towed body for collecting oceanographic data. Ocean Engineering, Vol.3: 99-111.

Hori, T., Nakamura, M., Koterayama, W., Honji, H., and Takahashi, M., 1988. A design of
the towed vehicle system for the acoustic Doppler current profiler. Transactions of the
West-Japan Society of Naval Architects, No.76: 97-112.

Joyce, T.M., Bitterman ,J.R. and Prada K.E., 1982. Shipboard acoustic profiling of upper
ocean currents. Deep-Sea Research, No.29: 903-913.

Kaneko, A., Koterayama, W., Honji, A., Mizuno, S., Kawatate, K and Gordon R.L., 1990.
Cross-streem survey of the upper 400m of the Kuroshio by an ADCP on a towed fish.
Deep-Sea Research, Vol.37, No.5: 875-889.

Koterayama, W., Kyozuka, Y., Nakamura, M., Ohkusu, M. and Kashiwagi, M., 1988. Motions of a depth controllable towed vehicle. Proc. of the 7th International Conference
on Offshore Mechanics and Arctic Engineering: 423-430.

Koterayama, W., Nakamura, M., Kyozuka, Y., Kashiwagi, M. and Ohkusu, M., 1990. Depth
and roll controllable towed vehicle DRAKE for ocean measurements. Proc. of the First
Pacific/Asia Offshore Mechanics Symposium: 257-264.

Lhermitte, R., 1982. Doppler sonar observation of tidal flow. Journal of Geophysical Research, No.88: 725-735.

Takematsu, M., Kawatate, K., Koterayama, W., Suhara, T., and Mitsuyasu, H., 1986.
Moored instrument observations in the Kuroshio south of Kyushu. Journal of the
Oceanographycal Society of Japan, Vol.42: 201-211.

A STUDY OF THE KUROSHIO IN THE EAST CHINA SEA AND THE CURRENTS
EAST OF THE RYUKYU ISLANDS IN 1988

YAOCHU YUAN, JILAN SU AND ZIQIN PAN
Second Institute of Oceanography, State Oceanic Administration,
P. O. Box 1207, Hangzhou 310012, China

ABSTRACT

    The inverse method is used to compute the Kuroshio in the
East China Sea with hydrographic data collected during early
summer and autumn 1988 and the currents east of the Ryukyu
Islands with data during early summer 1988 only.   In the East
China Sea the Kuroshio Countercurrent is located further to the
west and its transport is larger during autumn 1988 than in late
summer 1987 and early summer 1988.   The Kuroshio water also
intruded onto the shelf with a reduced width during autumn 1988.
The transport of the Kuroshio in the East China Sea is 23.4 and
24.3×10^6 m^3/s, respectively, during early summer and autumn
1988.   The seasonal change of the Kuroshio axis at the surface
does not correlate with the seasonal change of the position of
the centerline of the Kuroshio width.   There is a cyclonic gyre
on the shelf north of Taiwan during all the cruises.   A part of
the Taiwan Warm Current in the survey area has a tendency to
converge to the shelf break.   Currents east of the Ryukyu
Islands flow northward over the Ryukyu Trench in early summer
1988, of which the transport is 26.9×10^6 m^3/s. The core of this
flow is between 400 and 800 m depths.

1 INTRODUCTION

    There have been many studies on the current structure and
volume transport of the Kuroshio in the East China Sea (Guan,
1982, 1988; Nishizawa et al., 1982; Saiki, 1982; Yuan and Su,
1988; Yuan, Endoh and Ishizaki, hereafter referred to as YEI,
1990).   Dynamic computations give the overall mean volume
transport through section G(PN) as 21.3×10^6 m^3/s for the years
between 1955 and 1978 (Guan, 1982, 1988) or 19.7×10^6 m^3/s if data
from 1954 to 1980 are used (Nishizawa et al., 1982). Guan (1982,
1988) and Nishizawa et al. (1982) used the 700 dbar as the
reference level. However, a recent work by YEI (1990) using the
inverse method showed that there may still be a significant
current at 700 m depth. Therefore, the Kuroshio transport through
the East China Sea may have been underestimated by previous
authors using dynamic computation methods.
    Studies on the currents east of the Ryukyu Islands are

relatively few. Konaga et al. (1980) and Yuan and Su (1988) pointed out that there is a northward current east of the Ryukyu Islands and this may be partly responsible for strengthening the Kuroshio south of Japan. YEI (1990) found that the northward currents east of Ryukyu Islands flowed over the Ryukyu Trench during Sep.-Oct., 1987.

In this study the inverse method is used to compute the Kuroshio Current in the East China Sea, using hydrographic data collected during two cruises in 1988 (May to June and October to November). The currents east of Ryukyu Islands are also computed with early summer data only. The current structure and transport of the Kuroshio and the currents east of the Ryukyu Islands will be discussed.

2 NUMERICAL CALCULATIONS

Since the inverse method was described in detail in previous studies (Yuan, Su and Pan, 1990; YEI, 1990), we will not repeat the description of the method here. The bathymetry, hydrographic sections and stations, and computation boxes are shown in Fig. 1. Two boxes are set up for the region in the East China Sea while east of the Ryukyu Islands only one box is set up. Because the water depths are shallow within boxes 1 and 2, we divide the boundary sections of these boxes into two layers. The value $\sigma_{t,P}$ of the isopycnal level is taken to be 25. The water depths in

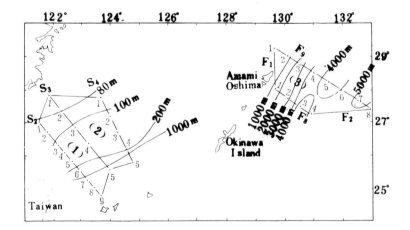

Fig. 1. (a) Depth (in m), hydrographic stations and computation boxes during early summer of 1988 (May and June, 1988).

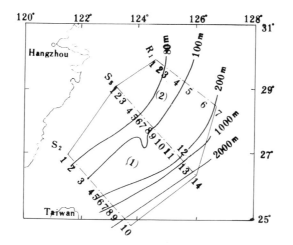

Fig. 1. (b) Depth (in m), hydrographic stations and computation boxes during autumn of 1988 (October and November, 1988).

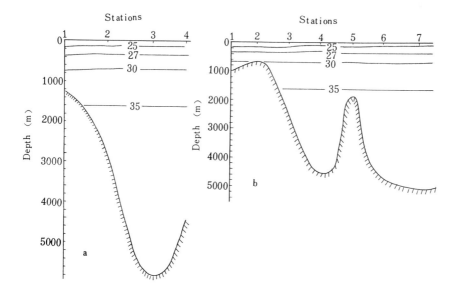

Fig. 2. Isopycnal levels along (a) section F$_8$ and (b) section F$_9$.

box 3 are deep, so that we divide each boundary section of box 3 into five layers according as the $\sigma_{t.p}$ values of four isopycnal levels 25, 27, 30 and 35. Figure 2 shows the isopycnal levels along sections F$_8$ and F$_9$, respectively. The depths of the $\sigma_{t.p} = 25$, 27, 30 and 35 levels lie between 125 to 200 m, 300 to

360 m, 700 to 730 m and around 1640 m, respectively.

TABLE 1

The balance of the transport in box 3 during early summer of 1988

layer	1st	2nd	3rd	4th	5th	total
$\triangle_1$	-2.03 $\times 10^{-3}$	-3.01 $\times 10^{-3}$	-7.61 $\times 10^{-3}$	9.73 $\times 10^{-3}$	1.46 $\times 10^{-3}$	-1.46 $\times 10^{-3}$
$\triangle_{1,D}$	-1.09	-0.68	1.50	-2.88	-2.14	-8.29

Note

$\triangle_1$: balance of the transport into the i-th layer by inverse method (positive into the layer).

$\triangle_{1,D}$: balance of the transport into the i-th layer by dynamic method (positive into the layer).

(units: 10^6 m^3/s).

Table 1 compares the balance of the transport for each layer of box 3 by the inverse method with those by dynamic method. The reference level of no motion is assumed at the bottom for the dynamic method. The dynamic method yields large unbalanced transport for each layer. The total unbalanced transport amounts to 8.3×10^6 m^3/s, which is too large to be ignored. This indicates that conservation of mass is not satisfied by dynamic method, when the reference level of no motion is assumed at the bottom.

3 THE KUROSHIO IN THE EAST CHINA SEA

In this section we first discuss the Kuroshio Current structure and its transport in early summer and autumn of 1988, and then compare these results with results from an earlier cruise.

3.1 Early summer (May-June) of 1988

In early summer of 1988 the Kuroshio has only one current core at section S_2 (Fig. 3). In Fig. 3 and following figures, the computation points are located at the mid-points between neighboring hydrographic stations. The core is located at point S_2-6 where the water depth is about 1260 m. The maximum velocity is found at the 50 m level with a magnitude of 97.3

cm/s. The velocities in the upper 150 m of this core are all greater than 80 cm/s. Its current speed at the 600, 700 and 800 m level is 17.3, 7.9 and 1.6 cm/s, respectively. In the deep layer there is a countercurrent. This countercurrent reaches the 600 m level at point S_2-7 whose maximum speed (6.6 cm/s) is at the 800 m level of point S_2-7. The transport of the Kuroshio through section S_2 in early summer is 23.4×10^6 m^3/s and its width is about 170 km.

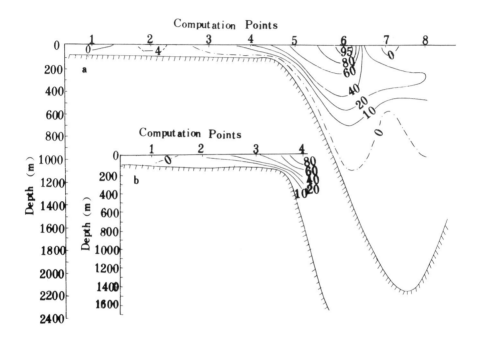

Fig. 3. Velocity distribution at (a) section S_2 and (b) section S_4 (positive value denotes northward flow). (units: cm/s).

The hydrographic stations at section S_4 does not reach deep waters. Figure 3(b) shows that the computed Kuroshio velocities in the upper 100 m of point S_4-4 are all greater than 80 cm/s.

There is a cyclonic cold gyre to the west of the Kuroshio on the shelf north of Taiwan (Fig. 4). A cyclonic gyre is often found here (Su and Pan, 1987; Yuan and Su, 1988; YEI, 1990). East of this cold gyre and west of the Kuroshio there is a northward current over the shelf which seems to be the offshore branch of the Taiwan Warm Current (TWCOB) (Su and Pan, 1987; Yuan et al., 1987) with a transport of about 0.11×10^6 m³/s. This may be an underestimate, because the distances between the shelf stations are too large to resolve the horizontal density gradient and because the data noise is not completely removed in our inverse method to make a reliable estimate of the small transport. In fact, the inverse method is better suited for the computation of the velocity field in a deep ocean than in a shallow ocean.

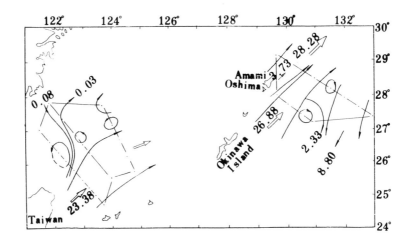

Fig. 4. Distribution of the transport in the computational region during early summer, 1988. (units: 10^6 m³/s).

3.2 Autumn (October-November) of 1988

In autumn 1988 the Kuroshio again has only one current core at section S_2. Its core is located at point S_2-7, as during the early summer cruise of 1988. The Kuroshio is stronger than during the early summer cruise. The velocities in the upper 250 m at the core are greater than 100 cm/s and the maximum velocity of 117.3 cm/s is at the 100 m level. Even in deep layers, its velocities are still quite strong. For example, at point S_2-7 its speed is 82.4, 27.6 and 10 cm/s at the 400, 800 and 1000 m level, respectively. The width of the Kuroshio is about 110 km, much narrower than during other cruises (September-October of 1987 and May-June of 1988). The transport of the Kuroshio through section S_2 is 24.3×10^6 m³/s (Fig. 6). There is a region of strong currents near the shelf break around point S_2-4. The maximum velocity in this region is 43.9 cm/s, while the maximum

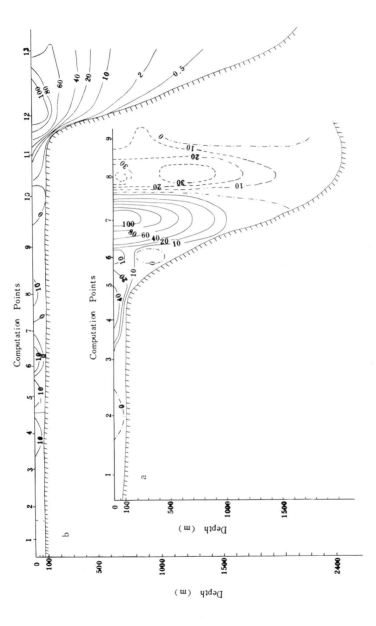

Fig. 5. Velocity distribution at (a) section S₂ and (b) section S₅ during the autumn cruise (positive value denotes northward flow). (units: cm/s).

312

velocity in this region during the early summer cruise is 23.9 cm/s.

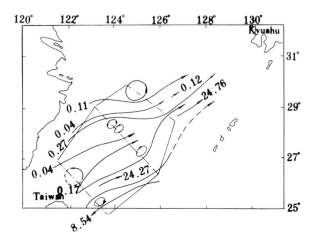

Fig. 6. Distribution of the transport in the computaitonal region during autumn 1988. (units: 10^6 m³/s).

There is a strong countercurrent at points S_2-8 and S_2-9. Its maximum speed is 42.1 cm/s at the 75 m level of point S_2-8 (Fig. 5a). The speed is 32.4, 17.8 and 6.9 cm/s at 1000, 1200 and 1500 m level of point S_2-8, respectively (Fig. 5a). The width of this countercurrent is about 55 km and its transport is about 8.5x10⁶ m³/s. A countercurrent is also found here during the September-October cruise of 1987, but its position was further to the east and both its width and volume transport were smaller (YEI, 1990).

At section S_5 (Fig. 5b) the core of the Kuroshio is located near the shelf break at point S_5-12. Its velocities in the upper 150 m of the core are all greater than 80 cm/s. The maximum velocity of 131.5 cm/s is at the 50 m level. The Kuroshio axis seems to move gradually eastward below 250 m. This characteristic was also found in previous studies (Yuan and Su, 1988; YEI, 1990).

Like the early summer cruise result there is also a cyclonic cold gyre to the west of the Kuroshio on the shelf north of Taiwan (Fig. 6). The position of this gyre is almost the same as that during early summer of 1988. The northeastward current which intrudes onto the shelf has a transport of about 0.17x10⁶ m³/s.

As pointed out before, this value may be an underestimate.
West of the gyre there is another northeastward current which
seems to originate from the Taiwan Strait. The total transport
of both northeastward currents into this computational region is
about 0.63×10^6 m³/s. It is worthy to note that this value is not
the total transport of the Taiwan Warm Current (TWC), because the
computational region does not include the coastal region. In
previous studies (Yuan and Su, 1988; YEI, 1990; Yuan, Su and Ni,
1990), a part of the TWC had a tendency to converge to the shelf
break and joins the Kuroshio, which is similar to the present
result.

3.3 Comparison of results from the three cruises

We shall discuss the computed results from the three
cruises, namely, September-October of 1987 (YEI, 1990), early
summer and autumn of 1988. Figure 7 shows the sketch of the
Kuroshio width and the position of the main Kuroshio axis for the
three cruise. The following main features are found:

1) The intrusion of the Kuroshio water onto the shelf during
the autumn of 1988 is close to the winter feature (Su and Pan,
1987). This means that the position of the Kuroshio is further
into the shelf and its width is narrower during the autumn of
1988 than its respective positions and widths in both late summer
1987 and early summer 1988. In addition, the countercurrent's
position is further to the west and its transport is greater
during autumn of 1988 than its respective positions and volume
transports in both late summer 1987 and early summer 1988 (Fig.
7a). The Kuroshio undergoes a slight meander near 26°30'N (Fig.
7), which may be related to the change of orientation of the
shelf-break line as seen from the position of the 200 m isobath
near 26° 30'N (Fig. 7). Finally, the transport of the Kuroshio in
the East China Sea at the section north of Taiwan is 25.8, 23.4
and 24.3×10^6 m³/s during the late summer 1987, early summer and
autumn 1988, respectively.

2) The Kuroshio Countercurrent's position is found to be
furthest to the west and its transport is the largest (8.5×10^6 m³
/s) during the autumn 1988 cruise among the three cruises.

3) In all the three cruises, the position of the main axis
of the Kuroshio (the position of the maximum surface speed)
is found immediately east of the 200 m isobath (Fig. 7b). At
section S_2 the depth at this position is 210, 1260 and 1260m

during late summer 1987, early summer 1988 and autumn 1988, respectively. During late summer 1987 shelf-intrusion of the Kuroshio Surface Water was weak so that the Kuroshio Subsurface Water was lifted to the shelf (Su and Pan, 1987). This results in large density gradient and hence the surface velocity of the Kuroshio is larger near the shelf break (≈ 200 m isobath) than at other positions. It is worthy to note that the seasonal change of the centerline position of the Kuroshio width (Fig. 7a) does not correlate with the seasonal change of its main axis position (Fig. 7b).

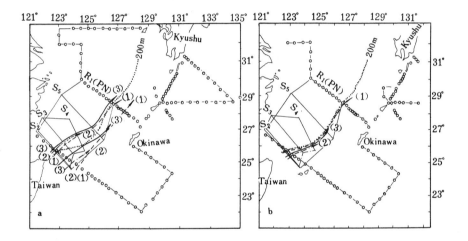

Fig. 7. The sketch of (a) the Kuroshio width and (b) the position of the main Kuroshio axis.
Sep.-Oct. of 1987: (1) - - - ,
May -June of 1988: (2) -·-·- ,
and Oct.-Nov. of 1988: (3) ———— .

4) During all the three cruises, there is a cyclonic cold gyre west of the Kuroshio on the shelf north of Taiwan. Its position in late summer 1987 is further to the southeast than during the two 1988 cruises. This result is in agreement with the fact that in summer 1987 the extent of the Kuroshio water intrusion onto the shelf is less than during the two 1988 cruises.

4 THE CURRENTS EAST OF THE RYUKYU ISLANDS DURING EARLY SUMMER
   1988
      This section presents the current structure and volume

transport at sections F_8 and F_9.

4.1 Section F_8

There is a large topographic variation along section F_8 (Fig. 8). The water depth at the west end of section F_8 is shallow (about 1200 m), but at the middle of the section there is a trench (the Ryukyu Trench) 5820 m deep. East of the Ryukyu Islands there is a weak northward current over the Ryukyu Trench. Its maximum surface speed is about 17.8 cm/s and its speed increases slightly with depth, reaching the maximum velocity of 38.2 cm/s at the 400 m level of point F_8-1. Below the 400 m level its maximum speed shifts gradually eastward to point F_8-2. The core of this flow lies between 400 and 800 m depths. This feature was also found in a previous study (YEI, 1990).

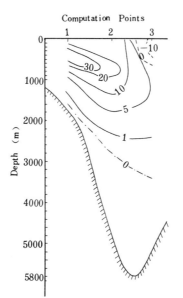

Fig. 8. Velocity distribution at section F_8 during early summer 1988 (positive value denotes northward flow). (units: cm/s).

4.2 Section F_9

The topographic variation is much larger along section F_9 than along section F_8. The topography of the western half of section F_9 resembles the topography of section F_8. In the eastern half of section F_9 there is a ridge and a deep basin.

West of the ridge there is a northward current, and east of the
ridge there is a southward current in the upper 700 m layer (Fig.
9). Most of the northward flow come through section F₈, and the
remainder comes from the Tokara Strait (Fig. 4). The surface
velocities of the northward current vary from 9 to 30 cm/s. The
velocity increases gradually with depth and reaches the maximum
of 32.1 cm/s at 600 m level of point F₉-3. At point F₉-3 its
velocity is 18.2, 10.9, 5.8 and 1.7 cm/s at 1000, 1200, 1500 and
2000 m levels, respectively (Fig. 9).

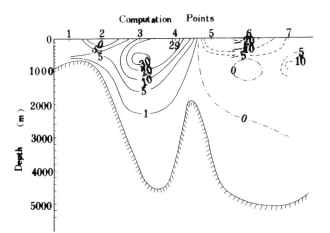

Fig. 9. Velocity distribution at section F₉ during early summer
1988 (positive value denotes northward flow). (units: cm/s).

Near the ridge there is an anticyclonic gyre between the
northward and southward currents (Figs. 4 and 9). The maximum
speed of the southward current, 25.9 cm/s, is at the surface of
point F₉-6. In the southward current there is a small core of
the northward current between 650 to 1300 m water depths at
point F₉-6, but its strength is rather weak with the maximum
velocity of about 4.1 cm/s. Below the southward current the flow
is again northward. Its speed is only about 0.1 cm/s or less.

4.3 The volume transport
Figure 4 shows that the northward current east of the Ryukyu
Islands flows through the western part of section F₈ and its
transport is 26.88x10⁶ m³/s. Most of this current continues to
flow northward through section F₉ with a transport of 24.55x10⁶

m³/s. The remaining 2.33x10⁶ m³/s of this current turns eastward
from the eastern part of section F₈ and makes an anticyclonic
meander to flow southward through section F₂ (Fig. 4). There is
also another weaker northward current coming through sections F₁
and F₉ near the Tokara Strait. Its transport is about 3.73x10⁶ m³
/s. The totals of the northward and southward transports at
section F₉ are 28.28x10⁶ m³/s and 8.80x10⁶ m³/s, respectively
(Fig. 4). In the next place, the transports of the northward
current through section F₈ in the first to fifth layers are 2.16,
4.18, 8.68, 10.29 and 1.57x10⁶ m³/s, respectively, while those
through section F₉ in the first to fifth layers are 2.82, 5.11,
9.41, 9.14 and 1.79x10⁶ m³/s, respectively. The transport of the
southward current through section F₂ in the first to fifth layers
are 2.69, 1.78, 1.67, 2.34 and 2.66x10⁶ m³/s, respectively.

The hydrographic structure in the upper water layer shows a
temperature decreases from the west to the east (Fig. 10a) and
does not reflect a northward current. However, the distribution
of temperature below the 400 m level (Fig. 10b) differs from that
in the upper layer, reflecting the existence of the northward
current. Thus, the measurement of hydrographic structure in the
deep layer is important to understanding the current feature east
of the Ryukyu Islands.

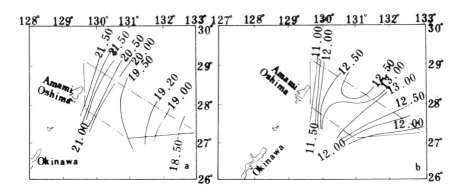

Fig. 10. Temperature distribution at (a) 200 m and (b) 500 m
levels east of the Ryukyu Islands in May-June, 1988.

5 SUMMARY

Based on hydrographic data from both the early summer and
autumn 1988 cruises, the Kuroshio in the East China Sea and the

currents east of the Ryukyu Islands are studied by the inverse method. It is found that:

1) The current pattern in the East China Sea during Oct.-Nov. 1988 is close to the winter pattern. The features are: (1) The positions of the main Kuroshio Current and countercurrent are both further to the west than during other cruises; (2) The countercurrent is stronger and the width of the Kuroshio is narrower during Oct.-Nov. 1988 than during other cruises.

2) The seasonal change of the Kuroshio axis (the position of the maximum surface current) does not correlate with the seasonal change of the position of the centerline of the Kuroshio Current width.

3) The transport of the Kuroshio in the East China Sea at the section north of Taiwan is 25.8, 23.4 and 24.3×10^6 m^3/s, respectively, during the late summer 1987, early summer and autumn 1988. During all these three cruises a cyclonic cold gyre is present west of the Kuroshio on the shelf north of Taiwan.

4) During autumn 1988 the transport of the Taiwan Warm Current (TWC) flowing into our computational region is about 0.63×10^6 m^3/s, which is not the total transport of the TWC because the computational region excludes the coastal region.

5) In the survey area the current east of the Ryukyu Islands flows northward over the Ryukyu Trench during early summer 1988, of which the transport at sections F_8 and F_9 is about 26.88 and 28.28×10^6 m^3/s, respectively. This northward current is not strong throughout the depths. Its core lies between 400 and 800 m.

6 REFERENCES

Guan Bingxian, 1982. Analysis of the variations of volume transport of Kuroshio in the East China Sea. In: Proceedings of the Japan-China Ocean Study Symposium. Oct., 1981, Shimizu, pp. 118-137.

Guan Bingxian, 1988. Major feature and variability of the Kuroshio in the East China Sea. Chin. J. Oceanol. Limnol. 6(1): 35-48.

Konaga, S., Nishiyama, K., Ishizaki, H. and Hanazawa, Y., 1980. Geostrophic Current Southeast of Yakushima Island. La Mer, 18: 1-16.

Nishizawa, J., Kamihira, E., Komura, K., Kumabe, R. and Miyazaki, M., 1982. Estimation of the Kuroshio mass transport flowing out of the East China Sea to the North Pacific. La Mer, 20: 37-40.

Saiki, M., 1982. Relation between the geostrophic flux of the Kuroshio in the Eastern China Sea and its large-meanders in south of Japan. The Oceanographical Magazine, 32(1-2): 11-18.

Su Jilan and Pan Yuqiu, 1987. On the shelf circulation north of Taiwan. Acta Oceanologica Sinica, 6 (supp. I): 1-20.

Yuan Yaochu, Su Jilan and Xia Songyun, 1987. Three dimensional diagnostic calculation of circulation over the East China Sea Shelf. Acta Oceanologica Sinica, 6 (supp. I): 36-50.

Yuan Yaochu and Su Jilan, 1988. The calculation of Kuroshio Current Structure in the East China Sea - Early Summer 1986. Progress in Oceanography, 21: 343-361.

Yuan Yaochu, Endoh Masahiro and Ishizaki Hiroshi, 1990. The Study of the Kuroshio in the East China Sea and Currents East of Ryukyu Islands. In: Proceedings of Japan China Joint Symposium of the Cooperative Study on the Kuroshio, Science and Technology Agency, Japan & SOA, China, pp. 39-57.

Yuan Yaochu, Su Jilan and Pan Ziqin, 1990. Calculation of the Kuroshio Current South of Japan During Dec., 1987-Jan., 1988. In: Proceeding of the Investigation of Kuroshio (II) (in press).

Yuan Yaochu, Su Jilan and Ni Jufen, 1990. A Prognostic Model of the Winter Circulation in the East China Sea. In: Proceedings of the Investigation of Kuroshio (II) (in press).

THREE NUMERICAL MODELS OF GUANHE ESTUARY

ZHANG DONGSHENG AND ZHANG CHANGKUAN
Hohai University
Nanjing, China

ABSTRACT

Three numerical models of Guanhe estuary have been set up for different practical motives. The first model, dimension combined model, is to study the circulation pattern and to evaluate the influence of the river discharge on the current pattern in the region arround the entrance. An explicit characteristic difference scheme is used for the inner part while a kind of triangular element method is used for the outer part. The second one is to simulate current field, sediment transport and the corresponding bed variation. This second model is discreted by ADI scheme in which the technique of non—uniform grids and movable boundary is introduced to improve the accuracy of the simulation in the vicinity of the river mouth. The third model has been developed with curvilinear orthogonal coordinate system to improve the computed current field close to the physical boundary.

The current fields obtained by the three models are similar and show their own advantages in representing the nature.

1 INTRODUCTION

The Guanhe estuary, with an irregular geometry and a seabed topography complicated by shoals at the entrance, is one of important estuaries entering the East China Sea, Jiangsu provence(Fig.1). Tides enter the estuary up to the top, 75km from the entrance. Freshwater enters the entrance through four sluice gates during flood season and mixes with sea water in the estuary. The numerical simulation of Guanhe estuary is of scientific significance on the studies about the interaction of sea water and the river discharge and about the formation of the entrance bar. The work so far carried out is for the purpose of predicting beach response and shoal change due to natural forces and constructed works. Three numerical models of Gaunhe estuary have been set up for different practical motives.

The first model, an one—two dimension combined model, is to study the circulation pattern, and to evaluate the influence of the river discharge on the current pattern in the region arround the entrance. In this model an explicit characteristic difference method is used for the inner part and a kind of triangular element method with three nodes is used for the outer part. The model is successful in simulating tidal current fields, especially in demonstrating the influence of Guanhe river discharge on the current pattern, but the current structure in the vicinity of the entrance is not accurate enough to represent the influence of complicated topography near the entrance bar. The second model is set up to simulate the tidal current, suspended sediment fields and corresponding bed variation. The modelled domain includes the outer part and an 11km long inner part. This model is discreted by ADI scheme. To improve the accuracy of the simulation in the vicinity of the entrance, a technique of non–uniform grids and movable boundary is introduced. The third model has been developed with curvilinear orthogonal coordinate system for the purpose of fitting the boundary better and considering the effect of the river bend on the current pattern in the vicinity of the entrance. Based on the solution of a set of Laplace equations, a curvilinear orthogonal grid system is generated by fixing a part of boundary points and moving the others. ADI scheme has been extended for the transformed shallow water equations in curvilinear coordinate system.

2 MODELS

2.1 Basic equations

2.1.1 Basic equations with Cartesian coordinate system

(i) Equation for one dimensional calculation (for the inner part of the first model)

$$\frac{\partial Q}{\partial x} + b\frac{\partial H}{\partial t} = 0$$

(1)

$$\frac{\partial u}{\partial t} + u\frac{\partial u}{\partial x} + g\frac{\partial H}{\partial x} + g\frac{u|u|}{C_1^2 R} = 0$$

where H is the total depth of flow ($H = D + h$, D: water depth with respect to the chart datum, h: water surface elevation above the chart datum); Q is section discharge; u is section mean velocity; b is water surface width; R is hydraulic radius; g is gravitational acceleration and C_1 is Chezy coefficient.

(ii) Equations for two dimensional calculation (for the outer part of the first model and the whole domain in the second model)

$$\frac{\partial h}{\partial t} + \frac{\partial}{\partial x}(Hu) + \frac{\partial}{\partial y}(Hv) = 0$$

$$\frac{\partial u}{\partial t} + u\frac{\partial u}{\partial x} + v\frac{\partial u}{\partial y} - fv + g\frac{\partial h}{\partial x} + \frac{g}{C_1^2}\frac{u\overline{V}}{R} = 0$$

(2)

$$\frac{\partial v}{\partial t} + u\frac{\partial v}{\partial x} + v\frac{\partial v}{\partial y} + fu + g\frac{\partial h}{\partial x} + \frac{g}{C_1^2}\frac{v\overline{V}}{R} = 0$$

where u and v are depth–averaged velocity components in x and y horizontal coordinate directions, respectively; f is Coriolis parameter; $\overline{V}$ is velocity modulus ($\overline{V} = \sqrt{u^2 + v^2}$); and the others are same as in the equation (1).

(iii) Equation of suspended sediment transport

$$\frac{\partial(HC)}{\partial t} + \frac{\partial(HuC)}{\partial x} + \frac{\partial(HvC)}{\partial y} - \frac{\partial}{\partial x}(HD_x\frac{\partial C}{\partial x}) - \frac{\partial}{\partial y}(HD_y\frac{\partial C}{\partial y}) - \alpha\omega(\beta S_* - \gamma C) = 0$$

(3)

where, C is depth–averaged sediment concentration, α is fall velocity of sediment particles, S_* is capacity carrying sand by flow, α is settling probability of sand particles, D_x and D_y are dispersion coefficients along x and y directions, respectively, β and γ are specified by the follows:

$$\beta = \begin{cases} 1 & \text{when} \quad u,v \geqslant u_c \\ 0 & \text{when} \quad u,v < u_c \end{cases}$$

$$\gamma = \begin{cases} 1 & \text{when} \quad u,v \geqslant u_f \\ 0 & \text{when} \quad u,v < u_f \end{cases}$$

where u_f is velocity for suspending sand particles and u_c is threshold velocity.

2.1.2 Basic equations with curvilinear orthogonal coordinate system

(i) Equations of unsteady flow in shallow water (for the third model)

$$C_\xi C_\eta \frac{\partial \zeta}{\partial t} + \frac{\partial [C_\eta (h + \zeta)u]}{\partial \xi} + \frac{\partial [C_\xi (h + \zeta)v]}{\partial \eta} = 0$$

$$\frac{\partial u}{\partial t} + \frac{u}{C_\xi}\frac{\partial u}{\partial \xi} + \frac{v}{C_\eta}\frac{\partial u}{\partial \eta} + \frac{uv}{C_\xi C_\eta}\frac{\partial C_\xi}{\partial \eta} - \frac{vv}{C_\xi C_\eta}\frac{\partial C_\eta}{\partial \xi}$$
$$+ gu\frac{\sqrt{uu + vv}}{C_1^2(h + \zeta)} - fv + \frac{g}{C_\xi}\frac{\partial \zeta}{\partial \xi} + \frac{v}{C_\eta}\frac{\partial B}{\partial \eta} - \frac{v}{C_\xi}\frac{\partial A}{\partial \xi} = 0 \qquad (4)$$

$$\frac{\partial v}{\partial t} + \frac{u}{C_\xi}\frac{\partial v}{\partial \xi} + \frac{v}{C_\eta}\frac{\partial v}{\partial \eta} + \frac{uv}{C_\xi C_\eta}\frac{\partial C_\eta}{\partial \xi} - \frac{uu}{C_\xi C_\eta}\frac{\partial C_\xi}{\partial \zeta}$$
$$+ gv\frac{\sqrt{uu + vv}}{C_1^2(h + \zeta)} + fu + \frac{g}{C_\eta}\frac{\partial \zeta}{\partial \eta} + \frac{v}{C_\eta}\frac{\partial B}{\partial \eta} - \frac{V}{C_\xi}\frac{\partial A}{\partial \xi} = 0$$

(ii) Basic Equations for generating curvilinear orthogonal coordinate

$$\frac{\partial}{\partial \xi}(\frac{C_\eta}{C_\xi}\frac{\partial x}{\partial \xi}) + \frac{\partial}{\partial \eta}(\frac{C_\eta}{C_\xi}\frac{\partial x}{\partial \eta}) = 0$$

$$\frac{\partial}{\partial \xi}(\frac{C_\eta}{C_\xi}\frac{\partial y}{\partial \xi}) + \frac{\partial}{\partial \eta}(\frac{C_\xi}{C_\eta}\frac{\partial y}{\partial \eta}) = 0 \qquad (5)$$

where ξ and η are variables in curvilinear orthogonal coordinate system and

$$C_\eta = \sqrt{(\frac{\partial x}{\partial \eta})^2 + (\frac{\partial y}{\partial \eta})^2} , C_\xi = \sqrt{(\frac{\partial x}{\partial \xi})^2 + (\frac{\partial y}{\partial \xi})^2}$$

and the corresponding boundary conditions are:

$$x = x_1 \quad (\xi = C_1, \eta)$$
$$y = y_1 \quad (\xi = C_1, \eta)$$

on the boundary of $\xi = C_1$

$$x = x_2 \quad (\xi = C_2, \eta)$$
$$y = y_2 \quad (\xi = C_2, \eta)$$

on the boundary of $\xi = C_2$

$$x = x_3 \quad (\xi, \eta = d_1)$$
$$y = y_3 \quad (\xi, \eta = d_1)$$

on the boundary of $\eta = d_1$

$$x = x_4 \quad (\xi, \eta = d_2)$$
$$y = y_4 \quad (\xi, \eta = d_2)$$

on the boundary of $\eta = d_2$

2.2 Numerical models

2.2.1 One--two dimension combined model (model 1)

In this model an explicit characteristic difference method is used for a 75km long inner part, which is divided into 30 segments about 2km long on the average. According to the Courant condition, the time step adopted is 150 sec. A kind of trianglar element method with three nodes is used for the outer part which has a size of about 60 * 40km. The shape of element is designed to fit the topographic feature of Guanhe estuary (Figs.1 and 2), and the size of elements is smaller for nearshore area than for offshore. The time step is 300 sec. Tidal levels are employed as boundary conditions at the open seaward boundaries. The measured water level variations are provided by the gauge stations located at Liu Wei and Zhong Shan, as shown in Fig.1. The water level variations at the rest part of the open boundaries are obtained throught linear interpolation considering some damping seaward (Zhang Dongsheng et al. 1987).

2.2.2 Two dimensional model of suspended sediment transport (model 2)

The modelled domain including a river section is about 60km * 25km and the upward boundary of the river section is located at Chen Gang, 11km from the entrance (Fig.1). The simulated area is discreted by a grid of 1200m * 1200m. For the sake of closely fitting the complicated topography near the entrance, where the river width is 800-1000m only, finner grid with spacing of 200m is adopted (Fig.3). The time step used is 150 sec for current field computation and 450 sec for the sediment transport computation.

Parameters such as S_*, ω, etc. are important for the sediment transport calculation. The capacity carrying sand by flow, S_*, is obtained by the following formula (Liu Jiaju, 1980) :

$$S_* = 0.0273 \rho_s \frac{(|V_1| + |V_2|)^2}{g H_1}$$

where ρ_s is the denisty of sand, V_1 is the combined velocity of tidal and wind induced currents and V_2 is orbital velocity of wave motion.

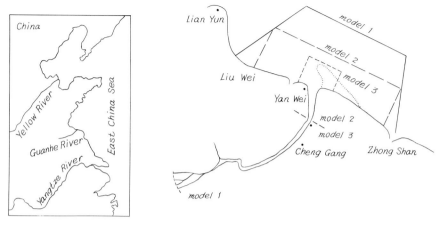

Fig.1　Map of Guanhe estuary showing the location of survey
area and boundaries of simulation domains

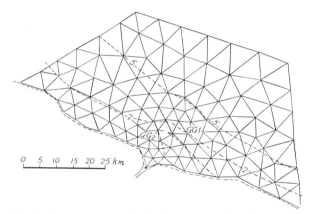

Fig.2　Simulation area divided into triangular elements of model 1
(Dash lines stand for bottom topography)

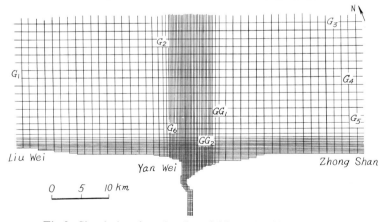

Fig.3　Simulation domain of model 2 resolved by ADI scheme

The fall velocity, ω, is calculated by (Zhang Ruijin, 1989) :

$$\omega = \frac{1}{24} \frac{\rho_s - \rho}{\rho_s} g \frac{d^2}{\gamma}$$

where d is the particle size of sands; γ is the kinematic viscosity of fluid. The threshold velocity, u_c, and the velocity for suspending sand particles, u_f, are calculated by (Qian Ning et al. 1983) :

$$u_c = (\frac{H}{d})^{0.14} (17.6 \frac{\rho_s - \rho}{\rho_s} d + 6.05 \times 10^{-7} \frac{1.0 + H}{d^{0.72}})^{\frac{1}{2}}$$

$$u_f = 0.812 d^{\frac{2}{5}} \omega^{\frac{1}{5}} H^{\frac{1}{5}}$$

where ρ ,ρ_* , are densities of water and sand, respectively.   The boundary condition for sediment transport is:

$$\frac{\partial(HC)}{\partial t} + \frac{\partial(HuC)}{\partial x} + \frac{\partial(HvC)}{\partial y} = 0 \quad \text{(in the case of outflow)}$$

or

$$C = C_* \quad \text{(in the case of inflow)}$$

where C_* is measured sediment concentration at G1,G3,G4 and G5 (refer to Fig.3 for their locations).

2.2.3 Two dimensional numerical model for unsteady flow with curvilinear orthogonal coordinate (model 3)

In consideration for the influence of river bend on the current pattern in the vicinity of the entrance, the modelled domain is limited to a smaller region. That is 15km * 10km. The upstream boundary is the same as that in the model 2. The grid has been formed according to Eq.(5) (Lau P C M, et al. 1979, Middlecoff J F, et al. 1979) and the spacing varies from 160m to 250m. Fig.4 shows the grid in physical domain and Fig.5 shows the grid in the calculating domain. The time step used is 100 sec.

3 MODEL VERIFICATION AND SIMULATED RESULTS

To collect the field data of tidal currents and water levels, four field measurements in Guanhe estuary were carried out in 1980,1981 and 1985 (two measurements), respectively. Parts of the field data obtained from the measurements have been employed to verify the above three models.

Fig.6 is the comparison of tidal currents between the measured and the calculated by the

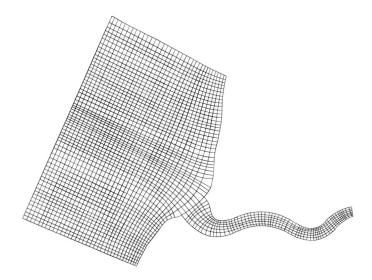

Fig.4 Grid of model 3 in physical domain

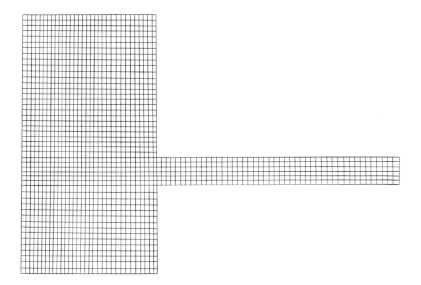

Fig.5 Grid of model 3 in calculating domain

328

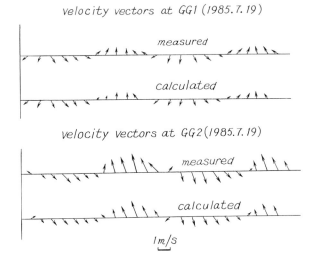

velocity vectors at GG1 (1985. 7. 19)

measured

calculated

velocity vectors at GG2 (1985. 7. 19)

measured

calculated

1 m/s

Fig.6 Comparison between field data and calculated results

water level at Yan Wei

----- measured

——— calculated by model 2

× × × × calculated by model 3

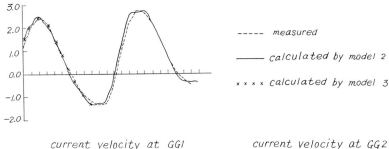

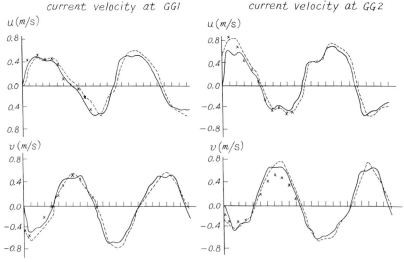

current velocity at GG1

current velocity at GG2

$u(m/s)$

$v(m/s)$

Fig.7 Comparison between field data and calculated results

model 1 at points GG1 and GG2 showing a good agreement between measured and calculated both in the magnitude of velocity and in the current direction. Fig.7 is the comparison of water levels at Yan Wei and tidal current velocities at GG1 and GG2 between field data and numerical results. In Fig.7 dot lines, solid lines and cross marks stand for observed values, calculated values by the model 2 and the model 3, respectively. The comparison shows that the simulated values, either the water levels or tidal current velocities, as well as the velocity phases agree well with the measured. Fig.8 gives measured tidal current vectors during a complete tidal period at different observation stations. Figs.9 and 10 demonstrate simulated tidal current vectors in Guanhe estuary during a complete tidal period obtained from the model 1 and the model 2, respectively. Figs.11 and 12 are tidal current vectors at flood tide and ebb tide obtained from the model 3. From the comparisons it can be concluded that either the current speeds and the directions at GG1 and GG2 or the general tidal current patterns in the whole estuary simulated by the three models are similar with the field observations.

Fig.6 shows that the tidal current is semidiurnal, counter clockwise rotatory, and the calculated tidal ellipses are essentially the same as that obtained from field observations. Comparing the Figs.9, 10, 11, and 12 with the Fig.8, it can be found that the dominant current directions at flood and ebb tides on the western side of the estuary are SE, as well as NW and SSE and WNW on the eastern side, respectively. The Fig.9 indicates that there is a small area near the entrance within which the tidal velocity is particularly smaller than that of adjacent area. The magnitude of velocity is about 0.2−0.5m／s, and at the position of shoals it reduces to 0.1m／s. Fig.9 shows the boundary of the small velocity area in dash line. It would be worth pointing out that the water depth in the small velocity area is so shallow that observation boats can not reach this area. The existence of this small velocity area interprets the formation of the entrance bar, at least from the viewpoint of hydrodynamic influence.

The computation results suggest that there is only a little difference in the current magnitude and direction between flood and dry seasons. The mean current velocities of flood and ebb tides during flood season are 0.01−0.05m／s larger than those during the dry season, and the region affected by the river discharge is limited to a small area near the entrance. This is because of the small river discharge released from four sluice gates, only 200m^3／s on the average during the flood season. Therefore, it can be concluded that river discharge variation under normal condition makes less contribution towards the current pattern of the outer part of the estuary.

Suspended sediment calculation has given acceptable results showing good agreement with field data. Fig.13 gives the comparison between measured sediment concentration and computed at two points G2 and G6(refer to Fig.3 for their positions). Figs.14 and 15 are the distrbutions of suspended sediment concentration in Guanhe estuary during the flood tide and the ebb tide, respectively, obtained from the model 2. This two figures show that a reasonable distribution of suspended sediment concentration has been achieved from the numerical model. The value of sediment concentration is 1.0kg／m^{3}on the average in the area around the entrance and decreases gradually seaward. During the period of flood tide, the suspended sedi-

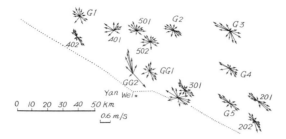

Fig.8   Measured tidal current vectors within a complete tidal period

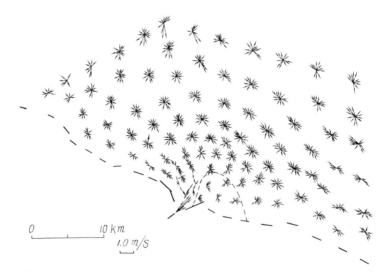

Fig.9   Tidal current vectors within a complete tidal period simulated by the model 1
(the area near the entrance)

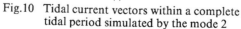

Fig.10   Tidal current vectors within a complete
tidal period simulated by the mode 2

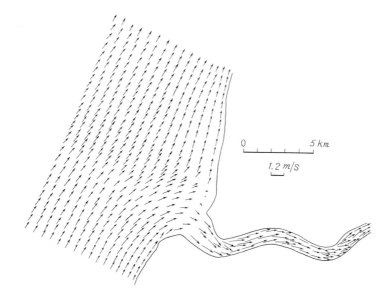

Fig.11 Tidal current vectors at flood tide by the model 3

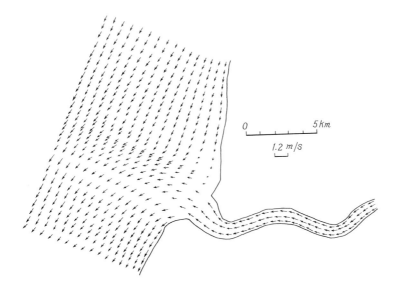

Fig.12 Tidal current vectors at ebb tide by the model 3

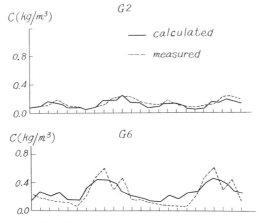

Fig.13　Comparison between the measured sediment
concentration and simulated by the model 2

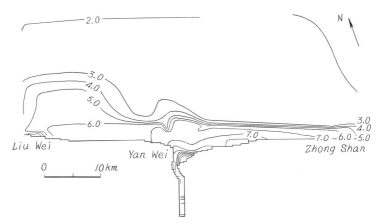

Fig.14　Distribution of sediment concentration at flood
tide obtained from the model 2(numbers in 0.1kg / m³)

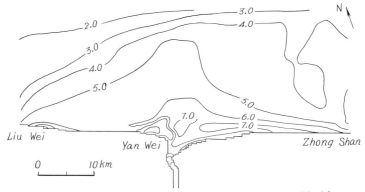

Fig.15　Distribution of sediment concentration at ebb tide
obtained from the model 2(numbers in 0.1kg / m³)

ment is pushed into the shore while pulled off during ebb tide. The range of moving of suspended sediments during a complete tide period is about 10km.

4 DISCUSSION

The verifications between numerical results from the three models 1 to 3 and field measured data indicate that all the three models are suitable for tidal current simulation of Guanhe estuary, and the model 2 can give an acceptable estimate and a reasonable distribution of the sediment concentration. For different practical motives, by judging computation cost, accuracy requirement, etc., we can choose one of them. If we are more interested in the influence of the river discharge on the current pattern in the estuary area, then the model 1 is preferable. If we want to know more about the effect of complicated topography near the entrance bar on the current structure in the estuary region, the model 2 is better. If we are more concerning the fitting of physical boundaries, especially in the case of treating problems involving coastal engineering in which the velocities close to the coast or constructed works become more interesting, certainly the model 3 should be on the top of list of candidates.

ACKNOWLEDGMENTS

The authors would like to thank Mr. Zheng Xiaoping, Mr. Jiang Qin and Mr. Liu Jinfang for writting computer programs and for performing computations. This work was supported by the National Foundation for Natural Scientific Research.

REFERENCES

Lan P C M, 1979. Curvilinear finite difference method for biharmonic equation. International Journal for Numerical Method in Eng. Vol. 14, No.6, PP. 791–812.

Lin Jiaju, 1980. Siltation calculation and prediction of the approach channel of Lianyun Harbour. Journal of Water Conservancy and Water Transpotation. No. 4. (in Chinese)

Middlecoff J F and Thomas P D, 1979. Direct control of the grid point distribution in meshes generated by elliptic equation. AIAA Computation Fluid Dynamics Conference, July.

Qian Ning and Wan Zhaohui ,1983. Mechanics of Sands. Science press (in Chinese)

Zhang Dongsheng, Xie Jingzhan and Zheng Xiaoping, 1987. The numerical simulation of hydraulic factors of Guanhe Estuary. Proceedings of Coastal and Port Eng. in Developing Countries, Vol.II, PP.2215–2224.

Zhang Ruijin, 1989. Dynamics of river sand. Water Conservency and Hydro power Press (in Chinese)

GEOSTROPHIC CURRENT AND ASSOCIATED VERTICAL MOTION OFF NORTHEASTERN TAIWAN

Hsien–Wen Li
Department of Oceanography, National Taiwan Ocean University, Keelung 20224, Taiwan, CHINA

ABSTRACT

The geostrophic velocity changes its direction with increasing depth by thermal wind relationship. The rate of rotation with depth is related to the vertical velocity by Hide's theorem. A method for calculating the three–dimensional absolute velocity by means of Hide's theorem and beta vorticity dynamics is developed. It is applied to hydrographic data in October 1987 and May 1988. The Kuroshio northeast of Taiwan flows nearly along the bottom contour. There exists a small cold core off northeastern Taiwan. The position of the Kuroshio's axis changes little in this period. The vertical velocity is mostly upward in the central part and west side of the Kuroshio and mostly downward east of the Kuroshio. Its typical magnitude is $10^{-4} \sim 10^{-6}$ m s^{-1}.

INTRODUCTION

In a rotating fluid, the Taylor–Proudman theorem implies that the horizontal velocity of a rotating homogeneous fluid is independent of the direction parallel to the rotation axis, and thus the motion is completely two–dimensional. However, the velocity field of a non–homogeneous fluid is three–dimensional.

Hide (1971) established a general theoretical relationship between the three–dimensional velocity field and density field for geostrophic flow of a non–homogeneous fluid. He found an equation for the rate of change with respect to the rotation axis of the direction of the horizontal flow. Schott and Stommel (1978) also derived an equation for the rate of turn of geostrophic velocity with height in the main thermocline, which they called the beta spiral by adding the beta vorticity constraint.

The atmosphere and the oceans are non–homogeneous. In the atmosphere strong vertical motion occurs in the neighborhood of a front (e.g., Holton, 1972). In regions of cold or warm advection, the geostrophic wind backs or veers with height. Although the fronts are of the nature of highly ageostrophic phenomena, they are necessary concomitants of highly geostrophic flow (Hide, 1971). Thus there exists a relationship between the vertical motion and the turning of geostrophic velocity with height or depth near the fronts in the atmosphere and the oceans.

The Kuroshio is an oceanic front in a sense that the temperature difference is remarkable across it. In this work a version of Hide's theorem is derived to depict the relationship between the rate of rotation of the horizontal velocity and the vertical distributions of temperature and salinity.

THEORY AND METHOD FOR DETERMINATION OF THE ABSOLUTE VELOCITIES

The thermal wind equation for the ocean is (e.g., Gill, 1982)

$$f \frac{\partial v}{\partial z} = g\alpha \frac{\partial T}{\partial x} - g\gamma \frac{\partial S}{\partial x} \tag{1}$$

$$-f \frac{\partial u}{\partial z} = g\alpha \frac{\partial T}{\partial y} - g\gamma \frac{\partial S}{\partial y} \tag{2}$$

where x, y and z are Cartesian coordinates with x to the east, y to the north and z upward. The velocity components in x, y and z direction are denoted by u, v and w, respectively, f is the Coriolis parameter, g the gravity acceleration, T the temperature, α the thermal expansion coefficient, S the salinity and γ the expansion coefficient for salinity.

We have steady–state equation for temperature and salinity,

$$u \frac{\partial T}{\partial x} + v \frac{\partial T}{\partial y} + w \frac{\partial T}{\partial z} = \frac{DT}{Dt} = \kappa_t \nabla^2 T \tag{3}$$

$$u \frac{\partial S}{\partial x} + v \frac{\partial S}{\partial y} + w \frac{\partial S}{\partial z} = \frac{DS}{Dt} = \kappa_s \nabla^2 S \tag{4}$$

where κ_t and κ_s are molecular kinematic diffusivities for temperature and salt, respectively.

In this study we consider only the large–scale geostrophic current far from the boundary layers, thus the eddy diffusivities due to turbulent small–scale flow may be neglected. We retain the molecular diffusivities according to physical nature of diffusion due to the existence of gradients of temperature and salinity.

If we combine (1) multiplied by u with (2) multiplied by v, we get

$$f(u \frac{\partial v}{\partial z} - v \frac{\partial u}{\partial z}) = g\alpha(u \frac{\partial T}{\partial x} + v \frac{\partial T}{\partial y}) - g\gamma(u \frac{\partial S}{\partial x} + v \frac{\partial S}{\partial y}) \tag{5}$$

Eqs.(3) and (4) are substituted into (5). Thus,

$$u \frac{\partial v}{\partial z} - v \frac{\partial u}{\partial z} = \frac{g\alpha}{f}(\kappa_t \nabla^2 T - w \frac{\partial T}{\partial z}) - \frac{g\gamma}{f}(\kappa_s \nabla^2 S - w \frac{\partial S}{\partial z}) \tag{6}$$

The velocity components can be expressed in polar form,

$$u = V \cos\theta \quad , \quad v = V \sin\theta$$

where V and θ are the magnitude and direction of the geostrophic velocity, respectively. Then (6) becomes

$$\frac{\partial \theta}{\partial z} = \frac{g\alpha}{fV^2} (\kappa_t \nabla^2 T - w \frac{\partial T}{\partial z}) - \frac{g\gamma}{fV^2} (\kappa_s \nabla^2 S - w \frac{\partial S}{\partial z}) \tag{7}$$

This is an equation for the rate of turn of the geostrophic current with depth.

Data in October 1987 and May 1988 are analyzed. The orders of magnitude of the calculated expansion coefficients for temperature and salinity are,

$$\alpha = 1.29 \text{ (at 1000m)} \sim 3.18 \text{ (at the surface)} \quad \times 10^{-4} (^{\circ}C)^{-1}$$

$$\gamma = 7.36 \text{ (at the surface)} \sim 7.63 \text{ (at 1000m)} \quad \times 10^{-4} (psu)^{-1}$$

Typical values for the diffusivities of temperature and salinity (e.g., Gill, 1982) are,

$$\kappa_t = 1.4 \times 10^{-7} \text{ m}^2 \text{ s}^{-1}$$

$$\kappa_s = 1.5 \times 10^{-9} \text{ m}^2 \text{ s}^{-1}$$

In the upper ocean where thermocline and halocline prevail, the vertical gradients of temperature and salinity are much larger than their horizontal gradients, respectively. Thus the three–dimensional Laplacians of temperature and salinity can be replaced by $\frac{\partial^2 T}{\partial z^2}$ and $\frac{\partial^2 S}{\partial z^2}$, respectively. Therefore, (7) can be approximated by

$$\frac{\partial \theta}{\partial z} = \frac{g\alpha}{fV^2} (\kappa_t \frac{\partial^2 T}{\partial z^2} - w \frac{\partial T}{\partial z}) - \frac{g\gamma}{fV^2} (\kappa_s \frac{\partial^2 S}{\partial z^2} - w \frac{\partial S}{\partial z}) \tag{8}$$

The numerical values of the first derivatives of temperature and salinity with depth in the upper ocean are larger than their second derivatives, respectively. Thus if the numerical value of vertical velocity is larger than the values of diffusivities as shown above, i.e., if

$$O(w) > 10^{-7} \text{ m s}^{-1}$$

then (8) can be approximated by

$$\frac{\partial \theta}{\partial z} = -\frac{gw}{fV^2} (\alpha \frac{\partial T}{\partial z} - \gamma \frac{\partial S}{\partial z}) \tag{9}$$

Eq.(9) gives the rate of change with depth of the direction of geostrophic velocity in relation to the vertical gradients of temperature and salinity, which is a version of Hide's theorem.

In the main thermocline, where

$$O(\alpha \frac{\partial T}{\partial z}) >> O(\gamma \frac{\partial S}{\partial z}) \tag{10}$$

Eq.(9) can be further simplified as follows,

$$\frac{\partial \theta}{\partial z} = -\frac{gw\,\alpha}{fV^2}\,\frac{\partial T}{\partial z} \tag{11}$$

Eq.(11) can be applied to the upper layer down to about 300 meters deep (upper thermocline) of the oceans in the low and middle latitudes, where the vertical gradient of temperature is stronger than the salinity gradient to satisfy (10). Thus (11) can be applied to the East China Sea.

The gradient $\frac{\partial T}{\partial z}$ is positive except in deep waters. The Coriolis parameter f is positive in the northern hemisphere, and the thermal expansion coefficient α is positive in normal seawater. Therefore,

$$w \gtrless 0 \ \ \text{if} \ \ \frac{\partial \theta}{\partial z} \lessgtr 0$$

Thus the upwelling or downwelling can be determined by the direction of rotation of geostrophic velocity with increasing depth in the upper thermocline through right—hand rule in the northern hemisphere.

Schott and Stommel (1978) added the following beta vorticity constraint to the above mentioned current spiral and called it beta spiral. A linearized steady vorticity equation is

$$\beta v - f\frac{\partial w}{\partial z} = 0 \tag{12}$$

We will use (9) and (12) to determined the best—fit level of no motion by least—squares method. In beta spiral method different density surfaces at difdetermined from an observed density field. The method presented here is quite simple, where no density surface determination is required.

We have N levels of temperature and salinity data for each station. The deepest level is designated by N. The relative velocities between different levels can be determined by dynamical method.

The procedure for determination of a best—fit level of no motion is as follows.

1) The deepest level N is assumed to be a reference level, where u, v and w are zero. The relative horizontal velocities in the whole water column are calculated, which are substituted into (9). The vertical velocities in the whole water column are determined.

2) The velocities v and w at many depths determined at the first step, are substituted into (12). The right—hand side would vanish at all the depths if the assumed reference level was a real level of no motion. Practically it will not vanish.

Deviations from zero (errors) are denoted by D_{N_i}, where the first subscript N refers to the reference level, and the second subscript i (i = 1,....,N−1) a level where D_{N_i} is calculated. The squares of the errors are summed up,

$$E_N = \sum_{i=1}^{N-1} D_{N_i}^2$$

3) The above steps (1) and (2) are iterated by assuming different reference levels from the deepest level to the surface through a water column. We then have

$$E_n = \sum_{\substack{i=1 \\ i \neq n}}^{N} D_{n_i}^2 \,, \qquad\qquad n=1,\ldots,N \quad .$$

The best–fit level of no motion is chosen in such a way that E_n becomes minimum.

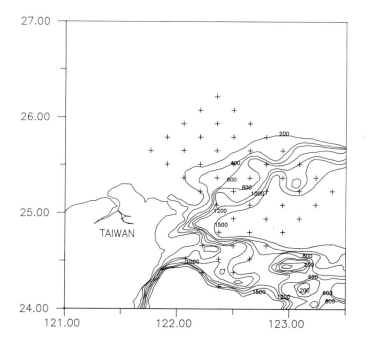

Fig. 1. Bottom topography off northeastern Taiwan. Symbols + denote the the station of Cruise 124 (8 to 11 October 1987) and 157 (13 to 17 May 1988) of R/V Ocean Researcher 1.

HYDROGRAPHY

Figure 1 shows the bottom topography of the sea northeast of Taiwan with CTD stations. The temperature and salinity distributions at 50m depth measured by CTD of R/V Ocean Researcher 1 on cruise 124 from 8 to 11 October 1987 are shown in dashed lines in Figs. 2 and 3, respectively. The region of maximum temperature gradient may mark the front of the Kuroshio. There exists obviously a cold core off northeastern Taiwan. The cold core extends from the sea surface to about 100m depth, where the water depth is approximately 200m.

The dashed lines in Figs. 4 and 5 show the temperature and salinity distributions at 50m depth measured by CTD on cruise 157 from 13 to 17 May 1988. The temperature gradient is apparently steeper in May 1988 than in October 1987. This indicates the Kuroshio is stronger in May 1988 than in October 1987. The cold core off northeastern Taiwan exists still there.

The motion described by (9) and (12) is basically a large–scale geostrophic flow. For eliminating the local effects on the temperature and salinity distributions which are mostly due to the tidal motion a smoothing operation is done basically by a 5–point scheme,

$$\overline{T}_0 = \tfrac{1}{2} T_0 + \tfrac{1}{8} (T_1 + T_2 + T_3 + T_4) \qquad \text{(Shuman, 1957)}$$

for inner stations, where T_1, T_2, T_3 and T_4 are data at the stations nearest to the station of T_0. Data at outer stations are smoothed by

$$\overline{T}_0 = \tfrac{1}{2} T_0 + \tfrac{1}{6} (T_1 + T_2 + T_3)$$

Data at the corner stations are smoothed by

$$\overline{T}_0 = \tfrac{1}{2} T_0 + \tfrac{1}{4} (T_1 + T_2)$$

Isolines of temperature and salinity by smoothed data are also shown in Figs. 2–5.

RESULTS

By means of the method described above the velocities at 50m depth are calculated for the cruises 124 and 157, as shown in Figs. 6 and 7. The calculated best–fit levels of no motion at different stations in these two cruises are different, as shown in Fig. 8. The Kuroshio off northeastern Taiwan changes little its direction in October and May. It is stronger in spring (May, 1988) than that in autumn (October, 1987).

There is a cyclonic eddy accompanied by the cold core northeast of Taiwan. The eddy is also stronger in May than in October. These figures show that upwelling exists in the central part and west side of the Kuroshio whereas downwelling on its east side. The calculated velocities are typically $10^{-4} \sim 10^{-6}$ m s^{-1}.

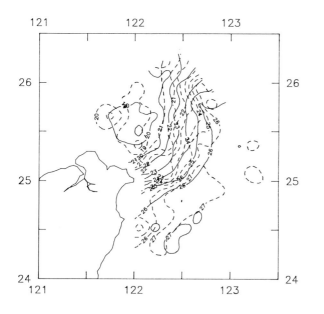

Fig. 2. Temperature distribution (——smoothed, --unsmoothed) at 50m depth from 8 to 11 October 1987.

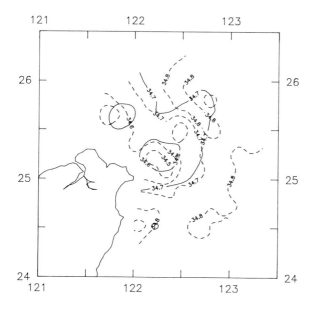

Fig. 3. Salinity distribution (——smoothed, --unsmoothed) at 50m depth from 8 to 11 October 1987.

342

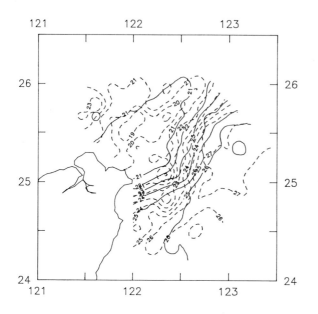

Fig. 4. Temperature distribution (——smoothed, --unsmoothed) at 50m depth from 13 to 17 May 1988.

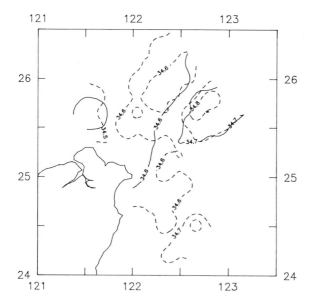

Fig. 5. Salinity distribution (——smoothed, --unsmoothed) at 50m depth from 13 to 17 May 1988.

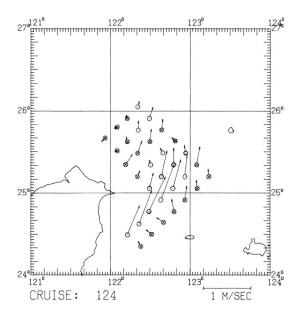

Fig. 6. The calculated velocity distribution at a depth of 50m from 8 to 11 October 1987. Symbols ⊙ denote upwelling, whereas symbols ⊗ downwelling.

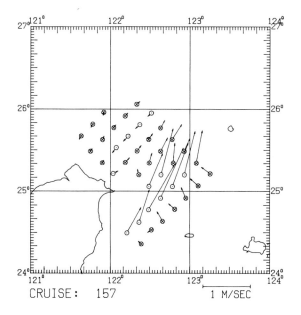

Fig. 7. The calculated velocity distribution at a depth of 50m from 13 to 17 May 1988. Symbols ⊙ denote upwelling, whereas symbols ⊗ downwelling.

344

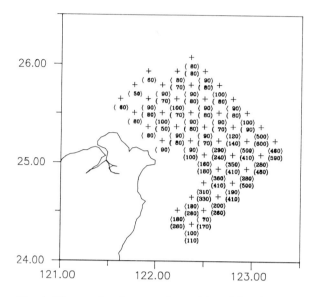

Fig. 8. The number in parentheses ( ) and brackets < > refers to the best–fit level of no motion at various stations on cruises 124 and 157, respectively.

CONCLUSION

The method for determination of the level of no motion presented here is quite simple. The magnitude of the calculated vertical velocities in and near the Kuroshio off northeastern Taiwan is typically $10^{-4} \sim 10^{-6}$ m s^{-1}, which may be larger than that in region far from the frontal area.

On the whole, the central part and west side of the Kuroshio is an upwelling area and its east side is a downwelling area.

ACKNOWLEDGEMENTS

The author is grateful to Mr. Yao–Tsai Lo and Mr. Ming–Jeng Chang for preparing diagrams and typing the manuscript. Special thanks are due to the crew and working team of R/V Ocean Researcher 1 for their assistance during the hydrographic survey. The referee's valuable comments are appreciated. This study is sponsored by the National Science Council of the Republic of China through grant NSC 78–0209–M019–01.

REFERENCES

Gill, A.E., 1982. Atmosphere–Ocean Dynamics. Academic Press, 662 pp.
Hide, R., 1971. On geostrophic motion of a non–homogeneous fluid. J. Fluid Mech., 49, 4: 45–751.
Holton, J.R., 1972. An introduction to Dynamic Meteorology. Academic Press, 319 pp.
Schott, F. and Stommel, H., 1978. Beta spiral and absolute velocities in different oceans. Deep–Sea Res., 25, 961–1010.
Shuman, F.G., 1957. Numerical methods in weather prediction: II. smoothing and filtering. Monthly Weather Rev., 85(11): 357–361.

WATER VOLUME TRANSPORT THROUGH THE TAIWAN STRAIT
AND THE CONTINENTAL SHELF OF THE EAST CHINA SEA
MEASURED WITH CURRENT METERS

Guohong FANG, Baoren ZHAO and Yaohua ZHU
Institue of Oceanology, Academia Sinica, 7 Nanhai Road, Qingdao, Shandong 266071, China

ABSTRACT

Historical current meter data from 24—hour anchored ships above the continental shelf of the East China Sea are analysed. The total (vertically integrated) flow averaged for each 2.2° latitude × 2.2° longitude square is clearly northeastward for all four seasons. The annual mean of the volume transport through the offshore area west of the 150 m isobath is estimated to be about 2 Sv. The anchored ship current meter data in the Taiwan Strait are analysed in combination with moored current meter data by Chuang (1985, 1986) and Wang and Chern (1988) for a rough estimate of the volume transport through the Strait. The annual mean is also about 2 Sv. The volume transport through the Taiwan Strait and the shelf of the East China Sea is comparable, in magnitude, with the transport through the Tsushima Strait (Korea Strait), which is about 2.8 Sv (Miita and Ogawa, 1984, 1985). These results suggest the existence of an extended current system including the Tsushima current system on the shoreside of the Kuroshio, which extends from the Taiwan Strait to the Tsugaru Strait. This current system is called "Taiwan—Tsushima—Tsugaru Warm Current System (T—T—T WCS)". The dynamic height of the sea surface relative to 1000 db in the northeastern South China Sea is higher than that in the area east of the Tsugaru and Soya Strait by about 0.7m. This sea—level difference is believed to be the main driving force of the T—T—T WCS.

INTRODUCTION

For a period of time it was believed that the current in the Taiwan Strait and the continental shelf of the East China Sea was driven by monsoon and thus went to the north in summer and to the south in winter. However, in the late 1950's and early 1960's direct current measurements showed that the current often flowed to the north even in winter, especially in the lower layers. Guan (1984, 1986) called it "Against—the—Wind Current" or "Counter—Wind Current", and Fang and Zhao (1988) called it "Upwind Current". Up to date, however, no quantitative estimate of the volume transport has been given. Since the flow in this area is mainly barotropic (Chuang, 1985, 1986; Fang and Zhao, 1988), and the conventional dynamic calculations provide no reasonable results, it is thus necessary to make use of direct current measurement data. Our calculation will show that the mean volume transport through the Taiwan Strait and the shelf of the East China Sea is comparable with that of the Tsushima Current, we shall propose existence of a current system on the shoreside of the Kuroshio, known as the Taiwan— Tsushima—Tsugaru Warn Current System (T—T—T WCS).

The driving mechanism of both the Taiwan Warm Current and the Tsushima Warm Current has been topics of many papers. In this paper we shall give an explanation of the main forcing mechanism for the T—T—T WCS, which was actually first proposed by Fang and Zhao (1988).

VOLUME TRANSPORT THROUGH THE TAIWAN STRAIT

Chuang (1985,1986) has made moored current measurements in the near—shore area west of Taiwan. Recently a moored current measurement lasting over one year has been reported (Wang and Chern, 1988). On the contrary, in the vast west part of the Strait no moored current measurement has been conduced . A few 24—hour current measurements have been made from anchored ships . The stations are shown in Fig. 1. The measurements at stations Nos. 1 to 9 were made in wintertime . Station 9 can be used to represent the channel between the Penghu Islands and Taiwan. The measurement was made by Chuang (1985,1986) with a moored current meter put at a level 20 m above the seabed. The mean northward component of the current was 24. 2 cm/s during the period of April 3 to May 5,1983 and 18. 0 cm/s from March 11 to May 25,1985. The width of the passage is about 35 km and the mean depth about 55 m. Thus the volume transport here is about 0. 4 Sv . The current data at stations 2 to 8 are quoted from Fu *et al* . (1989) and Guan (1980). These data are used to calculate the volume transport through the western Strait from the mainland to the Penghu Islands. The result is 0. 6 Sv as shown in Table 1. The total volume transport in wintertime is thus 1. 0 Sv.

TABLE 1

Calculation of volume transport through the Taiwan Strait

1. Winter (Dec. —May)

Channel	Section	Width	Mean depth	Mean speed of N—comp.	Transport
		(km)	(m)	(cm/s)	(Sv)
Mainland—Penghu Is.	P	250	45	5. 5	0. 6
	Q	190	45	8. 2	
Penghu Is. —Taiwan		35	55	21. 1	0. 4

2. Summer(June—Nov.)

Channel	Section	Width	Mean depth	Mean speed of normal comp.	Transport
		(km)	(m)	(cm/s)	(Sv)
Mainland—Taiwan	R	145	60	36	3. 1

3. Annual mean transport≈2 Sv.

The current records in the southern part of the strait are inadequate to determine the summer volume transport passing through the section across the Strait. There are a few stations, however,

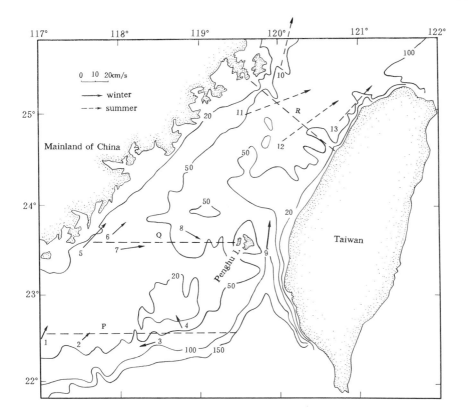

Fig. 1.   Measured depth—mean current vectors in the Taiwan Stuait.

in the northern Strait as shown in Fig. 1. Station 13 is a mooring station maintained by Chuang
(1985) and Wang and Chern (1988). Chuang obtained a mean current speed 28.5 cm/s during
April 3 to May 5, 1983 at a level 14 m above the seabed. Wang and Chern gave a progressive
current diagram for the period of July 19 to August 8, 1984 at a depth 15 m below the sea surface. From the diagram a mean current speed of 55 cm/s is obtained. It is likely that the average
of above two results gives an acceptable estimate for the vertically averaged current speed in summer. The current data at stations 10 to 12 are based on Wen *et al.* (1988) and Guan (1980).
These records show that the current in the northern Taiwan Strait in summer is rather stable. The
volum transport is calculated to be 3.1 Sv. Thus the annual mean volume transport through the
Taiwan Strait is about 2 Sv.

VOLUME TRANSPORT OVER THE SHELF OF THE EAST CHINA SEA

There are 566 24—hour anchored ship current measurements available in the East China Sea.
Most of these measurements were made with Ekman current meters. In addition to near—bottom
measurements at 2 to 5 m above the seabed, measurements were done in most cases at depths of

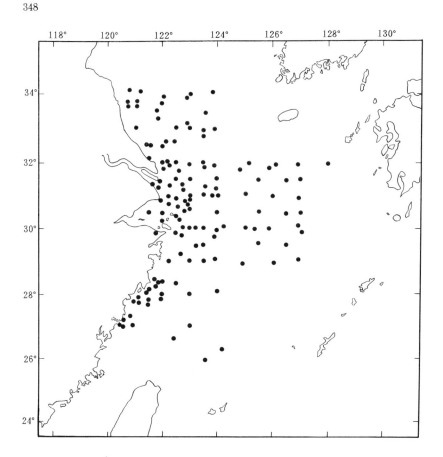

Fig. 2.   Position of anchored ship stations in the shelf part of the East China Sea.

5, 10, 15, 20, 25, 30, 40, 50, 75, 100, and 150 m below the sea surface, if these levels are above the near — bottom layer. Figure 2 shows the measuring sites with each dot representing a current station. The seasonal distribution of the number of measurements is listed in Table 2 by days and by stations. It can be seen that more measurements are carried out in spring and summer than in autumn and winter, but the difference is not very large.

TABLE 2

Seasonal distribution of the number of measurements

Season	Days	Stations
Spring	170	85
Summer	193	98
Autumn	108	65
Winter	95	58
Total	566	138

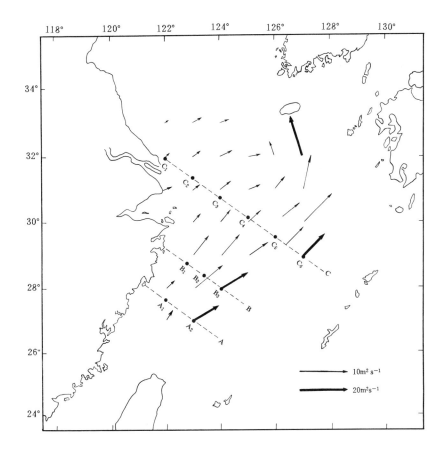

Fig. 3. Transport density in the shelf part of the East China Sea averaged for whole year. A, B and C are sections chosen to calculate stream function (see Fig. 8).

For each station and each measuring layer, we first make harmonic analysis of tidal current to get residual current (Fang, 1981). This approach is superior to 24—hour averaging in that the influence of the lunar tides M_2 and O_1 can be removed. Then for each 24—hour observation at a site we calculate the transport density Q (The quantity Q has many names, such as transport, total transport, or current amount. In this paper we use transport density) by the following formula:

$$Q = z_1 V_1 + (h - z_K) V_K + \sum_{k=1}^{K-1} (z_{k+1} - z_k)(V_{k+1} + V_k)/2$$

where K is the number of measuring layer, z_k and V_k the depth and residual current vector at the k—th layer, and h the water depth. The obtained 566 vectors of Q are grouped and averaged for

350

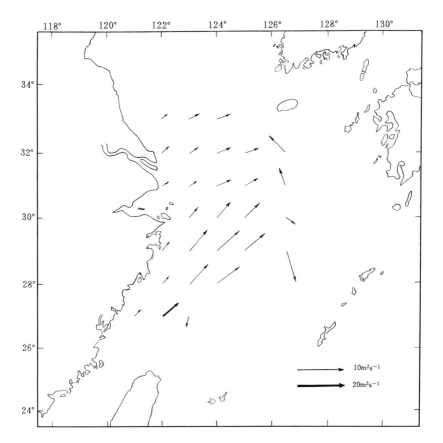

Fig. 4.   Transport density in the shelf part of the East China Sea averaged for spring.

each 2. 2° latitude × 2. 2° longitude square centered at the points 1° apart. The vectors of the averaged transport density are plotted in Fig. 3. Figure 4 to 7 show the averaged transport density for
four seasons. It can be seen that all vectors of annual mean transport density are directed northeastward except located at the northeast corner of the study area, where the current is likely to
have an anticlockwise deflection. As for the seasonal means, the overall tendency is the same as
that of the annual mean. It is especially noteworthy that the volume transport in winter is also
northeastward regardless of the southward blowing wind.

To give a quantitative estimate for the volume transport over the continental shelf of the East
China Sea we choose three sections in the area as shown in Fig. 3. At each section we take several
calculation points and calculate the transport density at these points by interpolation. Then the
components normal to the section, Q_y, are used to calculate the value of stream function defined by

$$\psi = \int_0^x Q_y \, dx$$

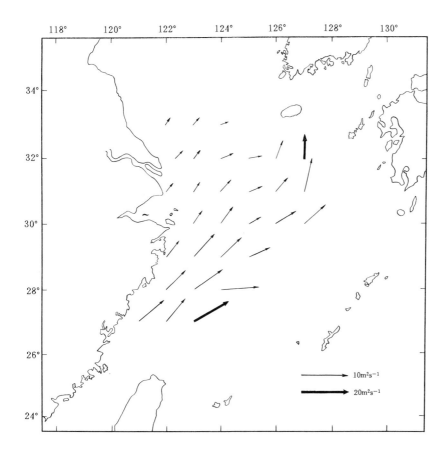

Fig. 5.   As in Fig. 4 but for summer.

where x is directed to the southeast along the section originating from the coast. The results are shown in Fig. 8. From these results we obtain the distribution of stream function over the shelf of the East China Sea as plotted in Fig. 9.

Figure 9 shows that the isopleth of 2 Sv is located slightly outside of the isobath of 100 m. Considering that in the above calculation more data are taken in summer and spring than in winter and autumn, and that most winter data were obtained under relatively calm weather condition, the transport in Fig. 9 should be slightly overestimated. Thus we can figure that the volume transport through the shelf part of the East China Sea west 150 m isobath is about 2 Sv.

DISCUSSION: TAIWAN — TSUSHIMA — TSUGARU WARM CURRENT SYSTEM AND ITS MAIN DRIVING FORCE

In the above sections we obtained the volume transport through the Taiwan Strait and the shelf of the East China Sea. Miita and Ogawa (1984) caluluated the volume tranport through the Tsushima Strait (Korea Strait) also from direct current measurements. The averaged volume transport is

352

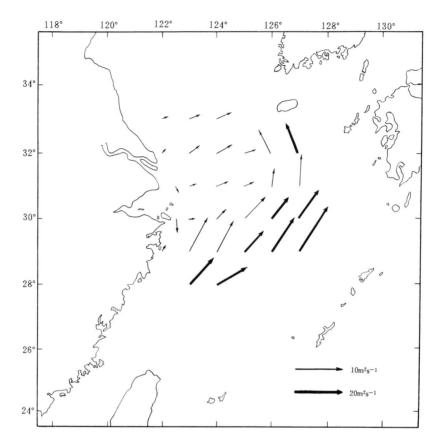

Fig. 6.　As in Fig. 4 but for autumn.

1. 8 Sv through the western channel and 1. 66 Sv through the eastern channel. But most of the data used in their calculation were taken in summer (62%) and spring (21%). Soon later Miita
(Coastal Oceanography Research Committee, The Oceanographical Society of Japan, 1985) got
an annual mean transport, 2. 77 Sv from the same data with seasonal correction. On the other
hand, recent studies indicated that the Tsushima Current did not directly come from the Kuroshio
near Kyushu (Rikiish and Ichiye, 1986). Fang and Zhao (1988) proposed that a current might
run parallel to the Kuroshio from the northeastern part of the South China Sea to the area east of
Hokkaido passing the East China Sea and Japan Sea by the Taiwan, Tsushima, Tsugaru and Soya
Strait. In the present study we have obtained the transport through the Taiwan Strait and the area
west of the 150 m isobath in the East China Sea to be about 2 Sv. The remaining transport (0. 8
Sv) of the Tsushima Current should also be from the Kuroshio but is likely not through the Taiwan
Strait. We propose here to call this current the Taiwan—Tsushima—Tsugaru Warm Current Sys-
tem, abbreviated to T—T—T WCS, which Ichiye (1989) called it T—T—T—S WCS to incorpo-
rate the Soya Strait. A schematic representation of the T—T—T WCS is given in Fig. 10.

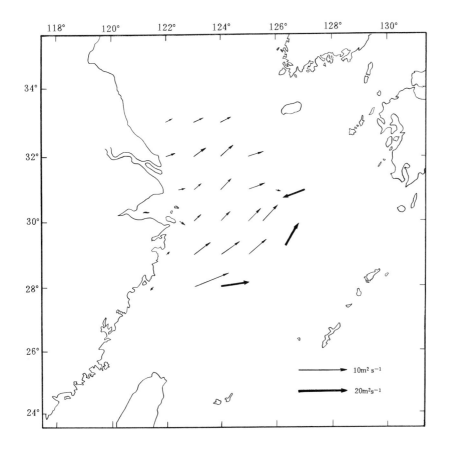

Fig. 7.   As in Fig. 5 but for winter.

The Tsushima Current in the Japan Sea is seperated from the Kuroshio by the Japanese Islands. But the current from the Taiwan Strait to the Tsushima Strait keeps in direct contact with the Kuroshio. Thus strong water exchange may occur between these two currents.

The driving mechanism of the Tsushima Current and the current in the Taiwan Strait has attracted many oceanographers' attention.

Chuang (1985) assumes that the current in the Taiwan Strait is driven by the sea surface slope and wind. After a regression analysis he concludes that there should exist a sea surface slope from the south down to the north with a slope varying from 1.5×10^{-7} to 2.33×10^{-7}. This sea surface slope was further verified based on land leveling data by Fang and Zhao (1988), who point out that the sea surface height along the southeast coast of China is higher in the southwest than in the northeast with a slope of 2.62×10^{-7}. The forces from the mean wind stress, atmospheric pressure and water density difference have effects opposite to the sea surface slope, but they altogether only amount to about one third of the force caused by the sea surface slope. The sea surface slope is certainly the main factor driving the northeastward current off the southeast coast of China. Ho—

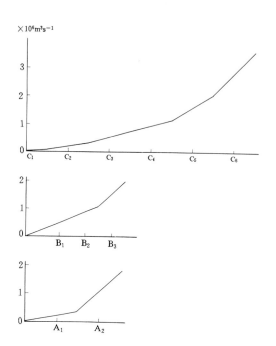

Fig. 8. Year—mean value of stream function along section A, B and C.

wever, there is still a question : what causes the sea surface slope?

It is quite evident that the current in the Tsugaru Strait is also driven by sea surface slope along the Strait. There are several tidal stations near the Tsugaru Strait (Nakano and Yamada, 1975). Iwasaki and Hachinohe are located on the coast of northern Honshu with the former by the Japan Sea and latter by the Pacific Ocean. The observed mean sea level at Iwasaki is 15 cm higher than at Hachinohe. Because both of them are on the southern side of the Tsugaru Current, this sea level difference can be regarded as the sea surface drop along the Strait. The mean sea level at Hakodate on the northern side of the Tsugaru Current is lower by 23 cm than the mean sea level at Iwasaki and Hachinohe. However, this sea level difference only reflects the geostrophic balance in the cross—strait direction and has nothing to do with the driving mechanism. Along the northern coast of Hokkaido there is also a sea surface drop of 14 cm from Wakkanai to Abashiri. This indicates that the current in the Soya Strait is also driven by sea surface slope. But there is still a question : why the sea level on the Japan Sea side is higher than on the Pacific Ocean side?

Many oceanographers have also attributed the Tsushima Current to the difference between the sea surface heights at the Tsushima and Tsugaru Straits. Minato and Kimura (1980) first considered the sea level variation along the western boundary of the ocean on the basis of Stommel's (1948) wind—driven ocean circulation model. Sekine (1988) improved Minato and Kimura's model by introducing the effect of density stratification. However, as pointed out by Ichiye (1984), the argument based on barotropic effect has two essential difficulties. First, the sea level

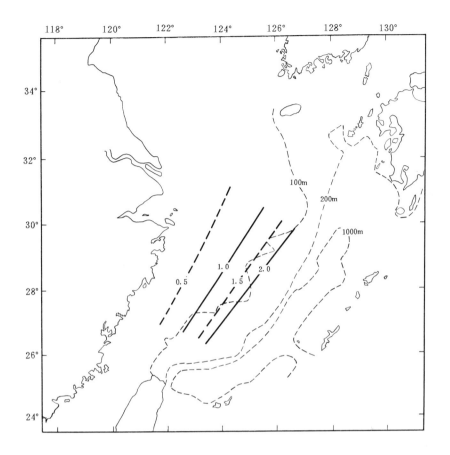

Fig. 9.   Year—mean stream function (in Sv=10^6m³s⁻¹).

difference derived from the wind—driven ocean circulation model is too small. The annual mean of the sea level drop from 30°N to 42°N is only 8 cm according to the calculation by Ichiye, which is even inadequate to account for an observed sea level drop along the Tsugaru or Soya Strait. Second, the seasonal variation in the sea level difference is almost opposite in phase to the observed seasonal variation in both observed sea level difference and volume transport of the Tsushima Current, and even negative in autumn (Ichiye, 1984; Sekine, 1988).

Fang and Zhao (1988) noticed a large difference of the sea level along the western boundary of the ocean resulting from the thermohaline effect. Figure 11 shows the sea surface dynamic height relative to 1000 db in the northwestern Pacific Ocean, redrawn from Wyrtki (1975). This figure indicates that the sea surface height in the northeastern South China Sea is about 70 cm higher than that east of the Tsugaru/Soya Straits provided that the 1000 db pressure surface coincides with the geopotential surface. This assumption should be basically true because the current in deep layers such as down to 1000 m is generally much weaker than the geostrophic current at the sea surface. Our diagnostic numerical model of the northwestern Pacific circulation reveals even

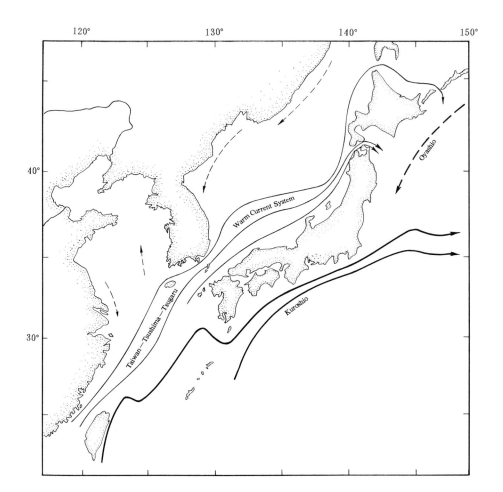

Fig. 10. Schematic representation of the Taiwan—Tsushima—Tsugaru Warm Current System.

a larger sea level difference along the west boundary. A sea level difference of 70 cm or so can produce a sea surface slope of order of 2×10^{-7}. This along—stream slope is comparable to that observed or derived in the Taiwan Strait (Fang and Zhao, 1988; Chuang, 1985) and is thus large enough to drive the T—T—T WCS. And furthermore, the sea surface drop along the southeastern coast of China ($\sim$30cm) and along the Tsugaru Strait ($\sim$15 cm)can be reasonably included in this sea level difference.

With this sea level difference as a basis, the seasonal variation of the transport of the T—T—T WCS can be more easily explained. Because the steric height in the deep ocean varies little from season to season, in summer the southerly monsoon and the wax of vertical stratification of sea water enhance this current, while in winter the northerly monsoon and the wane of vertical stratification weaken it.

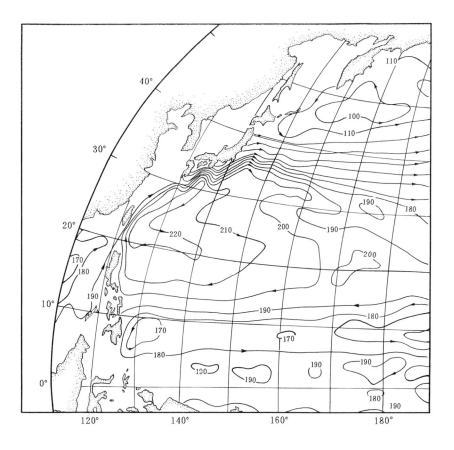

Fig. 11.   The dynamic height of the sea surface of the northwestern Pacific relative to 1000 db (in dynamic cm). Redrawn from Wyrtki (1975).

ACKNOWLEDGEMENTS
    The authors are especially grateful to Prof. Takashi Ichiye for his interest and valuable comments. Thanks are extended to Prof. Zilang Fu of Amoy University for making the current data in the Taiwan Strait available to the authors. The authors also thank the reviewers and the editor for their corrections and improvements to this paper.

REFERENCES

Chuang, W. —S. , 1985. Dynamics of subtidal flow in the Taiwan Strait. Journal of the O-
    ceanographical Society of Japan. 41:65—72.
Chuang, W. —S. , 1986. A note on the driving mechanisms of current in the Taiwan Strait.
    Journal of the Oceanographical Society of Japan. 42: 355—361.
Coastal Oceanography Research Committee, the Oceanographical Society of Japan, 1985. Coastal
    Oceanography of Japanese Islands (in Japanese). Tokai University press. 1106 pp.
Fang , G. , 1981. Quasi—harmomic constituent method for analysis and prediction of tides, Ⅲ.
    A practical procedure for analysing tidal streams and tidal elevations (in Chinese with English
    abstract). Studia Marina Sinica. 18: 19—39.
Fang , G. and B. Zhao, 1988. A note on the main forcing of the northeastward flowing current
    off the southeast China coast. Progress in Oceanography. 21: 363—372.

358

Fu, Z. ,J. Hu and G. Yu, 1989. Winter volume transport in the Taiwan Strait (Manuscript, in Chinese).

Guan ,B. , 1980. Oceanic current (residual current) in the Minnan (Southern Fujian) fishing ground (Manuscript,in Chinese).

Guan ,B. , 1984. Major features of the shallow water hydrography in the East China and Huang-hai Sea. In: T. Ichiye (Editor), Ocean Hydrodynamics of the Japan and East China Seas. Elsevier,Amsterdam,pp. 1—14.

Guan ,B. , 1986. Evidence for a counter—wind current in winter off the southeast coast of China. Chinese Journal of Oceanology and Limnology. 4:319—332.

Ichiye, T. , 1984. Some problems of circulation and hydrography of the Japan Sea and the Tsushima Current. In:T. Ichiye (Editor), Ocean Hydrodynamics of the Japan and East China Sea. Elsevier, Amsterdam, pp. 15—54.

Ichiye, T. , 1989. JECSS news No. 1: On JECSS V. Journal of the Oceanographical Society of Japan. 45: A79—82.

Minato,S. and R. Kimura, 1980. Volume transport of the western boundary current penetrating into a marginal sea. Journal of the Oceanographical Society of Japan. 36: 185—195.

Nakano, M. and S. Yamada, 1975. On the mean sea levels at various locations along the coasts of Japan. Journal of Oceanographical Society of Japan. 31: 71—84.

Rikiishi, K. and T. Ichiye, 1986. Tidal fluctuation of the surface currents of the Kuroshio in the East China Sea. Progress in Oceanography. 17: 193—214.

Sekine, Y. , 1988. On the seasonal variation in in—and outflow volume transport of the Japan Sea. Progress in Oceanography. 21:269—279.

Wang ,J. and C. —S. Chern,1988. On the Kuroshio branch in the Taiwan Strait during winter-time. Progress in Oceanography. 21: 469—491.

Wen,X. ,L. Huang,H. Lian and H. Li,1988. Hydrographic features of the middle and northern part of the Taiwan Strait (in Chinese). In: Report of Marine Comprehensive Survey of the Middle and Northern Taiwan Strait. Science Press, Beijing, pp. 138—188.

Wyrtki, K. , 1975. Fluctuations of the dynamic topography in the Pacific Ocean. Journal of Physical Oceanography. 5: 450—459.

A SUBSURFACE NORTHWARD CURRENT OFF MINDANAO IDENTIFIED BY DYNAMIC CALCULATION

D.X.HU, M.C.CUI, T.D.QU and Y.X.LI
Institute of Oceanology, Academia Sinica, 7 Nanhai Road, Qingdao, Shandong, China

ABSTRACT
On the basis of CTD data along a section perpendicular to Mindanao by R/V Science I in October 1987 to 1989, a subsurface northward flow is clearly identified by dynamic calculation, which is named the Mindanao Undercurrent (MUC). The MUC is of a double-core feature. The first core is located about 80 km away from the coast of Mindanao in 1987 and 1989 and close to the coast in 1988, and centered at a depth of about 500 m with a width of about 50 km, where the velocity is greater than 5 cm/s. The maximum velocity of the first core can be over 20 cm/s and the transport ranges from 3 to $9 \times 10^6 m^3/s$. The second is generally located about 200 km away from the coast, and is centered at a depth of about 370 m with a width of about 100 km. The maximum velocity can be over 30 cm/s and the transport ranges from 5 to $16 \times 10^6 m^3/s$. As a whole, the MUC transport ranges from 8 to $22 \times 10^6 m^3/s$ for the three years.

1 INTRODUCTION

The western boundary current (WBC) in the north Pacific (east of the Philippines) is distinctly different from WBCs in the other oceans, especially in terms of branching. There are two branches of the WBC east of the Philippines-the Kuroshio and the Mindanao Current, which should have significant effect upon the climate in East Asia and upon the formation and evolution of the warm pool in the western equatorial Pacific. So there have been many studies on them (Cui and Hu, 1989; Guan, 1986, 1989a and 1989b; Hu, 1989; Hu and Cui, 1989; Lukas, 1988; Masuzawa, 1969; Nitani, 1970 and 1972; Wyrtki, 1956 and 1961). However, most of those studies are concerned with the northward or southward surface currents only and few deal with return flows (undercurrents). The present study focuses on the subsurface northward current off Mindanao, based on geostrophic calculation.

2 DATA AND METHOD

Every October from 1987 to 1989, the Academia Sinica Institute of Oceanology (ASIO) carried out a cruise in the WBC area with R/V Science I, which provided us a number of CTD data by Neil Brown CTD or Sea-Bird CTD. In the following a CTD section along 7° 30′ N perpendicular to Mindanao (Fig.1) is geostrophically analyzed by dynamic calculation. All the CTD casts only reached down to 1500 m. For the choice of the level of no motion

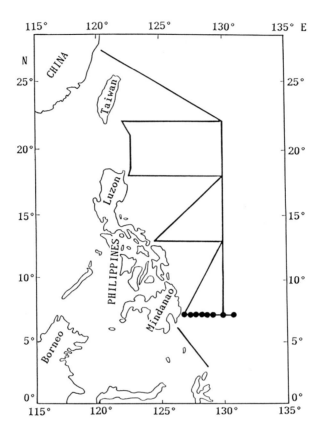

Fig.1   Study area in the Philippine Sea and 7° 30′ N section (thick line
with station dots)

the profiles of △D, the difference of dynamic depth between two adjacent
stations, are plotted in Fig.2. According to Defant (1965), the level of no
motion was chosen as 1200 m like Nitani (1970), 1300 m and 1500 m,
respectively.   The MUC transports for these levels of no motion are listed
in Table 1.

TABLE 1

Volume transport (× 10⁶m³/s) of the MUC for three levels of no motion

level of no motion	year 1987	1988	1989
1200 m	14.2	7.2	23.4
1300 m	15.6	7.4	23.4
1500 m	18.0	8.6	22.9

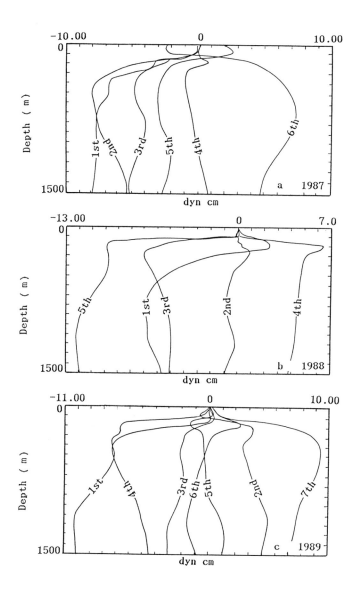

Fig.2. Vertical distribution of dynamic depth difference between adjacent stations in 1987(a), 1988(b) and 1989(c)

It is apparent from Tab.1 that different choice of the level of no motion results in little difference of volume transport of the MUC (for 1987, 1988, 1989, respectively); i.e., the maximum deviation from the mean is $2.1 \times 10^6 m^3/s$ (in 1987) . There is significant interannual variability, which is not caused by the choice of the level of no motion and is one of the MUC's features. In addition, $d(\triangle D)/dz$ is calculated for 1200, 1300 and

1500 m at all station pairs for the three years. If we take the 3-year average of d(△D)/dz at each depth of 1200, 1300 and 1500 m, respectively, the minimum will take place at 1500 m, which is 1.39×10^{-1} (dyn.cm) /m. So, a depth of 1500 m was chosen as the level of no motion for the following calculation.

3 RESULTS

The calculated geostrophic velocities are shown in Figs.3a,b,c, relative

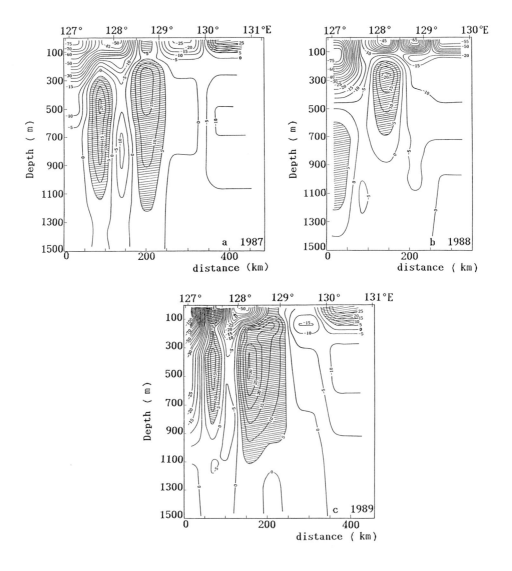

Fig.3. Geostrophic velocity section relative to 1500 db (shaded is northward velocity over 5 cm/s)

to 1500 db. It is evident from Fig.3 that the Mindanao Undercurrent generally consists of two cores. The first core is about 80 km away from the coast in general,with an exception of being close to the coast in 1988. It is about 50 km wide and about 700 m thick, centered at about 500 m depth with maximum velocity of 21 cm/s and maximum transport of $9.2 \times 10^6 m^3/s$ with average of about $6 \times 10^6 m^3/s$. The second one is about 200 km away from the Mindanao coast, about 100 km wide, about 900 m thick if it is defined by speeds greater than 5 cm/s. It is centered around 370 m in depth with maximum velocity over 30 cm/s and maximum transport over $16 \times 10^6 m^3/s$.

The characteristics of the Mindanao Undercurrent are summarized in Table 2.

TABLE 2

Characteristics of the Mindanao Undercurrent

charac-teris-tics	max. vel. (cm/s) and depth (m)		transport ($10^6 m^3/s$)		width(km) *		thickness (m) *		distance from coast (km) △	
year	core1	core2	core1	core2	core1	core2	core1	core2	core1	core2
1987	21 465	17 379	9.2	8.8	75	80	920 (230-1150)	1050 (170-1220)**	110	230
1988	10 724	20 258	3.2	5.4	50	80	600 (600-1200)	530 (170-700)	<30	180
1989	20 362	31 482	6.2	16.7	40	125	678 (172-850)**	1100 (0-1100)**	110	180
average	17 517	23 373	6.2	10.3	55	95	732	893	83	197

* : width and thickness are determined by an area of velocity over 5 cm/s.
**: thickness- (1) for core 2 (1987), a separate northward flow with velocity greater than 5 cm/s in the upper 100 m is excluded (Fig.3a); (2)for core 1 (1989),the upper core centered at 100 m depth with maximum velocity of 57 cm/s is excluded (Fig.3c). (3) for core 2 (1989), a northward flow over 5 cm/s reaches the surface. but the core is still below the surface (Fig.3c).
△ : a value of 30 km between the Mindanao coast and the station nearest to the coast is added.

The characteristics summarized in Tab. 2 are similar to those given by Hu and Cui (1989) with inverse calculation.

It should be noted that: (1) The MUC seems to exhibit a pronounced interannual variability in its strength, width, depth, and distance from the coast. Since the measurement was done only once a year, nothing could be known about the interannual variability, if the variability in seasonal or shorter time scales were large. In this context, more frequent measurements are really needed to explore more about MUC's nature; (2) Although the MUC is basically a subsurface current the northward flow associated with the MUC can reach the surface, as the second core of the MUC in 1987 and the two cores in 1989. (3) The interaction between the MUC and the southward surface Mindanao Current (MC) seems to be an interesting problem. Figure 3 shows that, when the MC was strong (over 100 cm/s in 1989, over 75 cm/s in 1987 and 1988) the MUC was strong in velocity and transport. When the first core of the MUC is closer to the coast, the MC is shallower and vice versa. The migration of the upper part of the northward flow associated with the MUC seems related to the location of the surface MC.

Finally, the mechanism of the MUC is still poorly understood, more works on which are to be conducted.

4 Acknowledgement

This work is jointly supported by the Natural Science Foundation of China (NSFC) under grant 4880230 and the State Commission for Science and Technology under grant 75-76-07-06. We are much grateful to Mr. Shaoying Bai for drawing the figures and Miss Ying Hu for typing the manuscript.

5 REFERENCES

Cui, M. and Hu, D., 1989. Inversion calculation of the western boundary current in the Pacific during October 1988. Chin. J. Oceanol. Limnol., 8(2):177-187.
Defant, A., 1960. Physical Oceanography. Pergamon Press. 1960, p.1319.
Guan, B., 1986. Current structure and its variation in the equatorial area of the western North Pacific Ocean. Chin. J. Oceanol. Limnol., 4(3):239-255.
Guan, B., 1989a. Variation of the Mindanao eddy and its relation with El Nino event. Oceanologia et Limnologia, 20(2):131-138.(in Chinese)
Guan, B., 1989b. Current structure and its variability in the area east of Taiwan and the Philippines in winter. Chin. J. Oceanol. Limnol., 20(5): 393-400.
Hu, D., 1989. A thought on the role of western Pacific Ocean circulation in climate change in southeast China. Chin. J. Oceanol. Limnol.,7(1):93-94.
Hu, D. and Cui, M., 1989. The western boundary current in far-western Pacific Ocean. Proceedings of Western Pacific International Meeting and Workshop on TOGA COARE, pp: 123-134, May 24-30, 1989, Noumea, New Caledonia, edited by Joel Picaut, Roger Lukas and Thierry Delcroix.
Lukas, R., 1988. Internannual fluctuations of the Mindanao Current inferred from sea level. J. Geophys. Res., 93:6714-6748.
Masuzawa, J., 1969. The Mindanao Current. Bull. Jpn. Soc. Fish. Oceanogr. Spec. No.(Prof. Uda's Commem Pap.)99-104.

Nitani, H., 1970. Oceanographic conditions in the sea east of the Philippines and Luzon Strait in summers of 1965 and 1966. In: The Kuroshio- A symposium on the Japan Current (edited by J. C. Marr), East-West Center Press, Honolulu, 213-232.

Nitani, H., 1972. Beginning of the Kuroshio. In: Kuroshio-its physical aspects. Tokyo Univ. Press, 129-163.

Wyrtki, K., 1956. The subtropical lower water between the Philippine Sea and Irian (New Guinea). Mar. Res. Indonesia, 1:21-52.

Wyrtki, K., 1961. Physical oceanography of the Southeast Asian Waters. Scientific results of marine investigations of the South China Sea and Gulf of Thailand, 1959-1961. Naga Report, 2:195 pp.

THE MECHANISM OF EXPLOSIVE CYCLOGENESIS OVER THE SEA EAST OF CHINA AND JAPAN

S.M.XIE[*], D.Y.WEI[*], C.L.BAO[*] and T.AOKI[**]
* National Research Center for Marine Environmental Forecasts, 8 Da Hui Si, Hai Dian Division, Beijing 100081, China
** Meteorological Agency, 1-3-4 Ote-machi, Chiyoda-ku Tokyo 100, Japan

ABSTRACT
By using 10-level data on a $2.5°$ latitude-longitude grid of March 14-15, 1988, a diagnostic analysis is made of the explosive intensification of a cyclone (pressure drop of 24hPa/24h) over the Kuroshio region. The cyclone experienced three development stages: energy storage stage, explosive development stage and continuous development stage. The explosive intensification of the cyclone is caused mainly by the heat transport from the Kuroshio into the cyclone, baroclinicity of the SST frontal zone and cumulus convection in the warm sector of the cyclone. The continuous explosive development is caused by a change in the atmospheric dynamic condition resulting from the development in the earlier stages and the release of baroclinic energy from the oceanic and atmospheric frontal zones.

INTRODUCTION

The Kuroshio is a strong current in the Pacific Ocean. It does not only transport a great deal of heat from south to north, but also supply a great deal of latent heat to the atmosphere. The Kuroshio region is also one of the most frequent occurrence region of explosive intensification of the cyclone in the Pacific Ocean (Sanders and Gyakum, 1980; Roebber, 1984). The explosive intensification of the cyclone over the North Atlantic and the Northeast Pacific is basically attributed to a kind of baroclinically unstable phenomena similar to CISK mechanism (Gyakum, 1983a,b; Bosart and Lin, 1984; Mellen 1983; Anthes and Kuo, 1983). However, very few studies were done on explosive deepening of the cyclone over the Northwest Pacific. This paper presents a case study on it over the Kuroshio region on March 14-15, 1988 for investigating the physical mechanism of this kind of event.

2 SYNOPTIC ANALYSIS

A depression circulation was observed near $30°N$, $120°E$ at 00UTC on March 14, 1988 with a cloud area on its northwest side. Both of them moved eastward. This depression moved to southeast of Japan ($32.5°N$, $132.5°E$) at 12UTC on March 14 and developed into an extratropical cyclone. On the surface map (Fig.1), a strong warm tongue extended from the southwest to the northeast of the cyclone with a strong warm advection at its eastern part. A cloud cluster

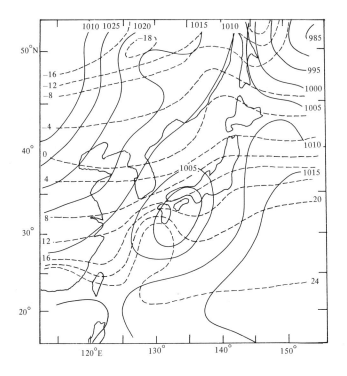

Fig. 1. Surface map at 12UTC on March 14, 1988. Heavy lines for the pressure, broken lines for the air temperature.

developed. The cloud system was thickest at the northeast part of four levels on an enhanced satellite image (Fig.2). On 500hPa chart (Fig.3), a weak trough was situated in the rear of the surface cyclone together with a weak temperature trough behind it. A weak cyclonic vorticity advection over the surface cyclone center provided a favorable dynamic condition for the deepening of the surface cyclone. But the vorticity advection was very weak. Moreover, it was just situated within zero temperature advection area on upper levels. Therefore, the environmental condition was far from the cause of explosive intensification of the cyclone in later stages.

Then, the cyclone moved to $37.5°$ N, $145°$ E at 00UTC on March 15. It began to develop rapidly with the central pressure fall at a rate of 10.8hPa/12h. It exceeded the critical pressure drop at that latitude (8hPa/12h) which is calculated from the critical value formula for the explosive intensification of the cyclone at latitude ϕ : 24hPa/24h*(sin ϕ /sin $60°$) proposed by Sanders and Gyakum (1980). At this moment, a meso-scale convective complex (MCC) being

Fig. 2. Infrared satellite image at 12UTC on March 14, 1988, replacing an enhanced image for convenience.

resolved into seven levels on an enhanced satellite image (Fig.4) was formed
to the southeast of the surface cyclone center which was a result of its
remarkable development. Following the appearance and development of the convective cloud system, the cyclone was intensified furthermore. At 12UTC on
March 15, the central pressure fell to 976.2hPa with a pressure drop rate of
13.2hPa/12h. The observed maximum surface wind speed was 26m/s and the calculated precipitation rate was 43.4mm/6h: the cyclone had an intensity of strong
tropical storm. From 12UTC on March 14 to 12UTC on March 15, the central pressure fell by 24hPa which was larger than the critical pressure-drop value
(16.9hPa/24h) by 7.1hPa.

The oceanic heating field can be calculated by Jacob's formula:

$$Q = 0.715 \times (5.97 - 0.006T_s) \times (e_s - e_a) +$$
$$0.475 \times (5.97 - 0.006T_s) \times (T_s - T_a) \times V$$

370

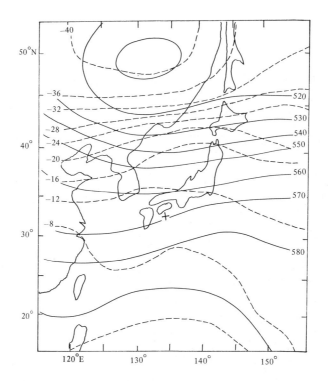

Fig. 3. Heights (solid lines) and isotherms (broken lines) on 500hPa for 12UTC on March 14, 1988. A mark + indicates the surface cyclone center.

Here, Q is the upward heat flux from the ocean to the atmosphere, T_s, T_a the sea surface temperature (SST) and air temperature at the sea surface, respectively, e_s, e_a the saturation vapor pressure (calculated from T_s) and sea surface vapor pressure, V the sea surface wind speed.

Figure 5A shows the oceanic heating field on March 14 , SST field at the second ten days of March, 1988, and the cyclone track. Figure 5B shows the heating field on March 15, SST field at the last ten days of March, and the cyclone track. In Fig.5A an oceanic frontal zone (SST isotherms concentration zone) extended from the northern area of Taiwan, China, to the northeast at an angle of about 60° from north direction. At the early stage before 12UTC on March 14, the cyclone moved from China mainland across the oceanic frontal zone at an angle of 20° to the Kuroshio. During this period, the SST beneath the cyclone increased from 12°C to 20°C. However, the cyclone development was not yet evident with a central pressure-drop of 1.8hPa. After the cyclone reached the Kuroshio region, its track became almost parallel to the oceanic frontal zone and was situated in uniformly high SST area to the south of the

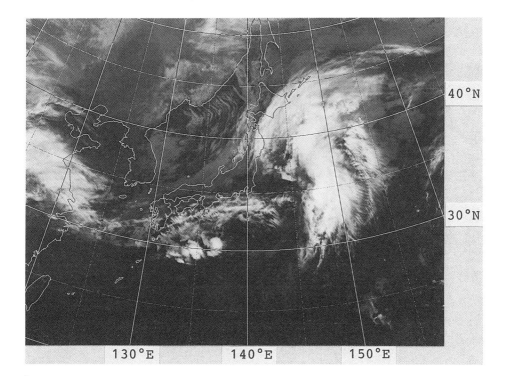

Fig. 4. Infrared satellite image at 00UTC on March 15, 1988, replacing an enhanced satellite image for convenience.

frontal zone. The cyclone's warm sector was just within a strong heating
central area. Just in this period (12UTC on March 14 to 00UTC on March 15),
the cyclone began to be intensified explosively. In a later stage from 00UTC
to 12UTC on March 15, crossing the oceanic frontal zone again at an angle of
30°, the cyclone moved from the Kuroshio region to a cold water area. The
cyclone continued to be intensified explosively. A synoptic analysis shows
that its explosive development from 12UTC on March 14 to 00UTC on March 15 was
due to the heating effect of the Kuroshio on the atmosphere. However, its subsequent explosive development from 00UTC to 12UTC on March 15 may be attributed to the baroclinic effect of the oceanic and atmospheric frontal
zones, while the oceanic heating effect could not be a principal cause because
of the SST lowering.

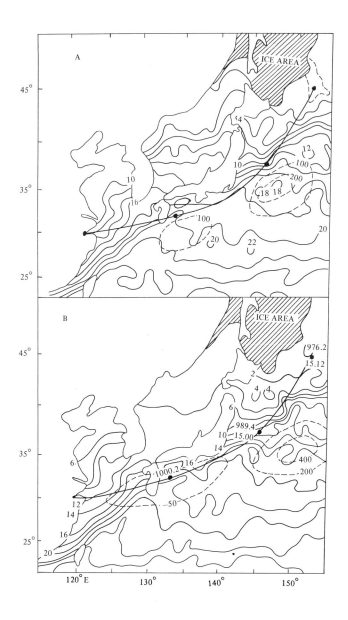

Fig. 5. Track of the surface cyclone (heavy lines with solid circles), SST(light lines) and oceanic heating (broken lines) fields. The central position of the surface cyclone is shown by a solid circle every 12 hours. A. Heating field on March 14, 1988, and SST for the second ten days of March. B. Heating field on March 15, and SST for the last ten days of March.

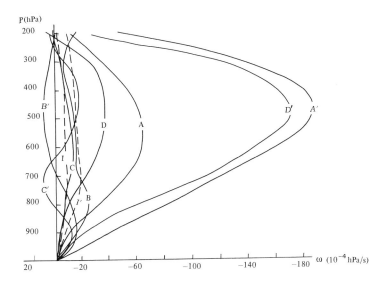

Fig. 6. Vertical motion profiles close to the surface cyclone center. Curves
A, B, C, D and I are ω, ω_1, ω_2, ω_3, and the upward motion caused by oceanic
sensible heating for 35°N, 135°E at 12UTC on March 14. Curves A', B', C', D'
and I' are ω, ω_1, ω_2, ω_3, and the upward motion caused by oceanic sensible
heating for 35°N, 147.5°E at 00UTC on March 15.

3 DIAGNOSTIC ANALYSIS

Vertical motion is an important factor to the cyclone's development. It is
closely associated with not only atmospheric energy transformation and balance
in weather system, but also the vertical distribution of water vapor, heat and
momentum in the atmosphere. In this context, we analyze the vertical motion in
the cyclone before and during the explosive development.

Under quasi-geostrophic assumption, the vertical velocity ω is given by
the following diagnostic equation:

$$\Gamma(\omega) = -\frac{1}{\sigma}\nabla^2 A_\theta - \frac{\partial}{\sigma\partial p} A_\xi - \frac{1}{\sigma}\nabla^2(\frac{RQ}{Pc_p})$$

where $\Gamma(\ ) = \nabla^2 + \frac{f^2\partial^2}{\sigma\partial p^2}$ is an operator, $A_\theta = -R/p\ V\cdot\nabla T$ is the temperature

advection, $A_\xi = -V\cdot\nabla(\zeta + f)$ the absolute vorticity advection, Q the diabatic
heating.

Since the equation is linear for ω, ω is written as:

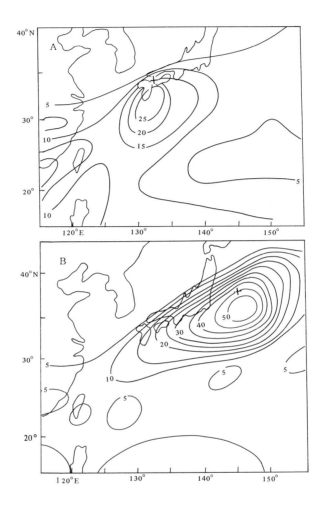

Fig. 7. Water vapor flux field on 850hPa. A at 12UTC on March 14, B at 00UTC on March 15. Units: g/s.hPa.cm. A mark + indicates the surface cyclone center.

$$\omega = \omega_1 + \omega_2 + \omega_3$$

Here, ω_1, ω_2, ω_3 are, respectovely, the vertical motion associated with the non-uniform horizontal distribution of temperature advection, with the vertical distribution of absolute vorticity advection and with diabatic heating. In the calculation, ω consists of oceanic heating of the atmosphere (calculated by Jacob's method by assuming a linear distribution below 700hPa) and condensation latent heat in the atmosphere (calculated by water vapor convergence scheme).

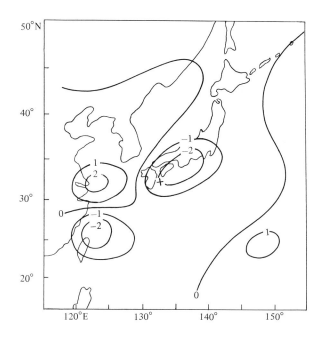

Fig. 8. Water vapor flux convergence field on 850hPa at 12UTC on March 14.
Units: 10^{-7}g/cm^2.s.hPa. A mark + indicates the surface cyclone center.

Using 10-level data on a 2.5^b latitude-longitude grid and SST data on a 1°
grid provided by Japan Meteorological Agency, we calculated the vertical motion before and during the explosive intensification. Figure 6 shows the
profiles of vertical motion at the most prosperous developing place of the
cloud system near the cyclone center: curves A, B, C, D are ω, ω_1, ω_2, ω_3
profiles at 35°N, 135°E at 12UTC on March 14, while curves A', B', C', D' are
for MCC (35°N, 147.5^{o}E) at 00UTC on March 15. The vertical motion during explosive deepening at 00UTC on March 15 was much more pronounced than before
(at 12UTC on March 14). The diabatic heating term (curves D and D') was much
stronger than the two other terms, and became more and more important with explosive development of the system. At 00UTC on March 15, diabatic heating
curve D' was qualitatively and quantitatively consistent with the total vertical motion (curve A'). However, the contribution of oceanic sensible heaing
(curve I') among diabatic heating ω_3 was rather small, indicating that
diabatic heating, especially latent heat, played a very important role during
the explosive development.

Figures 7A, B show the water vapor flux field at a level of 850hPa at 12UTC

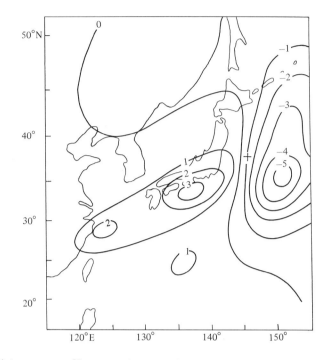

Fig. 9. Water vapor flux convergence field on 850hPa at 00UTC on March 15. Units: 10^{-7}g/cm^2.s.hPa. A mark + indicates the surface cyclone center.

on March 14 and 00UTC on March 15. A strong water vapor transportation took place in the warm sector of the cyclone with high water vapor flux extending along the Kuroshio. The concentration area of water vapor flux isopleths agreed with the oceanic and atmospheric frontal zones, which means that the water vapor flux was distributed asymmetrically with a strong shear in the cyclone. Instensities of the water vapor flux and shear at 00UTC on March 15 were much more remarkable than at 12UTC on March 14. Figures 8 and 9 give the water vapor flux convergence field on 850hPa at 12UTC on March 14 and 00UTC on March 15. The central and eastern parts of the cyclone were a convergence area of water vapor, while its western part was a divergence area. Similarly, water vapor convergence-divergence values at 00UTC on March 15 were larger than before with the largest value of $-5.5 \cdot 10^{-7}$g/cm$^2 \cdot$s·hPa to the southeast of the cyclone center. In other words, strong water vapor transportation, water vapor flux convergence and water vapor flux shear over the Kuroshio region were favorable to diabatic heating and also played an important role in the ex-plosive development.

Figure 10 shows the profiles of velocity divergence at the place of the

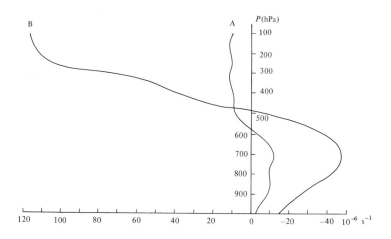

Fig. 10. Divergence profiles close to the surface cyclone center. A at 12UTC on March 14 at 35°N, 135°E, B at 00UTC on March 15 at 35°N, 147.5°E.

most pronounced development of the cloud system near the cyclone center; curve A at 35°N, 135°E at 12UTC on March 14 and curve B at 35°N, 147°E at 00UTC on March 15. Both show a convergence at middle-layers in the troposphere and a divergence at upper layers. It is worth to note that the value at 12UTC on March 14 was rather small, while the value at 00UTC on March 15 was very large, especially at upper layers; i.e., the divergence on 100hPa was as high as $120 \cdot 10^{-6}$/s. Moreover, the upper layer divergence at 00UTC on March 15 was much stronger than the lower layer convergence, which means that the cyclone was in two different development stages with very different dynamic condition in intensity, and that a potential of its furthermore rapid development was suggested at 00UTC on March 15.

Figures 11A,B show the vertical section of ageostrophic velocity along 37.5 °N and 145°E passing the cyclone center at 00UTC on March 15. Very strong supergeostrophic wind was observed at the upper troposphere. Strong supergeostrophic wind also appeared at lower layers but weak in middle layers. Comparison with Fig.6 implies that this kind of supergeostrophic wind was mainly caused by diabatic heating effect.

Based on the above analysis, we can infer the mechanism of explosive development of this cyclone as follows. First, when the cyclone moved into the Kuroshio region, strong oceanic heating, water vapor transportation and vapor convergence took place in lower layers. At the same time, the thermal-pressure structure of the cyclone including heat advection, vorticity advection and convergence in lower layers and divergence in upper layers became favorable to

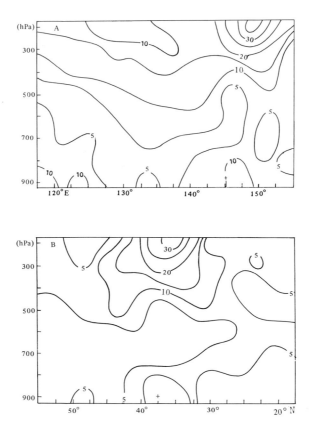

Fig. 11. Ageostrophic velocity distribution in the latitudinal and longitudinal sections passing the surface cyclone center at 00UTC, March 15. A
for 37.5° N, B for 145° E. A mark + indicates the surface cyclone center.

a large-scale upward motion which, in turn, brought about the condensation to
release latent heat, and then its feedback effect caused a strong upward motion. Consequently, the upper layer divergence became much intensified than
lower layer convergence, and then the surface pressure rapidly fell. Finally
the cyclone was intensified explosively.

4 DISCUSSION

The above analysis makes clear that the explosive development of the
cyclone was caused by a great deal of heat and water vapor supplied from the
Kuroshio. At 00UTC to 12UTC on March 15, the cyclone moved from the Kuroshio
region to the cold water area on the north side of the oceanic frontal zone.
During this period, the oceanic thermal effect decreased significantly so as
to be unfavorable to the cyclone's development. However, the cyclone was

still continuously in explosive intensification process with its central pressure drop rate even larger than before. It is, in our opinion, because the atmospheric dynamic condition was changed sugnificantly after its initial explosive intensification under the oceanic thermal effect. Both the existence of strong supergeostrophic wind in upper and lower layers (as shown in Fig.11) and the upper troposphere divergence much stronger than lower layer convergence (Fig.10) were favorable to the continuous explosive development.

Owing to the upward motion near the cyclone center, the latent heat of condensation was released to warm up the middle-layer troposphere, so that a warm-core structure on 500hPa was formed near the cyclone center. The feedback effect enhanced the upward motion. The warming up of the cyclone would introduce pessure falling in lower layers and prsure rising in upper layers, and then caused strong supergeostrophic wind in the upper and lower troposphere. Strong upward motion brought about a strong convergence in lower layers and a much stronger divergence in upper layers. When the cyclone moved northward across the oceanic frontal zone, the air temperature and SST contrast increased. Under this favorable dynamic condition, the baroclinic energy of the atmospheric and oceanic frontal zones was ready to be released. It was also very favorable to continuous explosive development of the cyclone.

5 CONCLUSION

The cyclone process on March 14-15, 1988 can be divided into three development stages.

1). Energy storage stage: The cyclone moved from China mainland across the oceanic frontal zone to the Kuroshio region in about 12 hours. The atmospheric dynamic environmental field was generally favorable but not enough for its full development. This was an energy storage stage. The cyclone got a lot of heat and water vapor from the ocean and possessed more baroclinic energy due to the existence of the oceanic frontal zone.

2). Explosive intensification stage: The cyclone reached the Kuroshio region. The latter continuously supplied heat and water vapor to the cyclone to increase its potential energy. The upward motion released the potential energy so that a much stronger upward motion was occurred. The baroclinic energy was released from the oceanic frontal zone. It turns out that the cyclone was explosively intensified.

3). Continuous development stage: After the cyclone was intensified explosively due to oceanic heating effect of the Kuroshio, the atmospheric dynamic condition was changed sugnificantly. The upper layer divergence became much larger than the lower layer convergence. At the same time, the oceanic and atmospheric frontal zones were going to release the baroclinic energy. In this way, the cyclone was continuously developed and intensified furthermore.

6 REFERENCES

Anthes, R.A. and Kuo, Y.-H., 1983. Numerical simulations of a case of explosive marine cyclogenesis. Mon. Wea. Rev., 111: 1174-1188.
Bosart, L.F. and Lin, S.C., 1984. A diagnostic analysis of the presidents' day storm of February 1979. Mon. Wea. Rev., 112: 2148-2177.
Gyakum, J.R., 1983a. On the evolution of the QE II storm. I: synoptic aspects. Mon. Wea. Rev., 111: 1137-1155.
Gyakum, J.R., 1983b. On the evolution of QE II storm. II: dynamic and thermodynamic structure. Mon. Wea. Rev., 111: 1156-1173.
Mellen, S.L., 1983. Explosive cyclogenesis associated with cyclones in polar air streams. Mon. Wea. Rev., 111: 1537-1553.
Roebber, P.J., 1984. Statistical analysis and updated climatology of explosive cyclones. Mon. Wea. Rev., 112: 1577-1589.
Sanders, F. and Gyakum, J.R., 1980. Synoptic-dynamic climatology of the "bomb". Mon. Wea. Rev., 108: 1589-1606.

THE EFFECT OF SWELL ON THE GROWTH OF WIND WAVES

HISASHI MITSUYASU
Research Institute for Applied Mechanics
Kyushu University 87, Kasuga 816, Japan
YOSHIKAZU YOSHIDA
Japan Port Consultant, Fukuoka 812, Japan

ABSTRACT
    A laboratory experiment has been made to clarify the effect of swell on the growth of wind waves, when the swell is propagating against the wind. It is shown that the growth of wind waves is not much affected by the swell when the swell steepness is small, but it is intensified by the swell when the swell steepness increases. This phenomenon is in clear contrast to the attenuation of wind waves by the swell propagating in the direction of the wind (Mitsuyasu 1966, Phillips & Banner 1974).

1 INTRODUCTION

    When the wind blows over still water surface in the ocean, wind waves are generated, develop with time and space, and finally become large ocean waves. In addition to such an idealized case, there are many cases where the wind waves are generated in the ocean with pre-existing swell. Although wave models presently used do not consider the interaction between the swell and wind waves, there have been no studies to make clear the effect of interaction, except for the studies on the interaction between the wind waves and swell both of which are propagating in the same direction.

    The purpose of the present study is to clarify experimentally the growth of wind waves under the existence of swell propagating against the wind. When the swell with large steepness is propagating to the same direction of the wind, wind waves are suppressed by the swell (Mitsuyasu 1966, Phillips and Banner 1974).

    The suppression mechanism of the wind waves by the swell has been presented by Phillips and Banner (1974), which gives fairly good explanation of the phenomenon. However, as far as we know, there have been no studies on the growth of wind waves under the existence of swell of the opposite direction.

2 EQUIPMENT AND PROCEDURE

2.1 The wind-wave flume

    The experiment was made in a wind-wave flume 0.8m high, 0.6m

382

wide, and with a usual test section 15m long. Water depth in the flume was kept at 0.353m throughout the experiment. The arrangement of the equipment is shown schematically in Fig.1. A centrifugal fan (outside left of the figure) sends the wind from the left to the right in the flume. The wind speed is monitored with a Pitot-tube installed 15cm above a transition plate. The wind speed measured at this point is used as a reference wind speed and denoted by Ur. The reference wind speed Ur in the flume was changed stepwise as Ur=5.0, 7.5, 10.0, 12.5 (m/s). A flap-type wave generator at the downwind side generates swells (regular oscillatory waves). The swell energy was absorbed by a beach at the upwind side, and the wind wave energy was absorbed by a filter in front of the wave generator. The swell energy passes through the filter, with small attenuation, due to its longer period than the wind wave period. Waves were measured simultaneously at eight stations (No.1-No.8) by using resistance-type wave gauges installed in the flume with 1m separation and recorded on a digital data recorder. The sampling frequency of the wave data is 200 Hz. Vertical wind profiles over the water surface were measured with Pitot-tubes at fetches, F_1, F_2, F_3 and F_4 in the flume (Fig. 1).

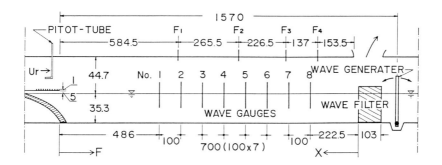

Fig. 1 Schematic diagram of wind-wave flume (units in cm).

The procedure of the experiment is as follows; Three minutes after the start of the wind blower wind-generated waves were measured for three minutes. Then swell was generated and sent into the generation area of wind waves and the co-existent system of the wind waves and the swell of opposite propagation direction was measured for the three minutes. In order to measure the attenuation of the swell without wind, the swell was measured independently for two minutes.

TABLE 1

Experimental conditions; wind speed and swell properties

Run No.	Wind speed Ur (m/s)	Wave period (sec)	Wave height (cm)
1 - 1	5.0		
1 - 2	7.5	wind waves	
1 - 3	10.0		
1 - 4	12.5		
2 - 1	5.0	1.024	3.0
2 - 2	7.5	1.024	3.0
2 - 3	10.0	1.024	3.0
2 - 4	12.5	1.024	3.0
3 - 1	5.0	0.620	1.3
3 - 2	7.5	0.620	1.3
3 - 3	10.0	0.620	1.3
4 - 1	5.0	0.620	2.6
4 - 2	7.5	0.620	2.6
4 - 3	10.0	0.620	2.6
5 - 1	0	1.024	3.0
5 - 2	0	0.620	1.3
5 - 3	0	0.620	2.6

TABLE 2

Constants for the spectral filter.

frequency band (Hz)	number of the spectral lines within the filter	effective band width (Hz)
$f \leqslant 3$	1	4.88×10^{-2}
$3 < f \leqslant 10$	11	3.66×10^{-1}
$10 < f \leqslant 40$	21	7.32×10^{-1}
$40 < f \leqslant 100$	41	1.46

The properties of the swell used in the experiment are summarized in Table-1 including the other parameters, where the wave period and wave height were measured at the station No.4.

2.2 Analysis of the wave data

The wave data of each run were divided into 6 samples, each of which contained $4096(=2^{12})$ digitized data. Power spectra of waves were computed through a fast-Fourier-transform procedure using each 4096-point data set. After taking a sample mean of the six spectra, the spectra were smoothed by the triangular spectral filters shown in Table 2.

3 RESULTS

3.1 Wind profile

Figure 2 shows some examples of the wind profiles over the water surface where the wind waves co-exist with the swell propagating against the wind. For comparisons, wind profiles over wind wave surface without the swell are shown with broken lines, where the individual data points are not shown. It can be seen that all wind profiles $U(z)$ near the water surface can be approximated by the logarithmic profile

$$U(z) = (u_*/K)\ln Z/Zo, \tag{1}$$

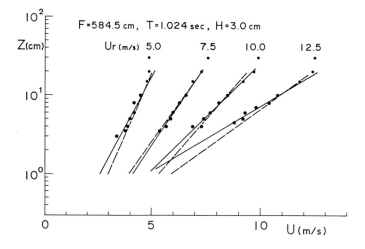

Fig. 2 Wind profiles over the water surface.
● data for the co-existing system; ——, regression line for the data; - - - , wind profiles over the pure wind waves.

where u_* is the friction velocity of the wind, K is the Karman's constant and Zo is the roughness parameter. It can be seen also from Fig.2 that the wind profiles are slightly affected by the existence of the swell propagating against the swell, though the trend is not clear.

3.2 Wave spectra

Figure 3 shows some examples of the wave spectra of the swell, i.e., monochromatic waves without wind action, which were observed at the stations No.2, No.4, No.6 and No.8. As usually observed, the spectra contain many higher harmonics due to their nonlinearity. Figure 4 shows the spectra of pure wind waves measured at the same stations No.2, No.4, No.6 and No.8. These data of the swell spectra and the pure wind wave spectra are used

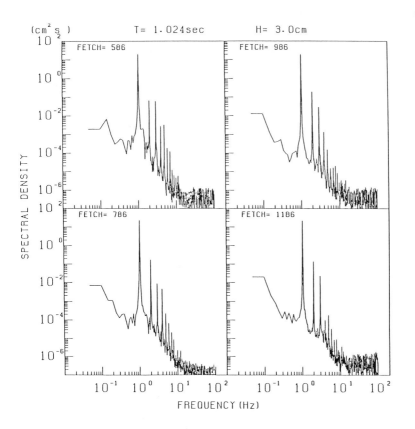

Fig. 3 Examples of swell spectra without wind action.
wave period, T=1.024sec; wave height, H=3.0cm.

as reference data to compare them with the spectra of co-existent systems of the wind wave and the swell.

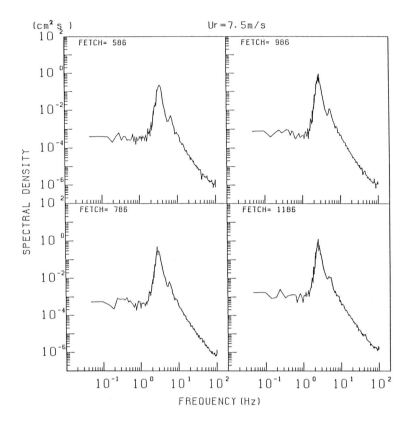

Fig. 4 Examples of wind wave spectra without swell.
wind speed, Ur=7.5m/s.

Figure 5 shows some examples of the wave records of the co-existent system of the wind waves and the swell. From the top to the bottom, the records correspond to those obtained at the stations, No.1, No.2,..., No.8; observation stations shift from the upwind side to the downwind side. Wind waves propagate and develop from the top to the bottom, and swell propagates and attenuates from the bottom to the top direction. Since the quantitative analysis of the wave records is difficult in this form, we computed the power spectra of the wave records of the co-existent systems. Examples of the wave spectra of the co-existent system of the wind waves and the swell are shown in Fig 6. A very sharp peak at low frequency side corresponds to the swell spectrum and a secondary peak at high frequency side corresponds to the wind wave spectrum. In our experimental conditions, frequency regions of the swell and the wind waves are different. Therefore we can separate their energies as schematically shown in Fig. 7, where the swell energy Es is computed as

$$Es = \int_{o}^{f_1} \phi \, (f) df, \tag{2}$$

and the wind wave energy Ew is computed as

$$Ew = \int_{f_1}^{f_2} \phi \, (f) df. \tag{3}$$

Characteristic frequencies f_1 and f_2 were selected respectively as $f_1 = 1.8 Hz$ and $f_2 = 40 Hz$ throughout the present analysis of the wave data after visually investigating all data of the spectra.

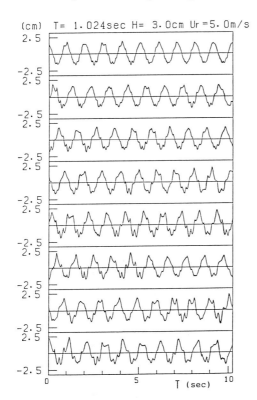

Fig. 5 Sample records of the co-existent system of wind waves and swell.
From the top to the bottom fetch F=4.86, 5.86,...,11.86cm
(c.f. Fig. 1).

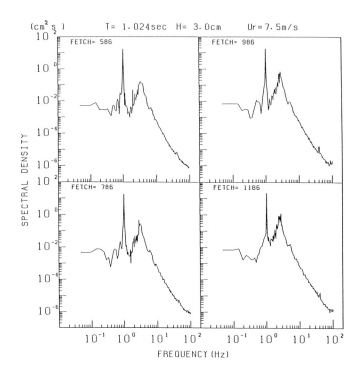

Fig. 6 Examples of wave spectra of the co-existent system; swell,
T=1.024sec, H=3.0cm; wind speed Ur=7.5m/s.

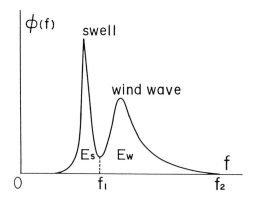

Fig. 7 Schematic representation of the wave spectrum of the
co-existent system.

3.3 The effect of swell on the growth of wind waves

In order to clarify the effect of swell on the growth of wind waves, wave spectra of the co-existent systems have been compared with pure wind wave spectra which were obtained in the same experimental conditions (wind speeds and fetches). Some examples are shown in Figs. 8a and 8b. It can be clearly seen that the

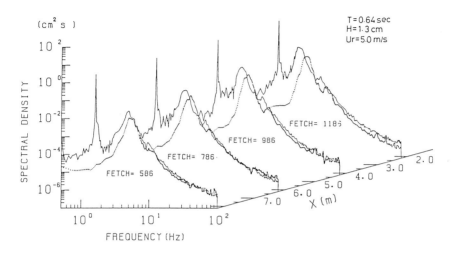

Fig. 8a Comparison of the wave spectra of the co-existent system (solid line) and those of the pure wind waves (dotted line). swell, T=0.64sec, H=1.3cm; wind speed Ur=5.0m/s.

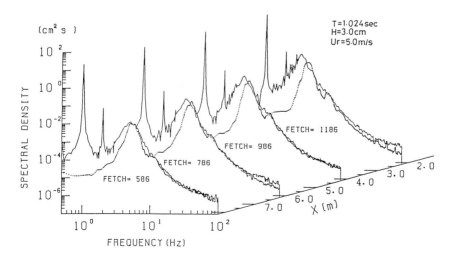

Fig. 8b The same as in Fig. 8a except for the different swell; swell T=1.024, H=3.0cm, Ur=5.0m/s.

growth of the wind wave spectra are much intensified by the existence of swell propagating in the opposite direction; the spectral energy increases greatly and the spectral peak frequency shifts to the low frequency side. Such phenomena are much different from those observed when the swell is propagating in the direction of the wind waves. In the latter case wind waves are suppressed by the swell when the steepness of the swell is large (Mitsuyasu 1966, Phillips and Banner 1974).

In order to show the change of the wind wave energy quantitatively, the ratio of wind wave energy of the co-existent system to the energy of pure wind waves is shown in Fig. 9 as a function of wave steepness of the swell, where H_I and λ are the wave height and wave length of the swell respectively. For

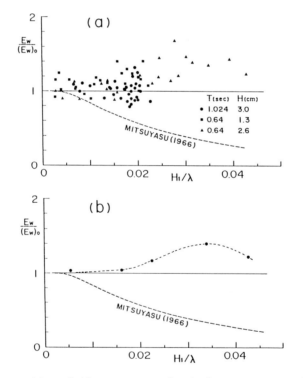

Fig. 9 The ratio of the energy of wind waves co-existing with the swell, Ew, to that without the swell, (Ew)o.
(a) original data, (b) sectional mean values.

comparison, Fig. 9 also shows the result of our previous study (Mitsuyasu, 1966) where the swell is propagating in the direction of the wind waves. Although the data for the present study scatter considerably, the relative wind wave energy is not much affected by the swell when the swell steepness is small, but it increases with the increase of the steepness of the swell, which shows clear contrast to the result of our previous study where the wind wave energy decreased with the increase of the steepness of the swell (Mitsuyasu, 1966).

The present results, in association with previous results (Mitsuyasu 1966, Phillips and Banner 1974), show that the effects of the swell on the growth of the wind waves depend, to a great extent, on the direction of the swell relative to that of the wind waves: wind waves are suppressed by the swell propagating in the same direction, while the wind waves are amplified by the swell propagating in the opposite direction. Such effects of the swell on the wind waves are not considered in the wave models which are presently used for the prediction of ocean waves. Practical application of the present results is one of the important subjects for future studies. It should be noted that the scatter of the data is very large in Fig. 9(a). This fact means that the swell steepness is not the only parameter controlling the phenomenon. The effect of other parameters need to be studied in future.

4 DISCUSSIONS

Phillips and Banner (1974) presented an attenuation mechanism of wind waves by the swell propagating in the direction of the wind waves. Essential points of the mechanism are as follows;

When long waves (swell) propagate on the sea surface under the action of the wind, there is a nonlinear augmentation of the surface drift near long-wave crests, so that short waves, superimposed on the longer ones, experience an augmented drift in these regions. The distribution of the surface drift q is given by

$$q = (C-u) - \left\{ (C-u)^2 - q_o(2C-q_o) \right\}^{1/2}, \tag{4}$$

where C is the phase speed of the swell, u is the orbital speed of the swell, and q_o is the surface drift at the point of the wave profile where the surface displacement $\eta=0$.

On the other hand, the maximum amplitude ηmax that the short wind waves can attain is given by

$$\eta\, max = (2g')^{-1} (c-q)^2, \tag{5}$$

where g' is the apparent gravitational acceleration which the short wind waves experience on the surface of the swell and c is the phase speed of the wind waves (Banner and Phillips, 1974).

The ratio γ of the maximum amplitude of the wind waves to their amplitude in the absence of the swell is given by

$$\gamma = g(c_c - q_c)^2 / g'(c_o - q_o)^2 \quad , \tag{6}$$

where the suffix "o" corresponds to the values in the absence of the swell and the suffix "c" corresponds to the values at the crest of the swell when the swell is superimposed on short wind waves. Laboratory measurements by Mitsuyasu (1966) and Phillips and Banner (1974) on the suppression of short wind waves by the swell gave good agreement with the above theoretical prediction.

The distribution of the surface drift q' can be obtained from similar inference to that of Phillips and Banner (1974), even when the swell is propagating in the opposite direction to the wind direction. That is given by

$$q' = -(C-u) + \left\{ (C-u)^2 + q_o(2C+q_o) \right\}^{1/2}, \tag{7}$$

which shows the augmentation of the surface drift at the crest of the swell, though the augmented values are slightly smaller than those in the presence of the swell propagating in the same direction as the wind. Therefore, the mechanism similar to that of Phillips and Banner (1974) can not explain the augmented growth of wind waves by the swell propagating against the wind waves. The investigation of other unknown mechanisms for explaining the present phenomenon is also a subject for future studies.

Acknowledgments The authors are indebted to Mr. K. Marubayashi and Miss M. Hojo for their assistance in preparing the manuscript.

REFERENCES

Banner, M.L. and Phillips, O.M., 1974. On small scale breaking waves. J. Fluid Mech., 65: 647-657.

Mitsuyasu, H., 1966. Interaction between water waves and wind (1). Rep. Res. Inst. Appl. Mech., Kyushu University, 14: 67-88.

Phillips, O.M. and Banner, M.L., 1974. Wave breaking in the presence of wind drift and swell. J. Fluid Mech., 66: 625-640.

THE THERMAL EFFLUENT PROBLEMS OF THREE NUCLEAR POWER PLANTS IN TAIWAN

K.L. FAN
Institute of Oceanography, National Taiwan University
Taipei, Taiwan, China

ABSTRACT
    There are three nuclear power plants in Taiwan. Each plant has
two units for generating power. The first two plants are located
along the northern coast of Taiwan. The third one is located
along the coast of southernmost Taiwan. The plants take sea water
for cooling, and discharge thermal effluents into the ocean
surface by the once-through cooling system. Fishermen living near
the power plants complain that the thermal water affect the
inshore fishery catch. Besides, the second plant has some
influence on the nearby swimming area. And the thermal water from
the third nuclear power plant bleaches or kills some corals in
shallow water near the outlet.

INTRODUCTION

    A total of six nuclear reactors installed in three power plants
along the northen and southern coasts of Taiwan (Fig. 1) have
started their operations one after another since October 1977.
Each plant has two units for generating power. Owing to the large
quantities of cooling water intake into and discharge from the
plants, some environmental factors such as water temperature,
environmental radioactivity and nearshore current may be
significantly changed. Variations of these abiotic environmental
factors may influence the biological activities in the ecosystem,
particularly may impart some kinds of damage toward marine
biological resources. Therefore, the possible environmental
impact upon the biological systems including the fishery resources
along the northern and southern coasts of Taiwan sould be studied
before and after the plant operation.
    The National Scientific Committee on Problems of the
Environment, Academia Sinica (SCOPE/AS), under the supports of
Atomic Energy Council and Taiwan Power Company, has started long-
term programs of biological, ecological, chemical (including
radionuclides) and hydrographical environmental monitoring along
the power plant sites in the northern and southern parts of Taiwan

394

since July 1974 and July 1979, respectively. The items monitored
include nonbiological factors such as ocean and nearshore currents
and physical and chemical properties of sea water, and biological
factors such as primary productivity, species compositions and
interspecific relationships among phyto- and zooplanktons, algae,
invertebrates, corals and fishes, radionuclides in sea water and
biological specimens, and fishery statistics. Each part of the
project has been directed by the best available specialist of the
field invited by SCOPE/AS. The data collected in each year were
documented and discussed in annual reports. Near the first
nuclear power plant the littoral zone is majorly composed of sand
stones. Near the second nuclear power plant, the littoral zone is
majorly composed of rocks with sandy beach for siwmming.

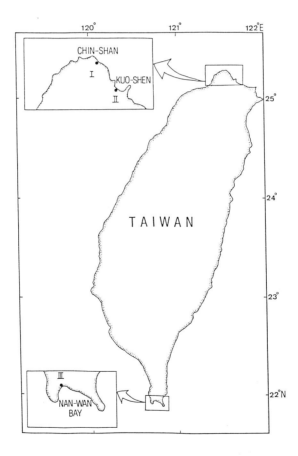

Fig. 1. The locations of three nuclear power plants in Taiwan.

There are plenty of corals along the coastline near the third nuclear power plant. Taiwan island is located between 20° 50'N and 25° 21'N in latitude. The mean surface water temperature around Taiwan is over 20°C. Nan-wan Bay, located at the southern tip of Taiwan, is a part of Ken-ting National Park. There are 179 coral species found in Nan-wan Bay. Of them, 155 species belong to scleractinians and 24 species to alcyonaceans (Yang et al., 1982). When the corals were exposed to a thermal stress of over 33°C water temperature, they would die within a few days (Yang et al., 1980).

The nearshore tidal current and the surface current adjacent to Taiwan affect the cooling efficiency of the nuclear power plants. The tidal currents control the distributions of thermal water plumed from the plants. And surface currents bring different water masses with different water temperatures to the power plant sites. Yao and Kao (1976) studied the current phenomena in Taiwan Strait. They found that the most predominant current component in the Strait is the tidal current of semidiurnal period. The tidal current flows along the coastline back and forth, and reverses its direction every 6 hours or so with average current speed of about 20 cm/sec. Then the thermal water nearshore moves in the direction parallel to the coastline about 4 km in both directions (Fan, 1987).

The surface current patterns in Taiwan Strait have been studied by Fan and Yu (1981). During winter time, the branch of the Kuroshio passes by Nan-wan Bay, and flows into Taiwan Strait. In summer time, South China Sea water flows into Taiwan Strait and Nan-wan Bay (Fig. 2). The surface water, either from the Kuroshio branch or South China Sea, is quite warm. It can get to about 24°C even in winter time, and to 29°C or higher in summer time (Su et al., 1981). This warm, clear water makes a well-developed coral community in the fringing reef in Nan-wan Bay. In 1979, the third nuclear power plant was constructed on Nan-wan Bay and began to operate in November 1984. The plant takes a large amount of sea water for cooling, and discharges to the sea surface from the outlet. Along the northern coast of Taiwan, the sea surface temperature is about 26°C in the summer time, and about 18°C in the winter time (Huang, 1989). The thermal effluent problems of the two nuclear power plants in northern Taiwan is not so serious as that of the third one in the southern tip of Taiwan.

The thermal water discharged from the third nuclear power plant is our major concern. Early in 1971, Jones et al. (1972) started to investigate the corals and fishes in Nan-wan Bay. Later on, Yang et al. (1976) collected biological data, and Liang et al. (1978) collected physical and geological data in the Bay area. If the discharged thermal effluents flew to the intake area, and the water was taken in for cooling again, then the thermal pollution will get serious. And there was also some evidence to indicate that the sea water near the outlet area might flow back to the intake area (Su et al., 1980). Then, after some model tests (Tang et al., 1980), TAIPOWER constructed a 120 m

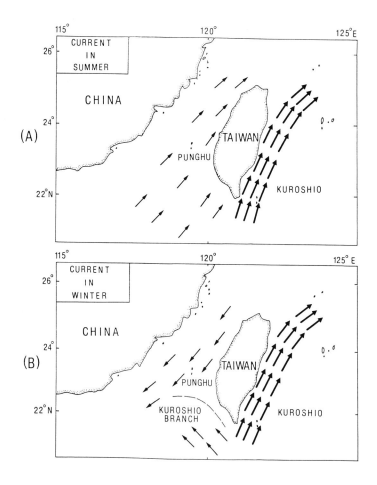

Fig. 2. Surface current patterns in the vicinity of Taiwan in summer (A) and winter (B).

offshore water channel into the ocean. This water channel forms a little bay on its west side. And this channel was finished before the operation of the plant. The second unit of the third nuclear power plant started to operate in May 1985. Unfortunately, the first unit caught fire and stopped operation until January 1987. In early July 1987, almost all the coral with the water depth less than 3 m in the little bay bleached. In this paper, current and water temperature data were employed to discuss the thermal effluent diffusion near the outlet area and the incident happened in early July 1987.

RESULTS AND DISCUSSION

The first nuclear power plant is located near Chin-shan in northern Taiwan. The No. 1 and the No. 2 units started to operate in October 1977 and October 1978, respectively. During flood tide, the current, parallel with the coastline, flows toward the WNW direction, and toward the ESE direction during ebb tide. Figure 3 shows the surface water temperature rises near the outlet area on Aug. 18, 1988. The data, supplied by the radioactive laboratory, TAIPOWER, were collected from 10:00 to 10:20 AM during flood tide. The surface water temperature at a background station near the intake is 28.3° C. The area of temperature rise larger than $4\,^\circ$C extends from the outlet to about 500-600 m away in the

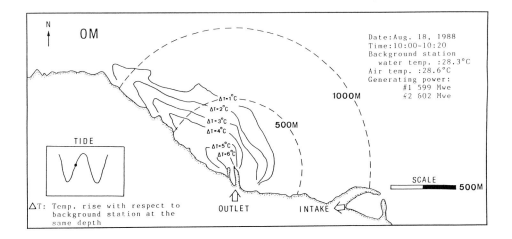

Fig. 3. Surface water temperature rises near the outlet area of the first nuclear power plant in the morning of Aug. 18, 1988.

WNW direction. The surface water temperature exceeds 32.3°C. It has a more or less important impact on the ecosystem in that small area.

The second nuclear power plant is located near Kuo-shen in northern Taiwan. The No. 1 and No. 2 units started to operate in December 1982 and March 1983, respectively. During flood tide, the current, parallel with the coastline, also flows toward the WNW directin, and toward the ESE direction during ebb tide. Figure 4 shows the surface water temperature rises near the outlet area on Aug. 30, 1988. The data, also supplied by the radioactive laboratory, TAIPOWER, were collected from 2:00-3:00 PM during ebb tide. The surface water temperature at a barkground station near the intake is 27.5°C. The area of temperature rise larger than 8°C extends from the outlet to even more than 700 m to the western side of the outlet. It affects not only the fishery catch in that small area, but also the swimming activities there.

Su et al. (1984) summarized and discussed the data collected from July 1974 to June 1984 for final assessment of ecological and environmental impact of the four units of the first and second nuclear power plants on the coastal environment of northern

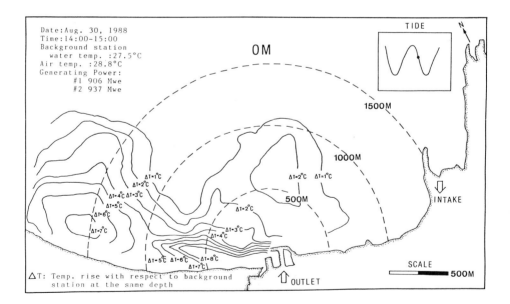

Fig. 4. Surface water temperature rises near the outlet area of the second nuclear power plant in the afternoon of Aug. 30, 1988.

Taiwan. The data indicated the major effects of these two power plants on the environment as follows:

1. The thermal plumes from the power plants affect the fishing activities and nearby swimming areas.

2. Although the contents of natural radionuclides, such as Tl-208, Pb-212, Pb-214, Bi-214 and the total beta activities in fishes remained at the background levels, the frequencies for the detection and the quantities of these radionuclies had been increasing during the operation of the second nuclear power plant. The results also indicate that there is no signficant difference in Sr-90, K-40, Co-60 and Zn-65 in fishes before and during the four reactor concurrent operation.

3. There is no obvious variation in biological, chemical and ecological parameters before and during the operation of the nuclear power plants.

The third nuclear power plant is located on the ocean side near Nan-Wan Bay in southern Taiwan. The No. 1 and No. 2 units started to operate in July 1984 and May 1985, respectively. Near the third nuclear power plant, the temperature and current data with RCM-4 current meters have been collected for more than ten years. The radioactive laboratory of TAIPOWER also supplied some water temperature data.

Liang et al. (1978) studied the current field in Nan-wan Bay. The current near the outlet area flows towards the SW most of the time during flood tide. During ebb tide, sometimes it flows to the SW, and sometimes to the NE, With the predominant SW direction. This flow feature can avoid the thermal effluent being used for cooling again to create recirculation problem. And this is also the reason why TAIPOWER designed to locate the intake to the northeast of the outlet (Fig. 5). They also found that the diurnal and semi-diurnal tides dominated the current patterns in Nan-wan Bay. Yip (1984)'s analysis of the tidal current in Nan-wan Bay also shows the similar features.

Figure 5 shows the temperature rises near the outlet area of the third nuclear power plant on July 3, 1987. The data, supplied by the radioactive laboratory, TAIPOWER, were collected from 8:10 to 8:30 AM, during flood tide. The surface water temperature at a background station near the intake is 27.9° C, the area of temperature rise larger than 4 °C extends from the outlet to 300-400 m away in the southern and western directions. In this case, the surface water temperature in the little bay, on the western

side of the outlet, exeeds 31.9°C. In some areas, the temperature
exceeds 34°C. And the thermal water majorly occupies the surface
3 m layer. It definitely will hurt or kill corals in shallow
waters near the shoreline. This incident was found in early July
1987 --the first serious incident ever happened in Taiwan (Fan,
1988). And, that also was the first summer to have two nuclear
reactors of the third nuclear power plant operated simultaneously.

   Su et al. (1988) summarized and discussed the data collected
from July 1979 to June 1988 for a preliminary assessment of
ecological and environmental impact of the two units of the third
nuclear power plant on the coastal environment of Nan-wan Bay.
The data indicated the major effects of this two power plant on
the environment as follows:

1. The thermal plume from the power plant affects the fishing
   activities.

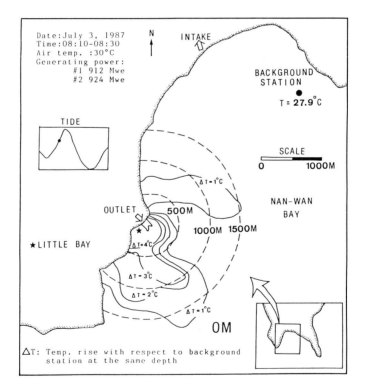

Fig. 5. Surface water temperature rises near the outlet area of
        the third nuclear power plant in the morning of July 3,
        1987.

2. The total alpha, beta and gamma activities of plankton and most of radionuclides detected in fishes collected from Nan-wan Bay showed no significant differences before and during the reactor operation. It should be noted, however, that the levels of Cs-137 (ranging from 6 to 91±20 pCi/Kg-flesh) and U-Series (31±6 pCi/Kg-flesh) in fish had been increasing during the period of two reactor concurrent operation although they still remained well below the regulatory levels.

3. The bleaching phenomenon of corals due to the thermal effluents was found only in the water layer of 0-5 m depth of a shallow bay near the cooling-water outlet in July 1987. The percentage of bleached coral colonies was 90-100% in the layer of 0-3 m and 50-70% in the layer of 3-5 m.

In the summer of 1988, the thermal plume from the third nuclear power plant even hurt more corals in a more extended area. In summer of 1989, near the outlet area, a few living corals in shallow waters left to be bleached.

CONCLUDING REMARKS

According to the government requirements, the water temperature rise is not allowed to exceed 4 °C beyond 500 m from the discharge structure of the power plant. The Little Bay area, where the incident happened in July 1987, is just on the western side of the outlet, and within 300 m from the discharge structure. The temperature rise may readily exceed 4 °C in the Little Bay. Even if the thermal effluent hurts or kills some corals in summer, the government regulations are not violated. However, on the bright side, after that incident for two consecutive years, people pay more attention to our environment in Taiwan, and the TAIPOWER try hard to improve the cooling systems of power plants.

ACKNOWLEDGEMENTS

The author wishes to express his appreciation to the radioactive laboratory, TAIPOWER. The study was supported partly by National Science Council, and partly by the Atomic Energy Council and TAIPOWER through the National Scientific Committee on problems of the Envionment, Academia Sinica.

REFERENCES

402

Fan, K.L., 1987. The pollutants movement along the southwestern coast of Taiwan. Proceedings of a Workshop on Ocean Outfalls, Taipei, pp. 12-23.

Fan, K.L., 1988. The thermal effluent incident of the third nuclear power plant in southern Taiwan. Acta Oceanogr. Taiwanica, 20: 107-117.

Fan, K.L. and Yu C.Y., 1981. A study of water masses in the seas of southernmost Taiwan. Acta Oceanogr. Taiwanica, 12: 94-111.

Huang, S.J., 1989. The Analysis of sea surface temperature in the seas adjacent to Taiwan and the applications of satellite remote sensing. Master Thesis, Dept. of Oceanogr., Nat'l Taiwan Ocean Univ., 48 pp.

Jones, R.S., Randall, R.H., Cheng, Y.M., Kami H.T. and Mak, S.M., 1972. A marine biological survey of southern Taiwan with emphasis on corals and fishes. Inst. of Oceanogr., Nat'l Taiwan Univ. Special Publ. No. 1, 93 pp.

Liang, N.K., Lien, S.L., Chen, W.C. and Chang, H.T., 1978. Oceanographic investigation in the vicinity of Ma-an-san --- Nan-wan Bay. Inst. of Oceanogr., Nat'l Taiwan Univ. Special publ. No. 18, 207 pp.

Tang, F.L.W., Hwung, H.H., Lin, C.P., Chen, Y.Y. and Chien, W.C., 1980. Studies of the improvement on the thermal diffusion of the Third Nuclear Power Plant. Tainan Hydraulics Laboratory, Nat'l Cheng Kung Univ. Bulletin No. 47, 40 pp.

Su, J.C., Hung, T.C., Chiang, Y.M., Tan, T.H., Chang, K.H., Yang, R.T., Jeng, I.M., Fan, K.L. and Chang, H.T., 1980. An ecological survey on the waters adjacent to the nuclear power plant in southern Taiwan: The progress report of the first year study. SCOPE/ROC, Acadamia Sinica, 7: 115 pp.

Su, J.C., Hung, T.C., Chiang, Y.M., Tan, T.H., Chang, K.H., Yang, R.T., Jeng, I.M., Fan, K.L. and Chang, H.T., 1981. An ecological survey on the waters adjacent to the nuclear power plant in southern Taiwan: The progress report of the second year study. SCOPE/ROC, Acadamia Sinica, 10: 118 pp.

Su, J.C., Hung, T.C., Chiang, Y.M., Tan, T.H., Chang, K.H., Lee, S.C. and Chang, H.T., 1984. The final report for the assessment of ecological impact on the operation of the nuclear power stations along the northern coast of Taiwan: The progress report from July 1974 to June 1984. SCOPE/ROC, Academia Sinica, 29: 151 pp.

Su, J.C., Hung, T.C., Chiang, Y.M., Tan, T.H., Chang, K.H., Shao, K.T., Huang, P.P., Lee, T.K., Huang, C.T., Huang, C.Y., Fan, K.L. and Yeh, S.Y., 1988. An ecological survey on the waters adjacent to the nuclear power plant in southern Taiwan: The progress report of the ninth year study (1987-1988). SCOPE/ROC, Academia Sinica, 59: 394 pp.

Yang, R.T., Huang, C.C., Wang, C.H., Yeh, S.Z., Jan, Y.F., Liu, S.J., Chen, C.H., Chen, S.J., Chang, L.F., Sun, C.L., Huang, W. and Tsai, C.F., 1976. A marine biological data acquisition program pertaining to the construction of a power plant in the Nan-Wan Bay area: Phase 1. a preliminary reconnaissance survey. Inst. of Oceanogr., Nat'l Taiwan Univ. Special Publ. 11, 134 pp.

Yang, R.T., Yeh, S.Z. and Sun, C.L., 1980. Effects of temperature on reef corals in the Nan-wan Bay, Taiwan. Inst. of Oceanogr., Nat'l Taiwan Univ. Special Publ. 23, 27 pp.

Yang, R.T., Dai, C.F., Yeh, S.Z., Sun, C.L., Su, F.Y., Liao, S.Y., Hsu, Y.K., Chang, C.L. and Chou, T.Y., 1982. Ecology and distribution of coral communities in Nan-wan Bay, Taiwan. Inst. of Oceanogr., Nat'l Taiwan Univ. Special Publ. 40, 74 pp.

Yao, N.C. and Kao, Y.C., 1976. Report on the computer and Analyses of the current meter data collected in the Taiwan Strait. A research project sponsored by Chinese Petroleum Corporation, 100 pp.

Yip, K.J., 1984. A study of low frequency motion in Nan-wan Bay. Master Thesis, Inst. of Oceanogr., Nat'l Taiwan Univ., 61 pp.

A SIMPLE DYNAMIC CALCULATION OF LAUNCHING A LINE

Kazuo KAWATATE, Takashige SHINOZAKI, Yoshio HASHIMOTO,
Tomoki NAGAHAMA, Hideo ISHII and Akimasa TASHIRO

Research Institute for Applied Mechanics, Kyushu University, Japan

ABSTRACT

Using a simple model consisting of anchor, buoy, and connection, we calculate motion and tension of a line in launch. Even after the anchor has hit the bottom, the buoy keeps descending, while it loses its speed and then it stops descending to start ascending. Repeating these series of motion the buoy reaches the position of its equilibrium. We derive conditions under which the slack in descending motion and the overshoot in ascending motion of the line take place.

1 INTRODUCTION

The authors have previously proposed a numerical calculation on the basis of a lumped mass method dealing with the motion of a mooring line (Kawatate, 1986; Kawatate et al., 1987). The proposed method of calculation has been applied to a problem pertaining to deploying a subsurface buoy, which was used for measuring the Kuroshio south of Tanegashima, as shown in Fig. 1, and the tension in line as well as the change of line shape were calculated. As a result of the calculation made by replacing the line with a model composed of 11 masses and 10 connections, we obtained variation of the tension in a nylon rope just above an anchor as shown in Figs. 2 and 3 together with change of the line shape as shown in Fig. 4. It has been made known that the whole line is almost straight when the top mass goes into the water and the line descends almost vertically hereafter. Based on the above findings, the present paper deals with an attempt to seek a simple solution by replacing the entire system with such a dual mass-one connection model as shown in Fig. 5, in order to simulate the tension arisen in the nylon rope. In replacing two lines having different length and stiffness with one line as shown in Fig. 6, we adopted a simplified method, where the length is the sum of the length of each line $l_1 + l_2$ and the stiffness is given by

$$\frac{l_1 + l_2}{EA} = \frac{l_1}{E_1 A_1} + \frac{l_2}{E_2 A_2}. \tag{1}$$

Incidentally, all symbols used in the present report are given in the appendix.

2 BASIC EQUATIONS

We write equations of motion of a buoy and an anchor

406

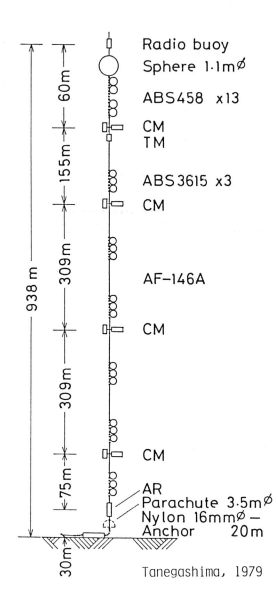

Radio buoy
Sphere 1·1m∅

ABS458 x13

CM
TM

ABS3615 x3

CM

AF–146A

CM

CM

AR
Parachute 3·5m∅
Nylon 16mm∅ —
Anchor   20m

Tanegashima, 1979

Fig. 1. A subsurface buoy used south of Tanegashima at a depth of 1000m October through November 1979.

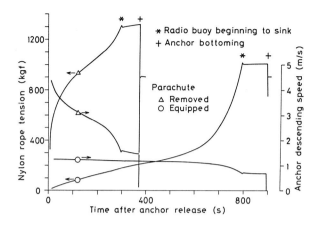

Fig. 2. Variation of tension on nylon rope just above an anchor.

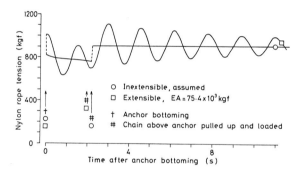

Fig. 3. Variation of tension on nylon rope just above an anchor after anchor bottoming.

$$\frac{W_B}{g}\frac{dv_B}{dt} = T - B - k_B v_B |v_B|, \tag{2}$$

$$\frac{W_A}{g}\frac{dv_A}{dt} = -T + W' - k_A v_A |v_A|, \tag{3}$$

where we have relations

$$x_A = x_B + l_0 + \Delta l, \tag{4}$$

$$\Delta l = \frac{l_0 T}{EA}. \tag{5}$$

In descending motion when the tension acting on a connecting line between the buoy and the anchor is constant, accordingly the elongation of the connecting line is constant, we have that

$$\Delta \dot{l} = 0, \tag{6}$$

$$\dot{x}_A = \dot{x}_B \quad \text{or} \quad v_A = v_B. \tag{7}$$

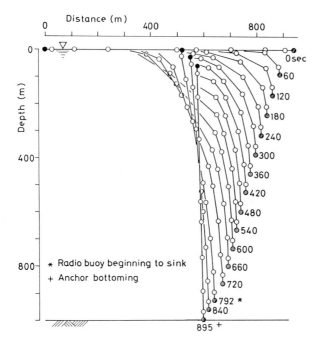

Fig. 4. Change of line configuration.

From the equations of motion we have that

$$\frac{W}{g}\frac{dv_B}{dt} = W' - B - kv_B|v_B|, \tag{8}$$

$$T = \frac{W_A(B + k_B v_B|v_B|) + W_B(W' - k_A v_A|v_A|)}{W}, \tag{9}$$

when we put

$$W = W_A + W_B, \tag{10}$$

$$k = k_A + k_B. \tag{11}$$

Expressing the acceleration in a form

$$\frac{dv_B}{dt} = \frac{dx_B}{dt}\frac{dv_B}{dx_B} = v_B\frac{dv_B}{dx_B} = \frac{1}{2}\frac{dv_B^2}{dx_B}, \tag{12}$$

we integrate the equation of motion in terms of the position x_B. Putting at the time $t = t_1$

$$v_B = v_1, \tag{13}$$

$$x_B = x_1, \tag{14}$$

we have a solution

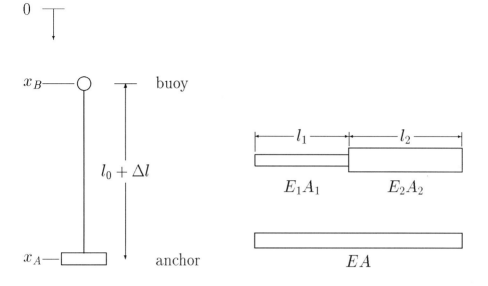

Fig. 5. Anchor, buoy, and connection.

Fig. 6. Conversion of extensional rigidity.

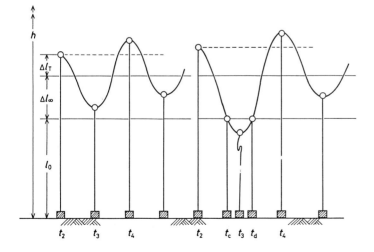

Fig. 7a. Descending and ascending motion of buoy after anchor bottoming   overshoot.

410

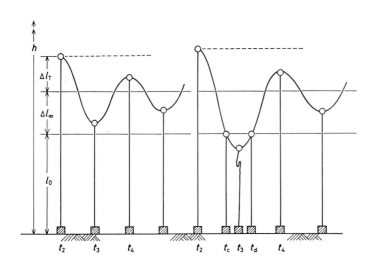

Fig. 7b. Descending and ascending motion of buoy after anchor bottoming   undershoot.

$$v_B^2 = (v_1^2 - V_T^2)\exp\left[-\frac{2gk(x_B - x_1)}{W}\right] + V_T^2, \tag{15}$$

where V_T is the terminal velocity of the system shown in Fig. 5,

$$V_T = \sqrt{\frac{W' - B}{k}}. \tag{16}$$

Time t is expressed in terms of the positions x_B as follows:

$$t - t_1 = \int_{x_1}^{x} \frac{dx'}{v_B(x')}. \tag{17}$$

Of the system shown in Fig. 1 we have that $W/2gk = 0.7164$m, and removing a parachute from the system we have $W/2gk = 0.8164$m. In both cases when the top float goes into water and moves down a few meters, the system gets the terminal velocity

$$v_B \simeq V_T, \tag{18}$$

when the tension is given by

$$T = B + k_B V_T^2 = W' - k_A V_T^2 = \frac{k_A B + k_B W'}{k}. \tag{19}$$

The tension is a sum of the buoyancy B of the buoy and the drag $k_B V_T^2$ acting on the buoy. Generally we have to check whether $v_B \simeq V_T$ by calculating $W/2gk$. However, we assume hereafter that at the time of the anchor bottoming $t = t_2$ the system gets the terminal velocity,

$$v_{B_2} = V_T. \tag{20}$$

We have that

$$x_{A_2} = h, \tag{21}$$

$$x_{B_2} = h - l_0 - \frac{l_0 B}{EA} - \frac{l_0 k_B V_T^2}{EA}. \tag{22}$$

Putting that

$$\Delta l_\infty = \frac{l_0 B}{EA}, \tag{23}$$

$$x_{B_\infty} = h - l_0 - \Delta l_\infty, \tag{24}$$

we express that

$$x_{B_2} = x_{B_\infty} - \frac{l_0 k_B V_T^2}{EA}. \tag{25}$$

After the anchor bottoming, the motion of the buoy and the tension in the connecting line are described by

$$\frac{W_B}{2g} \frac{dv_B^2}{dx_B} = T - B - k_B v_B |v_B|, \tag{26}$$

$$T = \begin{cases} \dfrac{EA}{l_0}(h - l_0 - x_B), & \text{if} \quad h - l_0 - x_B \geq 0; \\[2mm] 0, & \text{if} \quad h - l_0 - x_B < 0. \end{cases} \tag{27}$$

After the anchor has arrived at the ground, the buoy continues to descend $v_B > 0$, decreases its speed $\dot{v}_B < 0$, and gets to the lowest position at the time $t = t_3$, where the speed of the buoy becomes zero $v_{B_3} = 0$. Then the buoy turns to ascend $v_B < 0$. The absolute value of ascending speed increases $\dot{v}_B < 0$ for a while and decreases $\dot{v}_B > 0$. When the buoy gets to the highest position at the time $t = t_4$, its speed is zero $v_{B_4} = 0$. Repeating these motions the buoy diminishes its motion gradually and stops finally, when the tension of the line is equal to the buoyancy of the buoy $T = B$. We illustrate the motion schematically in Figs. 7a and 7b.

Putting that

$$L_B = \frac{W_B}{2g k_B}, \tag{28}$$

$$\omega^2 = \frac{EA}{l_0} \frac{g}{W_B}, \tag{29}$$

$$V_L^2 = \frac{EA}{l_0} \frac{L_B}{k_B} = 2\omega^2 L_B^2, \tag{30}$$

we define non-dimensional time, position, and speed

$$\tau = \frac{V_L}{L_B} t = \sqrt{2}\omega t, \tag{31}$$

$$\eta = \frac{x_B - x_{B_\infty}}{L_B}, \tag{32}$$

$$\sigma = \frac{d\eta}{d\tau} = \frac{v_B}{V_L}. \tag{33}$$

The tension is expressed by

$$T = \begin{cases} EA\dfrac{L_B}{l_0}(\eta_c - \eta), & \text{if } \eta \le \eta_c; \\[2mm] 0, & \text{if } \eta_c < \eta, \end{cases} \tag{34}$$

where

$$\eta_c = \frac{h - l_0 - x_{B\infty}}{L_B} = \frac{\Delta l_\infty}{L_B}. \tag{35}$$

Using the Heaviside unit step function and the signum function,

$$H(x) = \qquad\qquad 1,\ x > 0;\ \frac{1}{2},\ x = 0;\quad 0,\ x < 0; \tag{36}$$

$$\mathrm{sgn}(x) = H(x) - H(-x) = 1,\ x > 0;\ 0,\ x = 0;\ -1,\ x < 0; \tag{37}$$

we write the equation of motion in a form,

$$2\frac{d^2\eta}{d\tau^2} = \frac{d\sigma^2}{d\eta} = -[\eta_c - (\eta_c - \eta)H(\eta_c - \eta)] - \mathrm{sgn}(\sigma)\sigma^2. \tag{38}$$

We easily obtain a numerical solution of the above equation by use of, for instance, the Runge-Kutta-Gill method. On the other hand it is convenient to use an analytic solution, written bellow, for the purpose of treating a slack condition and a transition load in the line. We have a solution σ in terms of η, assuming $\sigma = \sigma_0$ when $\eta = \eta_0$, according to $\eta \lessgtr \eta_c$ and $\sigma \lessgtr 0$ as follows:

$$\sigma^2 = [\sigma_0^2 - (1 - \eta_0)]\exp[-(\eta - \eta_0)] + 1 - \eta, \quad \eta \le \eta_c,\ \text{descent } \sigma \ge 0, \tag{39}$$

$$\sigma^2 = \qquad (\sigma_0^2 + \eta_c)\exp[-(\eta - \eta_0)] - \eta_c, \qquad \eta_c < \eta,\ \text{descent } \sigma \ge 0, \tag{40}$$

$$\sigma^2 = \qquad (\sigma_0^2 - \eta_c)\exp[(\eta - \eta_0)] + \eta_c, \qquad \eta_c < \eta,\ \text{ascent } \sigma \le 0, \tag{41}$$

$$\sigma^2 = [\sigma_0^2 - (1 + \eta_0)]\exp[(\eta - \eta_0)] + 1 + \eta, \quad \eta \le \eta_c,\ \text{ascent } \sigma \le 0. \tag{42}$$

3 SLACK OF LINE

Putting that

$$r = \frac{V_T}{V_L}, \tag{43}$$

we obtain that

$$\sigma_2^2 = \quad r^2, \tag{44}$$

$$\eta_2 = -r^2. \tag{45}$$

Assuming that the tension is positive when the buoy reaches the lowest position and using

$$\sigma_3 = \sigma_4 = \cdots = 0, \tag{46}$$

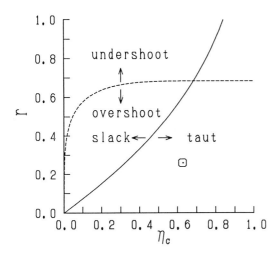

Fig. 8. Slack condition and overshoot condition.

we have from (39) and (42) that

$$\exp[-(\eta_3 + r^2)] + \eta_3 + r^2 - (1 + r^2) \quad = 0, \tag{47}$$
$$\exp[-(\eta_3 - \eta_4)] + (\eta_3 - \eta_4)/(1 + \eta_3) - 1 = 0, \tag{48}$$
$$\exp[-(\eta_5 - \eta_4)] + (\eta_5 - \eta_4)/(1 - \eta_4) - 1 = 0. \tag{49}$$

From the above equations we seek η_3, η_4, ... successively.

To solve the equation (47) we put

$$X = \eta_3 + r^2, \tag{50}$$

and seek X from the following

$$f(X) = \exp(-X) + X - (1 + r^2) = 0, \tag{51}$$

by use of the Newton-Raphson method. Then we determine η_3. Taking account of relations

$$f'(X) = -\exp(-X) + 1 \geq 0, \quad X \geq 0, \tag{52}$$
$$f(1 + r^2) > 0, \tag{53}$$

we know that

$$X < 1 + r^2, \tag{54}$$

accordingly

$$\eta_3 < 1. \tag{55}$$

To solve the equations (48), (49), ..., by taking account of relations

$$\eta_i < 0, \quad i = 2, 4, \ldots, \tag{56}$$

$$\eta_i > 0, \quad i = 3, 5, \ldots, \tag{57}$$

we put that

$$X = |\eta_{i+1} - \eta_i|, \quad i = 3, 4, \ldots, \tag{58}$$

$$\beta = 1 + |\eta_i|, \tag{59}$$

and seek X from

$$\exp(-X) + X/\beta - 1 = 0, \tag{60}$$

through the Newton-Raphson method, for instance. We have then that

$$\eta_{i+1} = \eta_i + (-1)^i X, \quad i = 3, 4, \ldots. \tag{61}$$

Let us consider the tension when the buoy stays at the lowest position. In the case of $\eta_c \geq 1$, since $\eta_3 < 1$ as shown in (55), we always have that $\eta_3 < \eta_c$. At the lowest position the tension is positive and the line is taut.

In the other case of $\eta_c < 1$, we write a condition that the tension is zero at the lowest position: $\eta_3 = \eta_c$ and $\sigma_3 = \sigma_c = 0$, and obtain from (47) that

$$\exp[-(\eta_c + r^2)] + \eta_c - 1 = 0, \quad \eta_c < 1. \tag{62}$$

Solving the above we have that

$$r^2 = -\ln(1 - \eta_c) - \eta_c, \quad \eta_c < 1. \tag{63}$$

We show this relation in Fig. 8 by a solid line. When the following condition holds

$$r^2 < -\ln(1 - \eta_c) - \eta_c, \quad \eta_c < 1, \tag{64}$$

we know that the line is taut at the lowest position. In order to determine the motion we use (51) and (60). In a range that

$$r^2 > -\ln(1 - \eta_c) - \eta_c, \quad \eta_c < 1, \tag{65}$$

the elongation accordingly the tension becomes zero before the buoy gets to the lowest position. The rope is slack. In order to determine the motion we seek the descending speed $\sigma = \sigma_c$ at the position $\eta = \eta_c$ where the elongation is zero, the lowest position $\eta = \eta_3$ where the speed is zero $\sigma = \sigma_3 = 0$, the ascending speed $\sigma = \sigma_d$ at the position $\eta = \eta_c$ after the buoy turns to move upward, and the highest position $\eta = \eta_4$ where the speed is zero $\sigma = \sigma_4 = 0$, using the equations given below, which are derived from (39), (40), (41), and (42),

$$\sigma_c^2 = -\exp[-(\eta_c + r^2)] + 1 - \eta_c, \tag{66}$$

$$0 = (\sigma_c^2 + \eta_c)\exp[-\eta_3 + \eta_c] - \eta_c, \tag{67}$$

$$\sigma_d^2 = -\eta_c \exp[\eta_c - \eta_3] + \eta_c, \tag{68}$$

$$0 = [\sigma_d^2 - (1 + \eta_c)]\exp[\eta_4 - \eta_c] + 1 + \eta_4. \tag{69}$$

Rewriting them in forms

$$\sigma_c^2 = -\exp[-(\eta_c + r^2)] + 1 - \eta_c, \qquad\qquad 0 < \sigma_c^2 < 1, \qquad\qquad (70)$$

$$\eta_3 = \eta_c + \ln[(\sigma_c^2 + \eta_c)/\eta_c], \qquad\qquad \eta_c < \eta_3, \qquad\qquad (71)$$

$$\sigma_d^2 = \eta_c \sigma_c^2/(\sigma_c^2 + \eta_c), \qquad\qquad \sigma_d^2 < \sigma_c^2, \qquad\qquad (72)$$

$$[1 + \eta_c^2/(\eta_c + \sigma_c^2)]\exp(-X) + X - (1 + \eta_c) = 0, \quad \eta_4 = \eta_c - X, \qquad (73)$$

we obtain σ_c, η_3, σ_d, and η_4 successively.

4   TRANSITION LOAD

As stated before, the buoy approaches to its equilibrium position by repeating descending and ascending motions after the anchor has attained the ground. The variable load during that time sometimes surpasses the tension at the moment of anchor bottoming and othertime not, as shown in Figs. 7a and 7b. Here, we will firstly evaluate such a condition as that the maximum value of transition load equals the tension at the time of anchor bottoming.

It is assumed here that the elongation of the line has attained exactly zero $\eta_3 = \eta_c$, at the moment when the buoy descends to the lowest position η_3, subsequent to the anchor bottoming. This condition is expressed as follows by using (62) which holds for $\eta_c < 1$,

$$1 - \eta_c = \exp[-(\eta_c + r^2)]. \qquad\qquad (74)$$

The buoy turns to ascend and gets the highest position η_4. It is assumed that the rope length at the highest position equals that of the anchor bottoming $\eta_4 = -r^2$. Using (48) we write the condition

$$\exp[-(\eta_c + r^2)] + (\eta_c + r^2)/(1 + \eta_c) - 1 = 0. \qquad\qquad (75)$$

From these two conditions of $\eta_3 = \eta_c$ and $\eta_4 = -r^2$ there will be induced

$$\eta_c^2 = -\ln(1 - \eta_c) - \eta_c. \qquad\qquad (76)$$

Solving the above we get $\eta_c = 0.6838$, where

$$r = \eta_c = 0.6838. \qquad\qquad (77)$$

In the case of

$$r > 0.6836, \qquad\qquad (78)$$

the tension at the highest position of the buoy does not exceed the tension at the time of anchor bottoming, $\eta_c - \eta_4 < \eta_c + r^2$, i.e., undershoot; whereas in the case of

$$r < 0.6836, \qquad\qquad (79)$$

the tension at the highest position of the buoy surpasses the tension at the anchor bottoming, $\eta_c - \eta_4 > \eta_c + r^2$, i.e., overshoot.

Also in the case of $\eta_c > 0.6838$, when $r = 0.6838$ the line is taut at the time when the buoy gets the lowest position η_3 and it is seen that the elongation of the line at the time when the buoy gets the highest position η_4 is equal to the elongation at the time when the anchor reaches the bottom. Accordingly, the above conditions are applicable.

Moreover, in the case of $\eta_c < 0.6838$, by putting $r = 0.6838$ it is known that the line is slack at the time when the buoy gets to the lowest position. We have to use the formula for the case where the line is slack at the lowest position of the buoy. Assuming $\eta_4 = -r^2$ that means the elongation at the highest position of the buoy is equal to the elongation at the anchor bottoming, then using equations (70) and (73) we induce

$$\sigma_c^2 = -\exp[-(\eta_c + r^2)] + 1 - \eta_c, \tag{80}$$

$$[1 + \eta_c^2/(\eta_c + \sigma_c^2)]\exp(-X) + X - (1 + \eta_c) = 0, \quad X = \eta_c + r^2. \tag{81}$$

From the above equations for a value of η_c we calculate X and r successively. The results are given in Fig. 8 with broken lines. The domains of undershoot and overshoot are also shown.

5 EXAMPLES OF CALCULATION AND CONSIDERATIONS

In considering the subsurface buoy system used for measuring the Kuroshio south of Tanegashima, as shown in Fig. 1, we carry out calculations by putting $B = 909\text{kgf}$, $W' = 1,437\text{kgf}$, $k_B = 200.6\text{kgfs}^2\text{m}^{-2}$, $k_A = 886.9\text{kgfs}^2\text{m}^{-2}$, $W_B = 1,882\text{kgf}$, $W_A = 13,387\text{kgf}$, $l_0 = 938\text{m}$, $h = 1,000\text{m}$, $EA = 2,829 \times 10^6\text{kgf}$, and we get $\eta_c = 0.623$ and $r = 0.260$, which is shown by a square in Fig. 8. From the figure it is known that the line maintains taut, while the tension of the line at the position to which the buoy has risen surpasses the tension of the line at the moment the anchor has bottomed. A phase plane trajectory and a time history of elongation and tension are respectively shown in Fig. 9 and 10, which have been drawn based on the Runge-Kutta-Gill method and by using PC9800.

Again, if we consider a system from which the parachute has been removed and assume $k_A = 52.4\text{kgfs}^2\text{m}^{-2}$, $W_A = 2,167\text{kgf}$ and the rest as same, we get $\eta_c = 0.623$ (the same) and $r = 0.538$. The phase plane trajectory and the elongation and tension history are given in Figs. 11 and 12.

While the tension in the actual mooring line is distributed, it is assumed as constant in the present calculation. Now, let us compare the present result of tension with that previously given in Fig. 3, where the lumped mass model of 11 masses-10 connections was used for calculation. The equilibrium convergent value is naturally the same. However, compared to the former model the present dual mass-one connection model gives higher value as for both the amplitude and the maximum value of tension. Therefore, the present method of calculation is justifiable as giving the value on the safe side.

The reason is accounted for by the concentration of weight. While with the previous model it occurs that the upper part, for instance, the part in the neighborhood of the radio buoy still keeps descending even when the part near the nylon rope has turned to move upward, with the present model the calculation has been made on an assumption that the part above the nylon rope starts rising uniformly. That is to say, while in the previous model, the movement at each point does not always act in concert, in the case of the present model

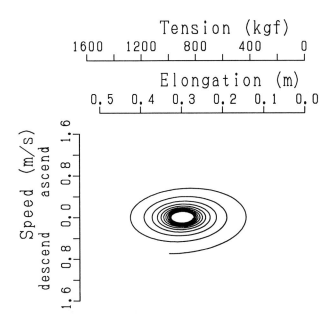

Fig. 9. Phase plane trajectory　　parachute equipped.

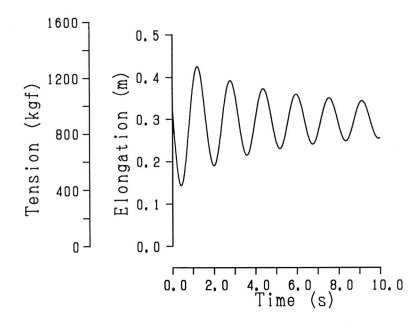

Fig. 10. Time history of elongation and tension　　parachute equipped.

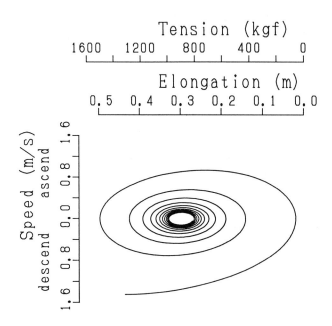

Tension (kgf)
1600   1200   800   400   0

Elongation (m)
0.5   0.4   0.3   0.2   0.1   0.0

Speed (m/s)

ascend   1.6   0.8   0.0

descend   0.8   1.6

Fig. 11. Phase plane trajectory   parachute removed.

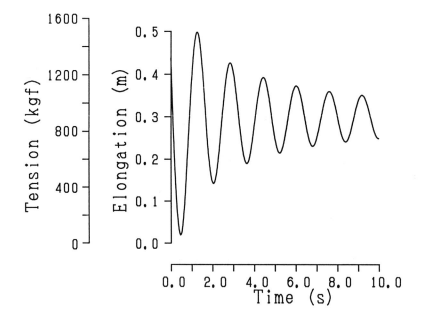

Tension (kgf)   Elongation (m)

1600 —   0.5 —

1200 —   0.4 —

800 —   0.3 —

400 —   0.2 —

       0.1 —

0 —   0.0 —

0.0   2.0   4.0   6.0   8.0   10.0
Time (s)

Fig. 12. Time history of elongation and tension   parachute removed.

each point is concentrated to the buoy and consequently each part is considered as to be acting simultaneously. For instance, when the buoy goes up and approaches the highest position, the acceleration acts in the direction toward which the tension increases. It is because of the concentration of weight that the tension given by the present method becomes high.

6 CONCLUSIONS

A mooring system is replaced with a model composed of a dual mass-one connection and calculation is carried out on this model at the time of its deployment. A simple method of estimation of the slack of the line and the transient tension subsequent to bottoming of the anchor is proposed.

Though, with an actual mooring line, tension is distributed, it is assumed as constant with the present model. Due to the concentration of the weight the calculated tension value becomes high. It is, however, considered to be appropriate for the estimation of the behavior of the rope right above the anchor.

7 REFERENCES

Kawatate, K., 1986. A dynamic two-dimensional calculation of underwater lines, Prog. Oceanog., Vol. 17, pp. 375-387.
Kawatate, K., Nagahama, T., Ishii, H., Shinozaki, T., and Tashiro, A., 1987. A dynamic calculation of underwater line in two-dimension (in Japanese), Bulletin of Research Institute for Applied Mechanics, Kyushu University, pp. 179-188.

APPENDIX

Symbols

B	buoyancy (kgf)
EA	extensional rigidity (kgf)
f	function
g	acceleration of gravity (ms^{-2})
h	water depth (m)
k	drag coefficient (kgfs^{2}m^{-2})
L	length (m)
l	length (m)
r	ratio of terminal velocity to characteristic velocity
T	tension (kgf)
t	time (s)
V	velocity
v	velocity
W	virtual weight (kgf)
W'	underwater weight (kgf)
X	variable
x	location (m)
η	non-dimensional location
η_c	non-dimensional location or elongation
σ	non-dimensional velocity
τ	non-dimensional time

Subscripts, Superscripts, etc.

A	anchor
B	float or buoy
T	terminal speed

420

0	origin
∞	equilibrium
Δ	elongation
.	d/dt

FREE TRANSVERSE VIBRATION OF AN ELASTIC CIRCULAR CYLINDER IN A FLUID

Guanghuan LIANG[*1] and Kazuo KAWATATE[*2]

[*1] Graduate School of Engineering, Kyushu University, Japan
[*2] Research Institute for Applied Mechanics, Kyushu University, Japan

ABSTRACT

In fluid-elastic problems it is always necessary to obtain the frequency of structural free vibration in a fluid. We use the separation of variable method to solve the coupling equations of elastic vibration and fluid motion. Considering a transverse vibration of an elastic circular cylinder piercing in a perfect fluid, we show the natural frequency equation. The present method can also be applied to the other fluid-elastic problems of the elastic cylindrical structures immersed in the fluid.

1 INTRODUCTION

When a structure vibrates in a fluid, the vibration of structure accompanies the motion of fluid. On the other hand the fluid motion affects the structural vibration. Thus, a coupling arises between the structure and the fluid. In order to estimate the natural frequency of the structural vibration in fluid, research workers have proposed several methods: empirical formulae, approximate solution by iteration, and numerical procedures on the basis of the finite-element or the boundary element (Clough, 1960; Kawatate, 1976; and Zienkiewicz et al., 1978).

Considering a transverse vibration of an elastic beam with circular section in the fluid, we try to obtain the natural frequency equation.

2 FORMULATION

We treat the transverse vibration of a circular section beam immersed in fluid, as shown in Fig. 1. We employ the Bernoulli-Euler beam (Timoshenko et al., 1974) and write the equation of motion in a form

$$\frac{\Gamma}{g}\frac{\partial^2 y}{\partial t^2} + EI\frac{\partial^4 y}{\partial x^4} = -F(x,t), \tag{1}$$

in the right hand side of which the fluid dynamic pressure term is involved. The equation of motion of the beam contains a fluid term.

We assume that the surrounding fluid is incompressible and irrotational. It is described by the velocity potential (Lamb, 1963), which satisfies the Laplace equation

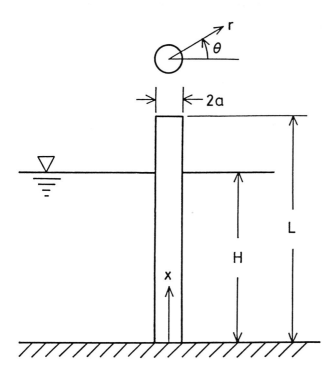

Fig. 1. A circular cylinder in a fluid.

$$\left(\frac{\partial^2}{\partial r^2} + \frac{1}{r}\frac{\partial}{\partial r} + \frac{1}{r^2}\frac{\partial^2}{\partial \theta^2} + \frac{\partial^2}{\partial x^2}\right)\Phi(r,\theta,x,t) = 0. \tag{2}$$

The fluid dynamic pressure on the beam is given by

$$F(x,t) = \frac{\gamma_w}{g}\int_0^{2\pi}\left(-\frac{\partial\Phi}{\partial t}\right)_{r=a}a\cos\theta d\theta. \tag{3}$$

The boundary conditions are written in the following. The kinematic and dynamic conditions at the mean free surface are combined into a form,

$$\frac{\partial^2\Phi}{\partial t^2} + g\frac{\partial\Phi}{\partial x} = 0, \quad x = H. \tag{4}$$

The vertical velocity component is zero at the bottom,

$$\frac{\partial\Phi}{\partial x} = 0, \quad x = 0. \tag{5}$$

The fluid at the surface of the beam moves with the beam,

$$\frac{\partial\Phi}{\partial r} = \frac{\partial y}{\partial t}\cos\theta, \quad r = a, \tag{6}$$

which also shows that the deflection of the beam influences the fluid motion. We take outward-traveling waves at infinity (Sommerfeld,1965),

$$\sqrt{r}\left(\frac{\partial}{\partial r} + \frac{1}{c(\omega)}\frac{\partial}{\partial t}\right)\Phi \to 0, \quad r \to \infty. \tag{7}$$

3 METHOD OF SOLUTION

We solve equations of coupling between the deflection of the beam and the velocity potential of the fluid by using the separation of variable method. We put that

$$y(x,t) = W(x)P(t), \tag{8}$$

and

$$\Phi(r,\theta,x,t) = \phi(r,\theta,x)\dot{P}(t). \tag{9}$$

We have the fluid dynamic pressure distribution, defined by $F(x,t)/\ddot{P}(t)$,

$$f(x) = \frac{\gamma_w}{g}\int_0^{2\pi}(-\phi)_{r=a}a\cos\theta\,d\theta. \tag{10}$$

We get from the equation of motion of the beam that

$$\ddot{P} = -\omega^2 P, \tag{11}$$

$$EI\frac{d^4W}{dx^4} - \omega^2\frac{\Gamma}{g}W = \begin{cases} \omega^2 f(x), & \text{if } 0 \le x \le H; \\ 0, & \text{if } H \le x \le L. \end{cases} \tag{12}$$

Taking $P = P_0e^{+i\omega t}$ we put

$$R(\alpha r) = \frac{H_1^{(2)}(\alpha r)}{H_1^{(2)\prime}(\alpha a)}. \tag{13}$$

We also put

$$R_j(\alpha_j r) = \frac{K_1(\alpha_j r)}{K_1(\alpha_j a)}, \tag{14}$$

and write that

$$\phi(r,\theta,x) = a\left\{R(\alpha r)d_0\cosh\alpha x + \sum_{j=1}^{N}R_j(\alpha_j)d_{0j}\cos\alpha_j x\right\}\cos\theta, \tag{15}$$

which is a solution of the Laplace equation (2). We easily find that ϕ satisfies both the bottom (5) and the radiation (7) conditions. Then the fluid dynamic pressure distribution is

$$f(x) = \frac{\gamma_w}{g}\pi a^2\left(d\cosh\alpha x + \sum_{j=1}^{N}d_j\cos\alpha_j x\right), \tag{16}$$

where

$$d = -d_0 R(\alpha a), \tag{17}$$

and

$$d_j = -d_{0j} R_j(\alpha_j a). \tag{18}$$

From the mean free surface condition (4) we have that

$$\frac{\omega^2}{g} = \alpha \tanh \alpha H, \tag{19}$$

and

$$\frac{\omega^2}{g} = -\alpha_j \tan \alpha_j H. \tag{20}$$

From the condition at the beam surface (6) we have that

$$\alpha a d_0 \cosh \alpha x + \sum_{j=1}^{N} \alpha_j a d_{0j} \cos \alpha_j x = W(x). \tag{21}$$

Using the relations

$$I = \frac{1}{H} \int_0^H \cosh \alpha x \cosh \alpha x \, dx = \frac{1}{2} \left(\frac{\sinh 2\alpha H}{2\alpha H} + 1 \right), \tag{22}$$

$$I_j = \frac{1}{H} \int_0^H \cos \alpha_j x \cos \alpha_j x \, dx = \frac{1}{2} \left(\frac{\sin 2\alpha_j H}{2\alpha_j H} + 1 \right), \tag{23}$$

$$\frac{1}{H} \int_0^H \cosh \alpha x \cos \alpha_j x \, dx = 0, \tag{24}$$

and

$$\frac{1}{H} \int_0^H \cos \alpha_i x \cos \alpha_j x \, dx = 0, \quad i \neq j, \tag{25}$$

we write that

$$d_0 = \frac{1}{\alpha a} \frac{1}{I} \frac{1}{H} \int_0^H W(x) \cosh \alpha x \, dx, \tag{26}$$

and

$$d_{0j} = \frac{1}{\alpha_j a} \frac{1}{I_j} \frac{1}{H} \int_0^H W(x) \cos \alpha_j x \, dx, \tag{27}$$

We also write that

$$\dot{d} = U(\alpha a) \frac{1}{I} \frac{1}{H} \int_0^H W(x) \cosh \alpha x \, dx, \tag{28}$$

and

$$d_j = U_j(\alpha_j a) \frac{1}{I_j} \frac{1}{H} \int_0^H W(x) \cos \alpha_j x \, dx, \tag{29}$$

where

$$U(\alpha a) = -\frac{R(\alpha a)}{\alpha a} = \frac{-J_1(\alpha a)J_1'(\alpha a) - Y_1(\alpha a)Y_1'(\alpha a) - i\frac{2}{\pi}\frac{1}{\alpha a}}{\alpha a\left[J_1'^2(\alpha a) + Y_1'^2(\alpha a)\right]},\tag{30}$$

and

$$U_j(\alpha_j a) = -\frac{R_j(\alpha_j a)}{\alpha_j a},\tag{31}$$

If the wave length is large compared with the diameter of the cylinder $\alpha a \ll 1$, the imaginary part of the above $U(\alpha a)$ is negligible to the real part. The imaginary part is proportional to the velocity of the beam motion and is related to damping, the real part to the acceleration and to added mass. Afterwards we consider the case of negligible damping and put

$$U(\alpha a) = \frac{-J_1(\alpha a)J_1'(\alpha a) - Y_1(\alpha a)Y_1'(\alpha a)}{\alpha a\left[J_1'^2(\alpha a) + Y_1'^2(\alpha a)\right]}.\tag{32}$$

The fluid dynamic pressure tends to vanish as the wave length becomes small (Patel, 1989) since

$$\alpha a \to \infty, \quad \alpha_j a \to \infty, \quad U \to 0, \quad U_j \to 0,\tag{33}$$

$$d \to 0, \quad d_j \to 0,\tag{34}$$

$$\phi(a, \theta, x) \to 0,\tag{35}$$

and

$$f(x) \to 0.\tag{36}$$

Defining that

$$\beta^4 = \frac{\omega^2 \Gamma}{g E I},\tag{37}$$

and

$$\epsilon = \frac{\gamma_w \pi a^2}{\Gamma},\tag{38}$$

we write the equation of the motion of the beam (12) that

$$\frac{d^4 W}{dx^4} - \beta^4 W = \begin{cases} \epsilon\beta^4\left(d\cosh\alpha x + \sum_{j=1}^{N} d_j \cos\alpha_j x\right), & \text{if } 0 \le x \le H; \\ 0, & \text{if } H \le x \le L. \end{cases}\tag{39}$$

This is a differential-integral equation (Ito and Kihara, 1972) in terms of W, since d and $d_j(j = 1, 2, \ldots, N)$ are expressed in the integration form of W.

Writing that

$$\eta = \frac{\beta^2}{\alpha^2 + \beta^2},\tag{40}$$

and

$$\eta_j = \frac{\beta^2}{\alpha_j^2 + \beta^2},$$ (41)

we have a solution in a form

$$W(x) = \begin{cases} W_1(x), & \text{if } 0 \le x \le H; \\ \\ W_2(x), & \text{if } H \le x \le L, \end{cases}$$ (42)

$$W_1(x) = C_1 \cos \beta x + C_2 \sin \beta x + C_3 \cosh \beta x + C_4 \sinh \beta x$$
$$+ \epsilon d\eta \left\{ \frac{(\cosh \alpha x - \cosh \beta x)\beta^2}{\alpha^2 - \beta^2} - \frac{\cosh \beta x - \cos \beta x}{2} \right\}$$
$$+ \sum_{j=1}^{N} \epsilon d_j \eta_j \left\{ \frac{(\cos \alpha_j x - \cos \beta x)\beta^2}{\alpha_j^2 - \beta^2} + \frac{\cosh \beta x - \cos \beta x}{2} \right\},$$ (43)

$$W_2(x) = D_1 \cos \beta(L - x) + D_2 \sin \beta(L - x)$$
$$+ D_3 \cosh \beta(L - x) + D_4 \sinh \beta(L - x),$$ (44)

where C_i and $D_i (i = 1, 2, 3,$ and $4)$ are arbitrary constants. In order to express d and $d_j (j = 1, 2, \ldots, N)$ in terms of $C_i (i = 1, 2, 3,$ and $4)$, we put that

$$A_i = \frac{1}{H} \int_0^H a_i(x) \cosh \alpha x \, dx,$$ (45)

and

$$A_{ji} = \frac{1}{H} \int_0^H a_i(x) \cos \alpha_j x \, dx,$$ (46)

where

$$a_1(x) = \cos \beta x,$$ (47)

$$a_2(x) = \sin \beta x,$$ (48)

$$a_3(x) = \cosh \beta x,$$ (49)

and

$$a_4(x) = \sinh \beta x.$$ (50)

We have from the beam surface condition that

$$\frac{Id}{U} = A_1 C_1 + A_2 C_2 + A_3 C_3 + A_4 C_4$$
$$+ \epsilon d\eta \left\{ \frac{(I - A_3)\beta^2}{\alpha^2 - \beta^2} - \frac{A_3 - A_1}{2} \right\}$$
$$+ \sum_{j=1}^{N} \epsilon d_j \eta_j \left\{ \frac{(0 - A_1)\beta^2}{\alpha_j^2 - \beta^2} + \frac{A_3 - A_1}{2} \right\},$$ (51)

$$\frac{I_k d_k}{U_k} = A_{k1}C_1 + A_{k2}C_2 + A_{k3}C_3 + A_{k4}C_4$$

$$+\epsilon d\eta \left\{ \frac{(0 - A_{k3})\beta^2}{\alpha^2 - \beta^2} - \frac{A_{k3} - A_{k1}}{2} \right\}$$

$$+ \sum_{j=1}^{N} \epsilon d_j \eta_j \left\{ \frac{(\delta_{jk} I_k - A_{k1})\beta^2}{\alpha_j^2 - \beta^2} + \frac{A_{k3} - A_{k1}}{2} \right\}, \tag{52}$$

where

$$\delta_{jk} = \begin{cases} 1, & \text{if} \quad j = k; \\[2mm] 0, & \text{if} \quad j \neq k. \end{cases} \tag{53}$$

Solving the above equations with respect to d and $d_k (k = 1, 2, \ldots, N)$, we express that

$$d = \sum_{i=1}^{4} d_{si} C_i, \tag{54}$$

and

$$d_k = \sum_{i=1}^{4} d_{sik} C_i, \tag{55}$$

where d_{s1} and d_{s1k} are, for instance, solutions of the equations when we put $C_1 = 1, C_2 = C_3 = C_4 = 0$.

We assume that the beam is fixed at the bottom $x = 0$ and free at the top $x = L$. The quantities of deflection, slope, moment, and shearing force are continuous along the beam. Thus, we have that

$$W_1(0) = 0, \tag{56}$$

$$W_1'(0) = 0, \tag{57}$$

$$W_2''(L) = 0, \tag{58}$$

$$W_2'''(L) = 0, \tag{59}$$

$$W_1(H) = W_2(H), \tag{60}$$

$$W_1'(H) = W_2'(H), \tag{61}$$

$$W_1''(H) = W_2''(H), \tag{62}$$

and

$$W_1'''(H) = W_2'''(H). \tag{63}$$

Putting

$$(p_i, q_i, r_i, s_i) = \epsilon d_{si}(p_s, q_s, r_s, s_s) + \sum_{j=1}^{N} \epsilon d_{sij}(p_{sj}, q_{sj}, r_{sj}, s_{sj}), \tag{64}$$

where

$$p_s = \eta \left\{ \frac{(\cosh \alpha H - \cosh \beta H)\beta^2}{\alpha^2 - \beta^2} - \frac{\cosh \beta H - \cos \beta H}{2} \right\}, \tag{65}$$

$$p_{sj} = \eta_j \left\{ \frac{(\cos \alpha_j H - \cos \beta H)\beta^2}{\alpha_j^2 - \beta^2} + \frac{\cosh \beta H - \cos \beta H}{2} \right\}, \tag{66}$$

$$q_s = \eta \left\{ \frac{(\alpha \sinh \alpha H - \beta \sinh \beta H)\beta}{\alpha^2 - \beta^2} - \frac{\sinh \beta H + \sin \beta H}{2} \right\}, \tag{67}$$

$$q_{sj} = \eta_j \left\{ \frac{(-\alpha_j \sin \alpha_j H + \beta \sin \beta H)\beta}{\alpha_j^2 - \beta^2} + \frac{\sinh \beta H + \sin \beta H}{2} \right\}, \tag{68}$$

$$r_s = \eta \left\{ \frac{\alpha^2 \cosh \alpha H - \beta^2 \cosh \beta H}{\alpha^2 - \beta^2} - \frac{\cosh \beta H + \cos \beta H}{2} \right\}, \tag{69}$$

$$r_{sj} = \eta_j \left\{ \frac{-\alpha_j^2 \cos \alpha_j H + \beta^2 \cos \beta H}{\alpha_j^2 - \beta^2} + \frac{\cosh \beta H + \cos \beta H}{2} \right\}, \tag{70}$$

$$s_s = \eta \left\{ \frac{\alpha^3 \sinh \alpha H - \beta^3 \sinh \beta H}{(\alpha^2 - \beta^2)\beta} - \frac{\sinh \beta H - \sin \beta H}{2} \right\}, \tag{71}$$

and

$$s_{sj} = \eta_j \left\{ \frac{\alpha_j^3 \sin \alpha_j H - \beta^3 \sin \beta H}{(\alpha_j^2 - \beta^2)\beta} + \frac{\sinh \beta H - \sin \beta H}{2} \right\}, \tag{72}$$

we write that

$$K_0 = p_1 - p_3 + r_1 - r_3, \tag{73}$$

$$L_0 = q_1 - q_3 + s_1 - s_3, \tag{74}$$

$$M_0 = p_2 - p_4 + r_2 - r_4, \tag{75}$$

$$N_0 = q_2 - q_4 + s_2 - s_4, \tag{76}$$

$$k_0 = p_1 - p_3 - (r_1 - r_3), \tag{77}$$

$$l_0 = q_1 - q_3 - (s_1 - s_3), \tag{78}$$

$$m_0 = p_2 - p_4 - (r_2 - r_4), \tag{79}$$

and

$$n_0 = q_2 - q_4 - (s_2 - s_4). \tag{80}$$

Using the above boundary and continuous conditions (56) through (63), we obtain a natural frequency equation in a form

$$
\begin{aligned}
&1+\cosh \beta L \cos \beta L \\
&+\frac{\cosh \beta L}{4}\{+(k_0+n_0)\cos \beta(L-H)+(l_0-m_0)\sin \beta(L-H)\} \\
&+\frac{\sinh \beta L}{4}\{-(l_0+m_0)\cos \beta(L-H)+(k_0-n_0)\sin \beta(L-H)\} \\
&+\frac{\cos \beta L}{4}\{-(K_0+N_0)\cosh \beta(L-H)-(L_0+M_0)\sinh \beta(L-H)\} \\
&+\frac{\sin \beta L}{4}\{+(L_0-M_0)\cosh \beta(L-H)+(K_0-N_0)\sinh \beta(L-H)\} \\
&+\frac{\cosh \beta H}{4}\{-(K_0+N_0)\} \\
&+\frac{\sinh \beta H}{4}\{L_0+M_0\} \\
&+\frac{\cos \beta H}{4}\{k_0+n_0\} \\
&+\frac{\sin \beta H}{4}\{-(l_0-m_0)\} \\
&+\frac{\cosh \beta(L-H)}{8}\{(-n_0K_0+m_0L_0+l_0M_0-k_0N_0)\cos \beta(L-H) \\
&\qquad +(m_0K_0+n_0L_0-k_0M_0-l_0N_0)\sin \beta(L-H)\} \\
&+\frac{\sinh \beta(L-H)}{8}\{(m_0K_0-n_0L_0-k_0M_0+l_0N_0)\cos \beta(L-H) \\
&\qquad +(n_0K_0+m_0L_0-l_0M_0-k_0N_0)\sin \beta(L-H)\} \\
&+\frac{1}{8}(K_0N_0-L_0M_0+k_0n_0-l_0m_0) \\
&=0.
\end{aligned}
\tag{81}
$$

When the depth of fluid becomes shallow $H \to 0$, no fluid affects the beam motion. In this case it is shown that we have the elastic frequency equation

$$
1+\cosh \beta L \cos \beta L = 0.
\tag{82}
$$

4   RESULTS

We calculated the natural frequencies by putting $L = 0.6090\mathrm{m}$, $H = L$, $a = 0.0381\mathrm{m}$, $\Gamma = 6.749\mathrm{kgf/m}$, $\gamma_w = 1000\mathrm{kgf/m^3}$ and $EI = 3.813\mathrm{kgf/m^3}$. We used $N = 20$ in (51) and (52). We compared the calculated results with Clough's experimental results.

We obtained that $\beta_1 L = 1.705$ and $\beta_2 L = 4.227$. We had in terms of period that $T_1 = 0.341\mathrm{s}$ and $T_2 = 0.0555\mathrm{s}$. Clough's results were that $T_1 = 0.333\mathrm{s}$ and $T_2 = 0.051\mathrm{s}$ (Clough, 1960).

The present calculated results in terms of period gave higher values compared to the measured results. However, the present method yields the practically acceptable values.

430

5 CONCLUSION

We presented a method to estimate the natural frequency of the transverse vibration of the beam with circular section immersed in the fluid.

We solved an interaction problem of beam vibration and fluid motion by use of the method of variable separation.

The present method can be applied to the other fluid-elastic problems, in which cylindrical structures are involved.

6 REFERENCES

Clough, R. W., 1960. Effects of Earthquakes on Underwater Structures, Proceedings of the 2nd World Conference on Earthquake Engineering, Vol. II, pp. 815-831.

Ito, K., and Kihara, T., 1972. Wave Making Resistance due to Oscillation of Circular Cylinder (in Japanese), Report of the Port and Harbour Research Institute, Vol. 11, No. 3, pp. 37-58.

Kawatate, K., 1976. A vibration analysis for underwater beam of circular section, Bulletin of Research Institute for Applied Mechanics (in Japanese), Kyushu University, No. 45, pp. 397-406.

Lamb, H., 1963. Hydrodynamics, Sixth Edition, Cambridge at the University Press, p. 38.

Patel, M. H., 1989. Dynamics of Offshore Structures, Butterworths, p. 207.

Sommerfeld, A., 1965. Partielle Differentialgleichungen der Physik, Sixth Edition, (translated into Japanese by H.Masuda), section 28.

Timoshenko, S., D. H. Young, and W. Weaver, Jr., 1974. Vibration Problems in Engineering, Fourth Edition, John Wiley and Sons, p. 416.

Zienkiewicz, O. C., R. W. Lewis, and K. G. Stagg (editors), 1978. Numerical Methods in Offshore Engineering, John Wiley and Sons.

ACKNOWLEDGEMENTS

We express our sincere gratitude to Professor Masayuki Oikawa for his stimulating discussions on the Sommerfeld radiation condition. We also extend our thanks to Professor Yusaku Kyouzuka for his valuable comments on the radiation problems associated with the added mass.

APPENDIX

Symbols

$A_i\ A_{ji}$	definite integral
a_i	functions
a	radius of beam
C_i	arbitrary constants
$c(\omega)$	phase velocity of fluid
D_i	arbitrary constants
$d\ d_0$	constants to be determined from beam surface condition
$d_k\ d_{0k}$	constants to be determined from beam surface condition
$d_{si}\ d_{sij}$	a set of solutions of the simultaneous linear equations
EI	flexural rigidity of beam
F	hydrodynamic force per unit length of beam
f	hydrodynamic force per unit length of beam per unit acceleration
g	acceleration of gravity
H	depth of fluid
$I\ I_j$	definite integral

J_1	the Bessel function of the first kind of order 1
K_1	the modified Bessel function of the second kind of order 1
L	length of beam
N	number of terms in series
P	transverse deflection of beam in terms of time
$p_i\ q_i\ r_i\ s_i$	a set of quantities defined by α and β
$p_s\ q_s\ r_s\ s_s$	a set of quantities defined by α and β
$p_{sj}\ q_{sj}\ r_{sj}\ s_{sj}$	a set of quantities defined by α and β
$K_0\ L_0\ M_0\ N_0$	a set of quantities defined by α and β
$k_0\ l_0\ m_0\ n_0$	a set of quantities defined by α and β
T	period
t	time
$R\ R_j$	quantity related to the Bessel functions
r	radial coordinate
$U\ U_j$	quantity related to the Bessel functions
W	normal (principal) function of axial coordinate
x	axial coordinate
Y_1	the Bessel function of the second kind of order 1
y	transverse deflection of beam
$\dot{}$	differentiation with respect to time $$\dot{f}_n(t) = \partial f_n(t)/\partial t$$
$'$	differentiation with respect to argument other than time $$f_n'(z) = \partial f_n(z)/\partial z$$
α	wave number
β	parameter related to frequency of beam
γ_w	fluid weight per unit volume
Γ	beam weight per unit length
δ_{ij}	Kronecker delta
ϵ	parameter related to influence of fluid to beam
Φ	velocity potential
ϕ	velocity potential, in terms of space coordinate
θ	angular coordinate
ω	circular frequency